Biochemistry of Carbohydrates

NIPA® GENX ELECTRONIC RESOURCES & SOLUTIONS P. LTD.
New Delhi-110 034

About the Authors

Dr. V.P. Vadodaria: The author has 40 years of teaching and research experience in the Veterinary Physiology and Biochemistry disciplines at the Veterinary Colleges at UG and PG levels. Guided various postgraduate students and published several National and International research papers. The author has received several silver medals and gold medals.

The author has participated in many seminars in India and abroad. The author is a co-recipient of the state-level award for research work of a student guided for Ph.D. research. The author has worked as a co-editor in one of the animal science research journals. The author is also working as a reviewer of research articles for research work in the field of Physiology and Biochemistry. The author was involved in technical screening for the suitability of many research projects at various Pharmacy colleges. The author has worked in various committees, at in different positions.

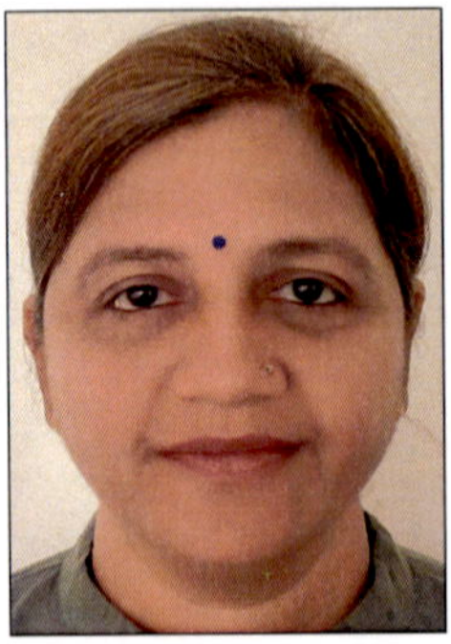

Smt. Dipali M. Sanandiya, Post-graduate in the subject of Biochemistry. She has experience in teaching, research and guiding PG students in the field of biochemistry. She has teaching experience in the subject of biochemistry. She worked as an assistant professor at one the constituent college of Sardar Patel University, Vallabh Vidyanagar, Gujarat.

Biochemistry of Carbohydrates

V.P. Vadodaria
Dipali M. Sanandiya

NIPA® GENX ELECTRONIC RESOURCES & SOLUTIONS P. LTD.
New Delhi-110 034

NIPA® GENX ELECTRONIC
RESOURCES & SOLUTIONS P. LTD.

101,103, Vikas Surya Plaza, CU Block
L.S.C. Market, Pitam Pura, New Delhi-110 034
Ph : +91-11-43860225, Mob.: +91 9717133558, 9540816132
E-mail: newindiapublishingagency@gmail.com
Website: www.nipaersources.com

Print ISBN: 978-93-58874-12-9

ebook ISBN: 978-93-58879-83-4

Composed and Designed by NIPA®.

With Eternal Gratitude

" In loving memory of my beloved parents —

the guiding stars of Dr. V. P. Vadodaria.

They planted the seeds of learning in my heart,

nurtured them with sacrifice, patience, and silent prayers.

Though their voices now echo only in the wind,

their love remains my compass, my light.

This book is born of the strength they gave me,

each word carrying the warmth of their dreams.

May these pages find them beyond the stars,

a humble tribute to the lives that shaped my own."

Preface

I am happy to complete the first edition of this book on the Biochemistry of carbohydrates to the level of expectations. Writing a textbook is time-consuming, and completing the book would take much longer if I covered all biomolecules. Therefore, I first thought of concentrating on carbohydrates, the most important energy-rich biomolecule, having a multifunctional role even as one of the components of nucleic acids, which are crucial for reproduction and expressing various genotypic and phenotypic characteristics.

Carbohydrates are found universally, are the primary energy source, and a very vast topic of biochemistry as they are found in making clothes, many industrial applications, and are modified to prepare medicines and many important substitutes. The textbook includes chemistry, structure, source, metabolism, energy, and carbohydrate-related disorders. The book is an extract of hard work over the years, what I learned from my teachers and mentors, discussions with the other teachers, interactions with many students, research scholars, and as a researcher publishing International and National research papers, and as a Guide to postgraduate research students, and reviewing many articles.

Understanding the importance of carbohydrates is taught at various colleges and universities offering bachelor's and postgraduate programs in Basic sciences, Pharmacy, Agriculture, Home Science, Nursing, Veterinary, and Medical branches. This book will also be helpful to students, faculty members, and research workers engaged in different fields. It also has precious references for research workers. Further, even nontechnical readers are eager to gain technical knowledge in straightforward language and understand the science of carbohydrates.

Attempts have been made to incorporate the technically and grammatically updated information. The figures in the book are simple to understand for all students and technical staff working in various disciplines. All suggestions from readers are welcome for improvement.

Authors

Acknowledgement

Behind every accomplishment lies the support of many individuals whose contributions make the task smoother and more meaningful. I wish to express my deep gratitude to all those who helped, both directly and indirectly, during the writing of this book on the Biochemistry of Carbohydrates.

First and foremost, I offer my adoration and thanks to the Almighty for granting me the health, strength, and courage to undertake and complete this project. Without this divine grace, the task of writing this book, especially at this stage in my life, would not have been possible.

I am sincerely grateful to my colleagues, students, and family members, whose encouragement, inspiration, and support played a vital role in motivating me throughout this journey.

I would like to extend my deepest appreciation to my son, Dr. Ketankumar V. Vadodaria, whose assistance with textproofing was invaluable. Without his help, this book could not have been presented in its final, polished form.

I also wish to acknowledge the full-time dedication and unwavering support of my family members, who devoted their time and efforts to creating a conducive environment for the completion of this work. Their presence has been a constant source of inspiration.

Finally, I express my heartfelt respect and gratitude to my wife, Smt. Sarojben V. Vadodaria, whose moral support, understanding, and sacrifices were crucial throughout the writing process. Her encouragement has been a steady source of strength, and for that, I am eternally grateful.

To all those mentioned and to others who have helped along the way, I extend my sincere thanks. This book would not have been possible without your support.

Dr. V.P. Vadodaria
Editor cum Author

Contents

Preface .... *vii*

Acknowledgement .... *ix*

1. Introduction 1

1.1 Carbohydrates Biochemistry 1

1.2 Modified Carbohydrates 2

1.3 Molecular Forces 2

1.4 Arrangement and Functions of Plasma Membrane Components 5

1.5 Transport Mechanisms Across Membranes 12

1.6 Mitochondria 25

Further Readings *29*

2. Chemistry of Carbohydrates 31

2.1 Introduction to Carbohydrates 31

2.2 Nomenclature and Classification of Carbohydrates 40

2.3 Monosaccharides 40

2.4 Oligosaccharides or Oligosacchroses or Low Molecular Weight Polysaccharides 42

2.5 Polysaccharides or Polysacchrose 43

2.6 Isomers and Isomerism 43

2.7 Racemic Mixture or Racemic Modification or Dl-Mixture or (±) Conglomerate 48

Further Readings *51*

3. Monosaccharides 53

3.1 Formulation of Monossacharides 53

3.2 General Properties/Reactions of Monosaccharides 59

3.3 Monosaccharides of Biological Importance 72

Suggested Readings *83*

4. Oligosaccharides...85

4.1 Disaccharides 85

4.2 Trisaccharides 93

4.3 Tetrasaccharides 97

Further Readings *102*

5. Polysaccharides (Glycans, Polyosides, Poly-Carbohydrates, Polyholosodes)...103

5.1 Homoglycans 103

5.2 Heteroglycans 112

Further Reading.. *127*

6. Metabolism, Metabolic Pathways, and Electron Transfer Chain..129

6.1 Metabolism 129

6.2 Metabolic Pathways for Carbohydrates.. 130

6.3 Mitochondrial Arrangement.. 131

6.4 Electron Transport Chain (ETC) or Respiratory Chain and ATP Synthesis 133

6.5 Inhibitors, Uncouplers, and Other Factors Influence the Electron Transport Chain.. 143

6.6 Carbohydrate Digestion in Non-Ruminants, Ruminants, Poultry, and Variation in Blood Glucose Levels. 147

6.7 Convergence of Metabolic Pathways... 151

Further Readings *159*

7. Glycolysis or Embden-Meyerhof-Pathway (EMP)........................161

7.1 Steps in EMP Cycle or Glycolytic Pathway 161

7.2 Utilization of Galactose and Fructose.. 167

7.3 Factors Influencing Glycolysis 167

7.4 Glycolysis in Erythrocytes or RBC—the Rapaport-Lumbering Cycle or Shunt (for Synthesis of 2, 3-Bisphosphoglycerate)......... 168

7.5 Points to Remember.. 169

Further Readings *174*

8. Tricarboxylic Acid Cycle (TCA)/ Citric Acid Cycle/ Krebs Cycle and Shuttle Mechanisms 175

8.1 Formation of acetyl-CoA from Pyruvic Acid 175

8.2 Steps of Tricarboxylic Acid Cycle (TCA)/ Citric Acid Cycle/ Krebs Cycle 178

8.3 Regulation of the TCA Cycle 184
8.4 Importance of the TCA Cycle 185
8.5 Shuttle Mechanisms 185
8.6 Plant Cycles 189
Further Readings *201*

9. Hexose Monophosphate (HMP) Pathway or Shunt (HMS)/ Pentose Phosphate Pathway (PPP)/ Pentose Cycle/ Phosphogluconate Pathway/ Warburg-Dickens-Lipman Pathway 203

9.1 Reactions of the Cycle 204
9.2 Important Points to Remember 210
9.3 Clinical Aspects 211
9.4 Important Points to Remember 212
9.5 Regulation of the Pathway 213
Further Readings *218*

10. Uronic Acid Pathway or Glucuronic Acid Pathway 219

10.1 Steps in Glucuronic Pathway 219
10.2 Important Points of Glucuronic Pathway 225
Further Readings *229*

11. Gluconeogenesis (Anabolic Pathway) 231

11.1 Gluconeogenesis 231
11.2 Cori Cycle or Glucose-Lactate Cycle 235
Further Readings *241*

12. Glycogenesis 243

12.1 Steps in Glycogenesis 244
12.2 Regulations of Glycogenesis 248
12.3 Glycogen Storage Diseases 249
Further Readings *253*

13. Glycogenolysis 255

13.1 Steps in Glycolysis 256
13.2 Significance of Glycogenolysis 258
13.3 Regulations of Glycolysis 258
Further Readings *265*

14. Enzyme Deficiencies and Carbohydrate Metabolism 267

14.1 Enzyme Deficiency and Disorders in Carbohydrate Metabolism . 267

Further Readings 279

15. Hormonal Disturbances and Carbohydrate Metabolism 281

15.1 The Hormones Involved in Homeostatic Regulation of Blood Glucose Level 281

15.2 Diabetes Mellitus 286

15.3 Bovine Ketosis/Acetonemia/Sweet Breath/ Hypoglycemia 292

Further Readings 300

Index 303

1

Introduction

1.1 Carbohydrates Biochemistry

Carbohydrates are essential for life, both in plants and in animals. They are organic chemical compounds made of C, H, and O. These elements get arranged in different ways so that several simple, and stored complex compounds (starch in plants and glycogen in animals) are available in abundance on the Earth. They are involved in enormous biological and nonbiological functions. Their biological functions, found in all living things, are the chief energy source for men and animals, even for herbivorous fishlike carp and rohu fingerlings. Carbohydrates are also essential structural components like nucleic acids, containing genetic information. One must recognize the protective function in cells, plants, and human health management. Even bacteria form a glycocalyx and biofilm for protection. Knowledge about their involvement in blood groups helped to identify universal donors, universal recipients, and genuine donors for the recipient to save a life. Carbohydrates are found as a storage structure for energy in emergency conditions.

Upon combustion, they yield carbon dioxide and water. Autotrophs (plants) synthesize them from carbon dioxide and water to grow, reproduce, and store, but heterotrophs (humans and other animals) use them, too. In plants, they occur as fibers, supporting structures, and simple sugars. The plant carbohydrates are used for preparing adhesives, fermentation, foods, and papers. Cotton fibers are used to make clothes to cover the body and protect humans and animals against adverse climatic conditions.

Through metabolic studies, carbohydrate biochemistry enlightens their utilization, the formation of intermediate products, and the involvement of various enzymes, coenzymes, hormones, and cofactors. This knowledge helped to find the reasons for defective metabolism, which can lead to either deficiency conditions or excessive unwanted accumulation of metabolites in the body, and the excretion of unwanted products formed in the body. Many of such conditions have been handled through the proper line of treatment. So, hopes have been created to handle some unknown conditions that may arise with knowledge of carbohydrate biochemistry.

However, carbohydrates are an essential part of the diet, and their abusive consumption is responsible for various pathophysiological conditions like obesity and type II diabetes.

Knowledge of these bindings and carbohydrate biochemistry has been used to manufacture specialized or modified carbohydrates.

1.2 Modified Carbohydrates

Ever-growing demand from health-conscious populations for modified carbohydrates all over the world has been targeted through carbohydrate biochemistry knowledge, which has been used to introduce modifications in their structure to make them specialized for their commercial use in various industries like nutraceuticals, cosmetics, food, beverage, and pharmaceuticals. Even these modified carbohydrates have entered into multiple sectors, e.g., animal nutrition in health management, sports nutrition, personal care products, and weight gain and loss products. Furthermore, people have started to create awareness for health concerns, such as consuming foods with high fiber content but for foods with low sweetness, digestibility, caloric values, glycemic index, cariogenic nature, and the sugar substitutes to improve their health.

This is possible if one has knowledge of carbohydrate chemistry and biochemistry and focuses on its structure and functions. To understand carbohydrates properly, one needs to understand their C, H, and O elements bonding and participation of related molecular forces, different transport mechanisms concerned with in and out movements of molecules across the plasma membrane, and knowledge of lipid bilayers in the plasma membrane in the transport of hydrophobic and hydrophilic molecules.

1.3 Molecular Forces

The linkages and arrangement of different groups formed by elements C, H, and O created various carbohydrates (simple form to complex form) which differ in their biochemistry". **Carl Neuberg,** a German chemist in 1903, introduced the word 'Biochemistry.' The meaning can be understood if it is split into **Bio (life or living thing)** and – **chemistry.** Simply, it means the **chemistry of life or living things.** It is a science dealing with the chemical structure, properties, and nature of the biomolecules (found in living matter), their transformations (anabolism and catabolism), and energy changes occurring in a biological system. The subject of biochemistry helps to understand the chemical basis of biological phenomena.

The knowledge of carbohydrate biochemistry helps to expand fields daily, like medical biochemistry (for human beings), veterinary biochemistry (for animals), and agricultural biochemistry (for plants and crops).

Various atoms and ions, etc., are held together by attractive forces, which are called chemical bonds. The various molecular forces that link elements C, H, and O create several different types of carbohydrates that differ in structure and function. Three primary types of bonding are ionic, covalent, and metallic. Atoms stick together to form molecules, including biomolecules.

Each atom has a nucleus and a shell. The nucleus possesses protons, which are positively charged, together with neutral neutrons. Outside the nucleus are shells containing negatively charged electrons that travel in shells around the nucleus of an atom. The number of electrons is the same as the number of proton present in the nucleus of an atom. Each shell can have a specific number of electrons. When the first shell is complete, the following shell, with a higher energy level, receives the additional electrons. Electrostatic forces form the bonds between positively charged atomic nuclei and negatively charged electrons. There are different types of chemical bonds, as listed below:

I. Strong interactions

A. Covalent bonds

These bonds are the most potent, and common chemical bonds in living organisms. The bonds are formed when participating atoms **share an electron to create electron pairs between atoms**. It is much stronger than ionic bonds. Carbon molecules interact primarily through covalent bonding commonly found in biomolecules. Long chains of covalently bonded carbon and hydrogen atoms occur in various arrangements in polymer structures.

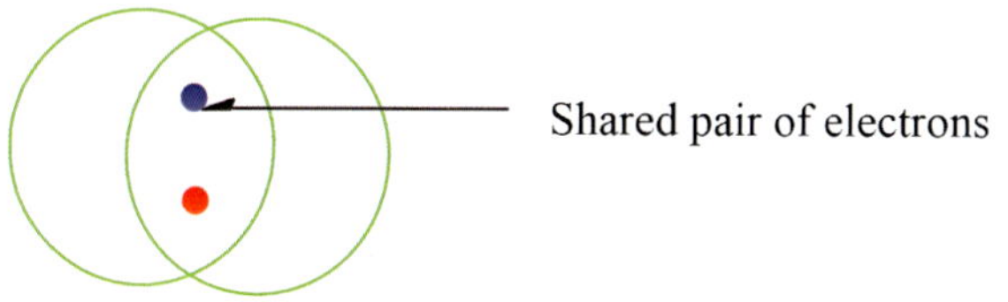

Fig, 1: Covalent Bond

The number of bonds formed between two atoms depends on the number of electrons they share. If they share one electron pair, they join by a single covalent bond; when they share two pairs of electrons, they participate in a double bond. Monosaccharides, when linked together, form polysaccharide chains with glycosidic bonds, which are covalent. Five hydrogen bonds (H-O) donors and six hydrogen bond acceptors (O) are within the glucose molecule. Both intramolecular (within) and intermolecular (between molecules) H-bonding exist in glucose. A glycosidic bond or glycosidic linkage joins a carbohydrate molecule to another group, which may or may not be a carbohydrate.

Lipids have non-polar covalent bonds between carbon and hydrogen to connect their atoms. Lipids (fatty acids, sterols, phospholipids, and glycosyl phosphatidylinositol can be covalently attached to proteins.

B. Ionic Bonds or Electrovalent Bonds

The bonds are formed when one atom gives up an electron to another atom to complete the outer electron shell. The atom losing electrons becomes a cation (a positively charged ion). The atom gaining an electron becomes an anion (a negatively charged ion). It is a type of linkage formed in a chemical compound from the electrostatic attraction between oppositely charged ions. Ionic bond occurs in sodium chloride. The sodium (Na) atom gives up its valence electron to complete the outer shell of the chlorine (Cl) atom.

Ionic bonds are formed as amino acids bearing opposite electrical charges are juxtaposed in the hydrophobic core of proteins.

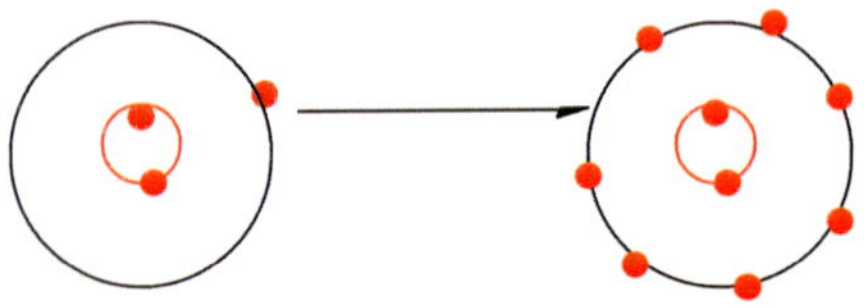

Fig. 2: Ionic Bond or Electrovalent Bond

II. Weak interactions

A. Hydrogen bonds

The bonds form between side chains in proteins. Linking of nitrogen bases for base pairing in nucleic acid structure is through hydrogen bonding, which is strong enough to hold the structure together and weak enough to break the DNA strands for its replication. They are formed between the hydrogen in one molecule and an electron-negative molecule in another. Hydrogen bonds can exist between atoms in different molecules or parts of the same molecule. Water is the perfect example of hydrogen bonds. Compared to an ionic or covalent bond, a hydrogen bond is weaker but more potent than Van der Waals forces. The bonds are constantly broken and reformed in liquid water.

The hydrogen bonds are essential in the confirmation of molecules of carbohydrates and in forming various structural formations. The carbohydrates have the elements C, H, and O in general, with a ratio of carbon to hydrogen to oxygen in proportion of 1:2:1 (CH2O). The simple carbohydrates resemble water molecules, and their hydrogen-bonding function comes from the hydroxyl group, which can have both donor and acceptor properties.

Hydrogen bond duplication. In lipids, there are carbon-hydrogen bonds, which are nonpolar.

B. Van der Waals

These forces are weak forces that contribute to bonding between molecules but still play an essential role in the properties of molecules. The attractive and repulsive electrical forces between atoms and molecules form the Van der Waals forces. Interactions between uncharged atoms/molecules give rise to these forces.

C. Hydrophobic Effect

It is a property of molecules that do not mix with water. "Hydro" means water, and "phobic" means fear. The molecules are called hydrophobes. Many biomolecules are amphipathic, which means they have hydrophobic (water-hating) and hydrophilic (water-loving) parts. The side chains of proteins may be hydrophobic or hydrophilic.

1.4 Arrangement and Functions of Plasma Membrane Components

Various biomolecules once enter the cell through their transport across the plasma membrane or cell membrane and are metabolized by the enzymes present in the cytoplasm or mitochondria; generated energy is captured as high-energy bonds when they undergo many metabolic reactions in the mitochondria. This requires understanding the structure and functions of the cell membrane and mitochondria.

I. Cell Membrane

All cells possess a cell membrane or plasma membrane. Cell walls protect cell membranes in plants, which are absent in animal cells. In multicellular organisms, the cell membrane performs many functions. One end of the molecules is hydrophobic (oily or non-polar), and the other end is hydrophilic (water-loving or polar) in nature._Several factors control the fluid properties of membranes. Examples: a. Long-chain saturated fatty acids, due to their strong interaction with each other, lead to decreased fluidity; oppositely, b. unsaturated fatty acids increase fluidity as they disrupt the interaction of fatty acyl chains due to double bonds. Cholesterol decreases fluidity. Higher temperature increases fluidity. The cell membrane, being semipermeable in nature, can perform various functions, which are as follows:

- Protective in nature.
- Regulate transport in and out of the cell.
- Allow selective receptivity and signal transduction.
- Allow cell recognition (intercellular recognition).
- The cell maintains its shape and moves to distant sites.
- It helps to compartmentalize subcellular domains (microdomains). Many cell organelles have single-layer membranes; some have two membrane layers (Mitochondria and nuclear envelope), and some are naked as they have no cell membrane cover.
- It provides a site for the binding and catalysis of enzymes.
- Regulates the fusion of the membrane with other membranes in the cell via specialized junctions.
- Provide a passageway across the membrane for specific, selective molecules.
- Allow directed cell or organelle motility.
- Response to external signals (hormones and neurotransmitters).

A. Composition of Cell Membrane

It is well-known that the cell is life's structural and functional unit. The membrane of a cell is composed of organic (proteins, carbohydrates, lipids) and inorganic components. The cell is highly organized, with many functional units or organelles. The cell membrane possesses a non-polar hydrophobic end impermeable to water-soluble substances. The polar hydrophilic end allows the movement of water soluble substances The cell membrane thickness varies from 7.5 to 10 microns. The general composition of the cell membrane includes Proteins -55% and Carbohydrates- 3%. Phospholipids- 25%, Cholesterol – 13%, and other Lipids – 4%. However, the proportion of these varies depending on the membrane.

Myelin contains 18% protein and 76% lipid. The mitochondrial inner membrane contains 76% protein and only 24% lipid.

The membrane is asymmetrical because the two sides of a membrane are very different in both the lipid and protein composition.

Cell membrane proteins. The proteins give protection to the central lipid layer. There are five main types of proteins in the cell membrane. Protein molecules are found within the lipid bilayer and are mobile-like lipids: integral proteins, peripheral proteins, glycoproteins, receptor proteins, and transport proteins.

Integral, intrinsic, or transmembrane proteins. Integral proteins are embedded within the membrane, extend deep into the lipid bilayer, and form channels through which water-soluble substances pass. They interface with the cytosol and the external environment. Examples: Transport proteins (glucose transporters), receptors for peptide hormones, and energy-transducing proteins, e.g., adenosine triphosphate [ATP] synthase. When integrated proteins are separated or removed, disruption of the bilayer occurs. Detergents, nonpolar solvents, or denaturing agents can separate the integral membrane proteins. There are three types of proteins, e.g., Monotopic proteins, which are embedded in the membrane and exposed to only one surface and do not span the lipid bilayer even for a single time. They are attached to the membrane from only one side. Bitopic proteins are exposed on both sides (many receptors) and span the lipid bilayer only once. In the case of Polytopic proteins, their polypeptide chains cross the membrane several times.

Peripheral proteins. They are not embedded in lipid bilayers, but attached to the hydrophilic portion of lipids or integral proteins at the surface. Their association with the membrane is temporary and they bond to integral proteins by ionic bonds. Many of them are enzyme proteins. They have a role in signal transduction and cytoskeletal organization. Peripheral proteins have loose affinity and can be removed without disrupting the membrane, with salt and pH changes. There are ionic and non-ionic detergents. Ionic detergents denature the proteins, whereas non-ionic detergents do not cause the denaturation of proteins.

Ionic detergent is Sodium dodecyl sulfate (SDS), used for the PAGE (polyacrylamide gel electrophoresis) technique to study membrane proteins. Non-ionic detergent Triton X-100 is used when active protein is required, as they do not denature the proteins.

Glycoproteins. They are the proteins covalently attached to carbohydrates. They are found on the extracellular side of cell membranes, performing a wide variety of functions.

Receptor Proteins

Receptors are integral membrane proteins. They recognize certain specific signal molecules, such as hormones, neurotransmitters, or growth factors.

Transport Proteins

They are involved in facilitating the movement of ions, small molecules, and macromolecules across the cell membrane.

The lipids. The lipid layer is fluid in nature. Membrane lipids are amphipathic or amphiphilic, which means they have hydrophilic and hydrophobic ends. They have hydrophilic globular (polar) heads that point outside, and the hydrophobic portion is a straight region (non-polar) that points inward. The lipid is arranged in a bilayer, making the membrane relatively impermeable to water-soluble molecules (glucose, urea, and electrolytes), allowing fat-soluble substances to pass through. The oxygen, carbon dioxide, and alcohol can pass through the lipid bilayer. Most contain phosphate groups except cholesterol and glycolipids. Phospholipids are one of the primary/significant membrane lipid types. They have a polar head group and two hydrocarbon tails. One of the tails is a straight-chain fatty acid (saturated), and the other has a kink in the tail because of a cis bond (unsaturated). The presence of a cis double bond prevents tight packing. The hydrophilic portion of the phospholipids is oriented facing outward (aqueous environment), and the hydrophobic portion is oriented facing inward (toward the center of the bilayer). The most lipid part is phospholipids (phosphatidylcholine, sphingomyelin, phosphatidylserine, and phosphatidyl ethanolamine, and in a much smaller quantity are inositol phospholipids). Phosphoglycerides are made of glycerol esterified to two fatty acids and one phosphate group. The phosphate group usually has choline, ethanolamine, serine, or inositol. Sphingolipids have sphingosine, a fatty acid. When phosphorylcholine attaches to the terminal alcohol of sphingosine, it forms sphingomyelin. In the case of glycolipid, a carbohydrate attaches to the terminal amino group. Cerebroside contains a simple sugar. If a carbohydrate is an oligosaccharide, it forms a ganglioside. Cholesterol is missing in most plants and bacterial cell membranes. The central lipid layer is a bilayer, and major lipids are phospholipids and cholesterol. The phospholipids are lipids with phosphorus and fatty acids. The phospholipids are amino phospholipids, sphingomyelins, phosphatidyl ethanolamine, phosphatidyl serine, and phosphatidyl inositol. The cholesterol molecules arranged in between phospholipid (soft and oily structure) molecules help in packing and maintaining the structural integrity of the cell membrane.

Membrane carbohydrates. The principal carbohydrates in cell membranes are galactose, mannose, fucose, galactosamine, glucosamine, glucose, and sialic acid. Carbohydrates are essential in cell-to-cell recognition and as a receptor site for many molecules (hormones). The surface layer covering the cell membrane is the glycocalyx. The glycocalyx is a sticky, hydrophilic layer comprising a fibrous meshwork of carbohydrates (glycoproteins, glycolipids, and proteoglycans) projecting out and covering the membrane. Glycoprotein is formed when a protein (lectins are carbohydrate-binding proteins) is attached covalently to a carbohydrate molecule. The cell surfaces are mostly bound to proteins as glycoproteins. Proteoglycans are different from glycoproteins. Proteoglycans have a protein with glycosaminoglycans chain linkages (glycosaminoglycans chain, heparan sulfate, dermatan sulfate, hyaluronan, and keratan sulfate). The sugar molecules, when covalently bound to lipids, form glycolipids. Glycolipids in plants and bacteria are glycerol-based, whereas in animals, they are ceramide-based. The myelin sheath's main glycolipid is galactocerebroside.

Bimolecular Organization of the Membrane

The membrane forms close compartments around the protoplasm of the cells and separates one cell from another. The membrane also makes compartments inside the cell, as it surrounds many organelles. Their selective permeable nature keeps the differences between intracellular and extracellular fluids. Gates, pumps, and specific receptors for enzymes, substrates, and hormones regulate this property of selective permeability. The membrane surrounds various cells (Prokaryotic, Eukaryotic, and eukaryotic –viruses).

Models of the Plasma Membrane

Various models have been proposed to explain the structural arrangement of the cell membrane. The different models proposed were as follows:

1. Overton, Gorion, and Grendel model (1925) or the Lipid and Lipid Bilayer Model

According to this model, biological membranes are thin polar membranes made of two layers of lipid molecules (or phospholipid bilayer). This became the foundation of future models of membrane structure.

2. Danielli and Davson Model or Lipo-protein Sandwich Model or The Trilaminar Model or Unit Membrane Model

This model was the first proposed basic model in 1935 by Davson and Daniell. Their suggested model was a bilayer of phospholipids located within two layers of globular protein, proteins on the outer surface of the lipid bilayer. Therefore, this model is a three-layered (trilaminate) and is lipo-proteinous. It was also called a "lipo-protein sandwich" model because the layer of phospholipids (hydrophobic) tails is sandwiched between two layers of proteins (hydrophilic). There is a tail-to-tail arrangement of the phospholipid layers. Out of the two layers of proteins (hydrophilic), one layer of heads faces the outside of the cell membrane, and another layer faces the inside.

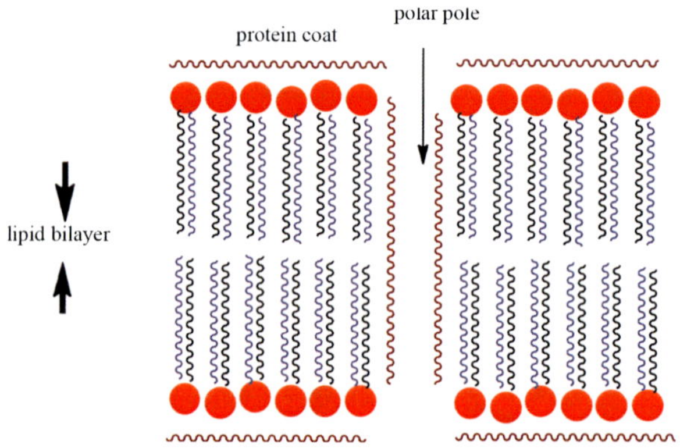

Fig. 3: Danielli and Davson Model

3. Robertson model

He suggested the structure of a membrane, which is similar to Danielli and Davson's model, according to which the unit membrane has a central core of a bimolecular leaflet of lipids on either side by a single-layered fully spread-out hydrophilic protein or nonlipid material. Later, he amended the concept and suggested that the layers on one side of the lipid core are made up of proteins, and the other end comprises carbohydrates, perhaps in the form of mucopolysaccharides.

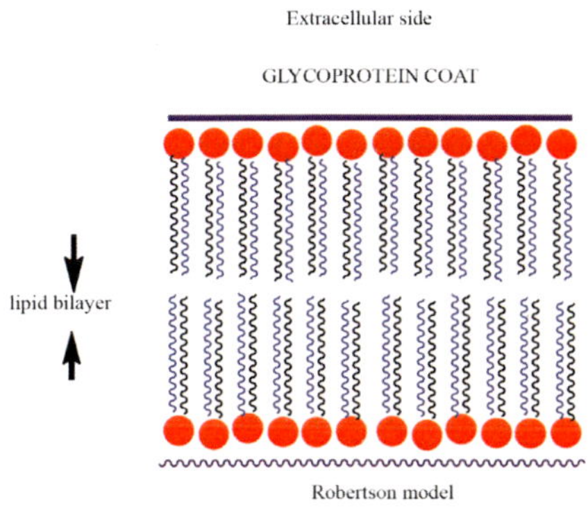

Fig. 4: Robertson Model

4. Singer and Nicolson proposed the Fluid-Mosaic Model (1972)

This model existed after the Davson-Danielli model was falsified through the membrane freeze-fracturing technique. It is a universally accepted model for the plasma membrane. The components of animal cells' fluid-mosaic model of the plasma membrane are phospholipids, proteins, cholesterol, and carbohydrates.

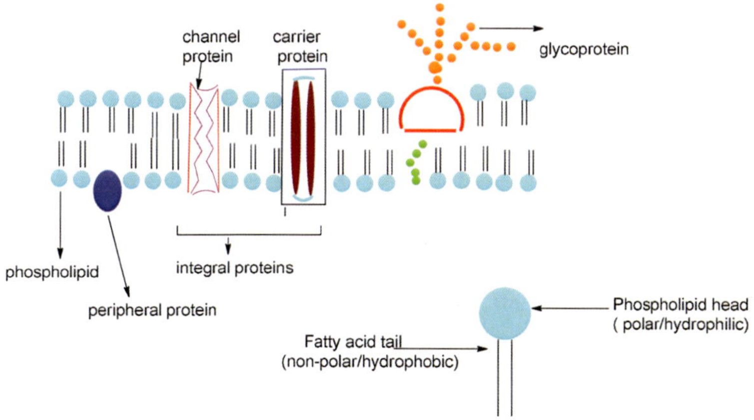

Fig. 5: Fluid Mosaic Model

Lipids form a fluid bilayer, and the proteins do not form a sandwich covering of a hydrophilic bilayer lipid covering. Lipids and proteins are mobile. A homogenous distribution of proteins in the lipid bilayer provides a mosaic nature to the cell membrane. Protein components are 1. Intrinsic (integral) proteins, and 2. Extrinsic (peripheral) proteins. Intrinsic globular protein molecules float free. They penetrate the membrane's interior and the fatty acids side chain. They are tightly bound to the lipids and not easily separable. The extrinsic or peripheral proteins are located on the lipid bilayer's exterior; many are enzyme proteins. They have a loose affinity and can be easily displaced.

The fluid character of the cell membrane is due to its lipid components, which give it a fluid character. Due to fluid characteristics, the cell membranes can adopt amorphous shapes as they are not fixed in position. Shorter chain fatty acids of phospholipid tails increase the membrane's fluidity as they are less viscous. In contrast, fatty acids with double bonds (unsaturated fatty acids) have kinked hydrocarbon tails that are harder to pack. Carbohydrates get linked to either proteins (glycoproteins) or lipids (glycolipids). Both lipid and protein components perform transitional movement within the lipid bilayer (dynamic structure).

- **Disorders of Membranes**

A change in the protein constituents of the membrane may cause many diseases. Proteins in the membrane are found in the receptors, transporters, ion channels, enzymes, and structural components. Membrane proteins are affected by mutation or by conditions other than mutation. Some conditions associated with the disorder of membranes are myasthenia gravis, ischemia, multiple sclerosis, Wilson's disease, Familial hypercholesterolemia, and cystic fibrosis.

1.5 Transport Mechanisms Across Membranes

The complex lipoprotein structure of the cell membrane is also selectively permeable in nature. The basic mechanisms involved in transporting substances across the cell membrane are 1: passive transport mechanism and 2. Active transport mechanism. Oxygen, carbon dioxide, and alcohol are lipid-soluble substances permitted by the cell membrane's lipid layer. Oxygen is highly soluble in lipids and diffuses rapidly. Electrolytes, which are water-soluble, diffuse through the protein layer of the cell membrane.

I. Passive transport mechanism

A mechanism of transport of substances without energy use and along the concentration gradient (from higher level to lower level), electrical gradient (electrochemical gradient), or both is known as a passive transport mechanism. It is a nonselective process, also known as downhill movement or diffusion. Any hydrophobic molecules get dissolved in the phospholipid bilayer to cross the plasma membrane, enter the other side, and dissolve in the present aqueous solution. Hydrophobic molecules, gases, and water can diffuse across the plasma membrane. Glucose, a larger polar and uncharged molecule, cannot cross the plasma membrane by passive diffusion.

A. Passive Diffusion –Simple and Facilitated Diffusion

Passive diffusion can be of two types, namely, 1. Simple diffusion, and 2. Facilitated diffusion.

1. Simple diffusion. It occurs either through the lipid layer or the protein layer. It involves the diffusion of hydrophobic molecules across the lipid membranes and those of small hydrophilic molecules through aqueous pores. The movement of molecules from an area of greater concentration to an area of lower concentration is called diffusion. However, if such movement of water molecules occurs through a semipermeable plasma membrane between the intracellular and extracellular compartments, is called osmosis.

Osmosis helps to maintain an osmotic balance between the two compartments (extracellular and intracellular) separated semipermeable membrane. If the extracellular compartment is hypertonic, then the water from the cells moves out, leading to the creation (shrinkage) of cells. On the other hand, if the extracellular compartment is hypotonic, then hydration (swelling) of cells will occur. Excessive hydration due to the entry of more water inside the cell may lead to the rupture of the cells, which is called hemolysis in the case of red blood cells. This is kept in mind during fluid transfusion given to the patient.

Osmosis is a special type of diffusion. which occurs in the direction opposite to that in which diffusion occurs. Osmotic pressure is the excess pressure applied to a solution to prevent the flow of solvents with low osmotic pressure when a perfectly semi-permeable membrane separates them. The cell membrane is semi-permeable in nature.

Osmosis is characterized by the movement of water molecules from a hypotonic (dilute-more water molecules) solution to a hypertonic (concentrated-less water molecules) solution when a semi-permeable membrane separates the two solutions.

The osmotic pressure depends on the number of osmotically active particles, i.e., molar concentration, and not on mass. The unit of measuring osmotic pressure is Pascal (Pa). There is a difference between the terms osmolarity and osmolality. An osmolarity solution has one osmol of the solute dissolved in a liter of the solution. A solution with one osmol of the solute dissolved in 1 Kg of solvent is called ***osmolality.*** One osmol has an osmotic pressure of 193000 mm of Hg. It does not depend upon the size of molecules. The solutions can be categorized into 1. Hypotonic 2. Hypertonic, and 3. Isotonic depends upon a solution's osmotic pressure compared to a cell's.

A. Hypotonic Solution

A solution with lower solutes but more water molecules than a cell is called a hypotonic solution. If the cells are placed in a hypotonic solution, the water will flow from outside to inside the cell, causing first swelling and then lysis of the membrane. The phenomenon is called plasmolysis. If it is red blood cells, it is called hemolysis; as a result, hemoglobin from red cells will come out and form a red color solution.

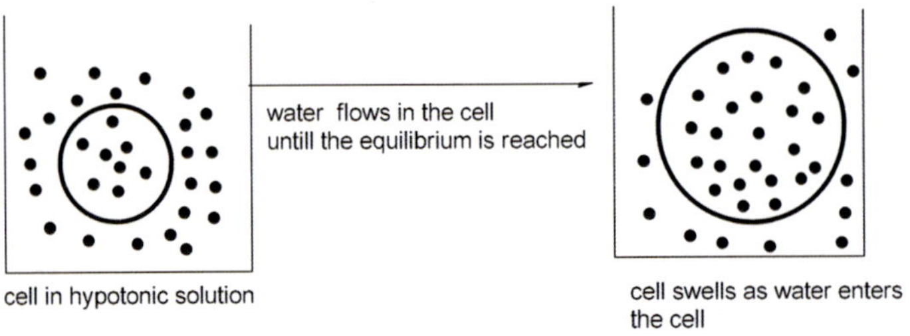

Fig. 6: Cell swelling in hypotonic solution.

B. Hypertonic Solution

A solution with more solute and less water molecules than a cell, is called a ***hypertonic solution.*** When placed in such a solution, the cells will lose water from within and come out into the solution. This will cause shrinkage (crenation) of cells.

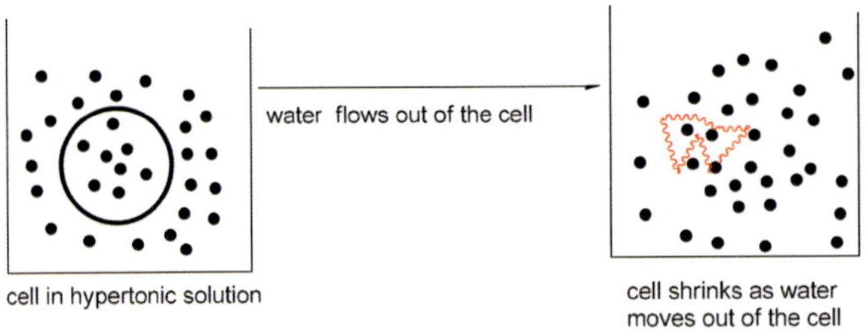

Fig. 7: Cell crenation when placed in a hypertonic solution.

C. Isotonic Solution

If the concentration of two solutions on either side of the cell membrane is the same, it is called an ***isotonic solution.*** When two solutions have the same osmotic pressure, they are called ***isosmotic.*** Normal saline (0.89% NaCl) is isotonic with plasma. Similarly, a 5% glucose solution has a similar osmotic pressure to that of plasma. For intravenous injection, only isotonic solutions are preferred.

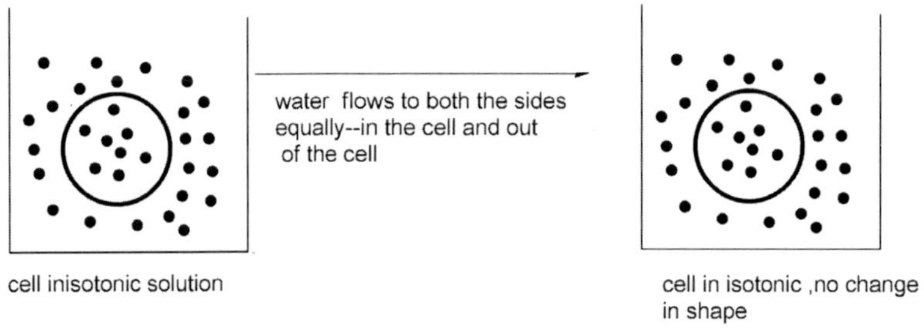

Fig. 8: No effect on cell volume in an isotonic solution

In the case of hydrostatic pressure differences, the water and solutes move toward the area of low hydrostatic pressure, which is known as filtration, which occurs from the arterial capillaries to the interstitial fluid. The filtration is also seen at the glomeruli of the kidneys.

2. Facilitated diffusion. It does not require any energy input. The facilitated diffusion is affected by various factors like temperature, concentration, diffusion distance, and size of the molecules.

Here, hydrophobic molecules do not have to dissolve in the phospholipid bilayers. The electrical potential across the membrane helps the transport of charged molecules. Facilitated diffusion takes the help of channel proteins and carrier proteins, which are membrane integral proteins.

It involves the spontaneous passive transport of polar and charged molecules (carbohydrates, amino acids, and nucleosides) and ions to cross a plasma membrane (into or out of a cell).

i. **Channel proteins or Ion channels.** Some pores are throughout the cell membrane's central lipid layer, forming a channel. These pores or channels are called protein channels as they are lined with invaginated protein molecules from both cell membrane surfaces. These channels have selective permeability and allow diffusion of water, electrolytes, and other water-soluble substances. They allow the passage of water, electrolytes, and other substances.

   There are two protein channels; some are always open and ungated, and others are permanently closed, known as gated protein channels. Gated channels are open as and when required.

- **Open-channel proteins.** They are protein-lined open pores passageways that form a continuous pathway across the lipid bilayer to transport ions and water-soluble substances of appropriate size and charge, and flow rapidly when the channel is open. Channels are substance-specific and are hydrophilic channels. One of the channel proteins is porin, which permits ions and small polar molecules to move freely. Aquaporin (protein) is for water channels that allow water to move more quickly. Channel proteins also facilitate the movement of molecules between the cells connected through gap junctions.

  Many channels, which are usually closed, open in response to specific signals. Ion channels permit the fast transport or fast movement of ions, which are present in all cells. Each protein channel allows only one type of ion to pass through it (selective permeability), e.g., Na^+, K^+, Ca^{2+}, and Cl^- channels. Some channels are continuously **open,** whereas others are **gated** (gates that open or close).
- **Gated -ion channels**. The gated channels are of three types:
- **Voltage-gated ion channels**. These channels operate when there is a change in the membrane's electrical potential and are called voltage-gated channels—for example, the neuromuscular junction. Calcium channels open when an action potential reaches axon terminals. At this time, calcium ions diffuse from cisternae into the sarcoplasm.
- **Ligand-gated ion channels:** They are operated when there is a binding of a specific molecule called ligands such as hormones. The hormonal substances are called ligands, and the channels are called ligand channels. The ligand may be external (neurotransmitters or hormones) or internal (intracellular calcium, cAMP).

  When the impulse arrives at the neuromuscular junction, acetylcholine in the vesicles is released to reach the synaptic cleft. These acetylcholine molecules open the sodium channels of the postsynaptic membrane. As a result, sodium ions diffuse into the neuromuscular junction from the extracellular fluid.
- **Mechanically gated channels.** They are open or closed in response to the mechanical vibration (sound) or pressure. Examples are the channels present in the pressure receptors and the receptor cells (hair cells) found in the internal ear for hearing. The sound waves cause the movement of the hairs of the receptors in the ear. The hair's movement causes ionic channels to open, leading to a receptor potential.

ii. **Carrier proteins**. Transport by carrier proteins can be either active or passive; transport of ions by channel proteins is always passive. Carrier proteins are transmembrane proteins and are also known as permeases, facilitated or carrier-mediated diffusion is faster than simple diffusion.

The carrier proteins help to transport large molecules like glucose, nucleosides, and amino acids. When a particular molecule (glucose) to be transported binds to the carrier protein, reversible conformational changes in shape occur so that molecules can cross the lipid bilayer membrane. As a result, the site with the bound molecule will move from one side of the membrane (outside) to the other side (inner side) of the membrane to release the molecules (inner side). Once molecules are released, the transporter proteins regain their original shape and become accessible to bind with the new molecule (glucose) to re-transport to the other side. The transported solute is delivered unchanged by the carrier proteins.

In other words, molecule (glucose) binding sites face alternatively to the outside and inside of the cell surface. Glucose can be transported from outside to inside and from inside to outside of the cell (liver), depending upon the need. Glucose, once entered into the cell, is immediately metabolized to allow the entry of the following molecule.

Glucose Transporters

Glucose transporters are carrier proteins. A group of carrier proteins, GLUT is a glucose transporter that moves glucose and hexose sugars, and does not require ATP. There are four types of glucose transporters, and they are found in different tissues. GLUT1 is the major glucose transporter in the brain, placenta, and erythrocytes; GLUT2 is found in the pancreas, liver, and kidneys; GLUT3 is neuronal and placental, while GLUT4 is the insulin-responsive transporter found in skeletal muscle, heart, and adipose tissue. Insulin increases the GLUTs on cells so that more glucose is taken up by the cells. Suppose glucose in the kidney filtrate is more than the capacity of the GLUTs present in the kidneys. In that case, glucose is excreted in the urine, as in diabetes mellitus conditions.

Facilitated diffusion of glucose by the Na^+-independent glucose transporter (GLUT) is reversible. The sodium-glucose cotransporters are found in the small intestine and renal tubules to transport glucose against the concentration gradient.

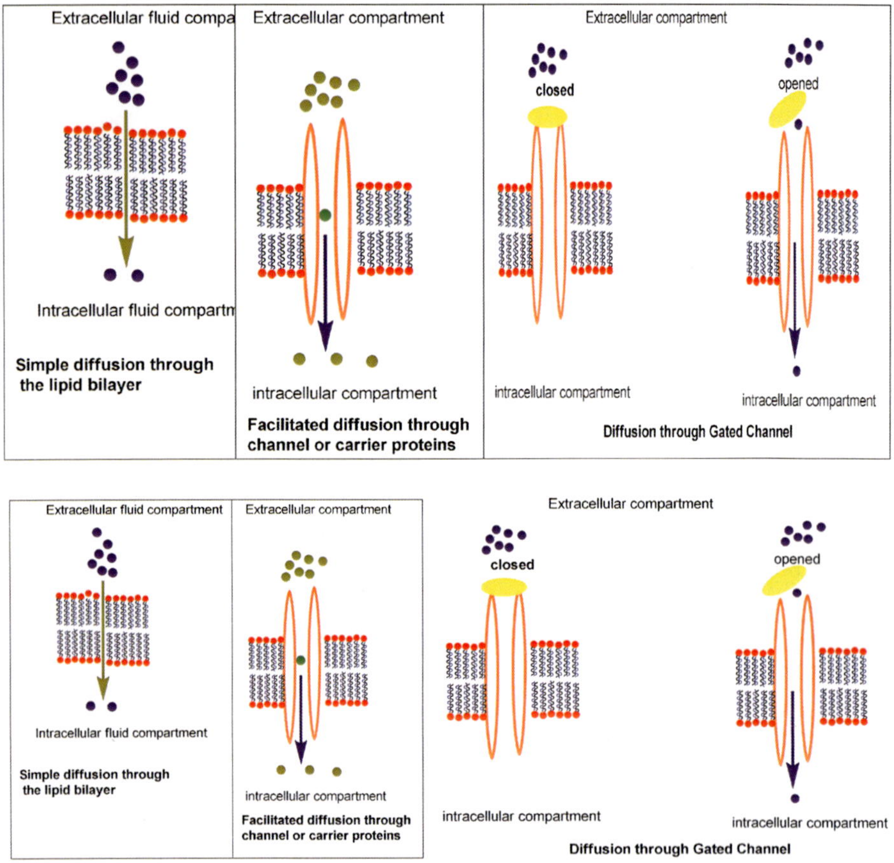

Fig. 9: Simple Diffusion and Facilitated Diffusion

There are different types of carrier-mediated transport, such as a. Uniport and b. coupled transport or cotransport (symport and antiport transport)

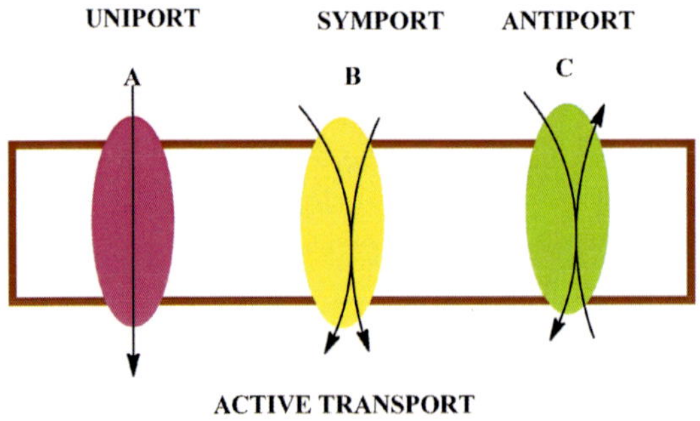

Fig. 10: Uniport, Symport and Antiport Transport

a. **Uniport transport.** It transports only a single substance by a carrier protein, such as a glucose transporter. The direction of transport is either inside or outside. Uniport carrier proteins facilitate the diffusion of a single substance (e.g., glucose, a particular amino acid).

- Na-independent glucose transporters (GLUTs) are uniporters that passively transport glucose, galactose, or fructose from the blood into most cell types down a steep concentration gradient.
- Cycling of the uniporter between alternative conformations allows the binding and release of the transported molecules.
- The direction of transport by the uniporter depends on the direction of the concentration gradient for the transported molecule.

b. **Coupled transport or cotransport.** Cotransport or coupled carrier proteins help transport two different types of molecules simultaneously across the plasma membrane using energy (stored in electrochemical gradients). The movement of molecules may be by facilitated diffusion or secondary active transport. If two molecules are the same or different, it decides the transport, which may be antiport or symport.

- **Antiport transport or counter transport.** It occurs when the antiporters move or transport two molecules or ions in opposite directions across the membrane. Here, the exchange of one molecule occurs for another. When 3 Na^+ move into the cell down the concentration gradient, there is an export of 1 Ca^{2+}. Therefore, sodium-calcium (Na+/Ca2+) transport is antiport or counter transport. Sodium-calcium antiport is present in the cardiac muscle cells. An antiporter that facilitates the diffusion of chloride-bicarbonate (Cl and HCO_3) functions in the transport of CO_2 from tissues to the lungs. Sodium-hydrogen counter transport, sodium-magnesium counter transport, sodium-potassium counter transport, calcium-potassium counter transport, chloride-sulfate counter transport.
- **Symport transport.** It is the transport of two distinct molecules or substances in the same direction at a time by transporter proteins. Along with the sodium, another substance is carried by the carrier proteins with the same energy. The substances carried along with sodium for co-transport are glucose, amino acids, chloride, iodine, iron, phosphate, and urate ions. Glucose passes through the small intestine apical membrane into an epithelial cell against its gradient using a Na^+/glucose symporter, also designated SGLUT1. Similarly, in the kidney's proximal tubules, 2 Na+ ions and one glucose molecule are simultaneously transported into the cell across the plasma membrane.

$2Na^+/1$ glucose, $2Na^+/1I^-$, $3Na^+/1Pi$, $Na^+/K^+/2Cl^-$, Na^+/Cl^-, K^+/Cl^-.

Factors Affecting the Rate of Diffusion

1. The rate of diffusion in the body is directly proportional to the temperature, concentration, electrical gradient across the cell membrane, surface area of the membrane, partial pressure, and solubility of the substance (lipid solubility). It is inversely proportional to the thickness of the cell membrane, the size of the molecules and the size of the ions.

II. Active Transport or Uphill Transport Mechanism

Transport of substances with the use of energy (ATP) and against the concentration gradient, electrical gradient, or both (electrochemical gradient) is known as active transport. It is also known as uphill transport.

A substance to be transported, when comes near the cell membrane combines with the carrier proteins and forms a substance-protein complex. This complex formed on the outer surface is released on the inner surface of the membrane, and the released carrier protein moves back to the cell membrane's outer surface. The mechanism is different than that of facilitated diffusion.

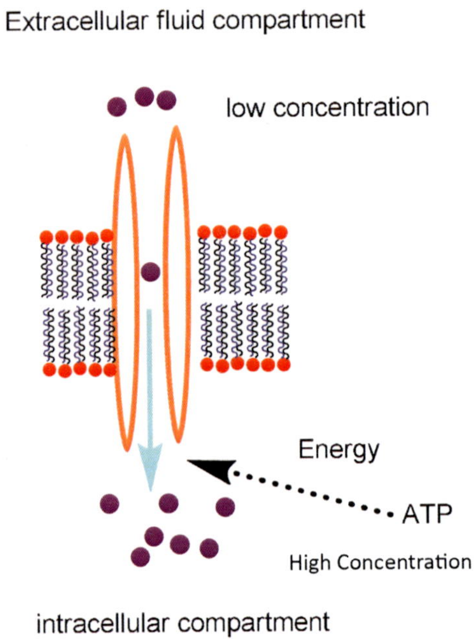

Fig. 11: Active transport

A. Primary or Direct Active Transport

In this method, the cell membrane transport of molecules from a lower to a higher concentration (against a concentration gradient) needs ATP energy. Pumps mediate the unidirectional movement of each molecule transported. Examples are the sodium-potassium pump, the calcium pump, glucose uptake in the intestine, hydrogen, and chloride.

• Na-K-ATPase Pump

Three sodium ions from the cell cytoplasm (inside the cell) get attached to their receptor sites. Two potassium ions from the extracellular fluid (outside the cell) get attached to their respective sites on carrier protein. After sodium and potassium ions bind, ATPase gets activated to break ATP into ADP and release the energy required to make conformational changes. As a result of the changes, the outer surface of the molecule with potassium ions now faces the inner side of the cell (cytoplasm), and the inner surface of the molecule with sodium ions now faces the outer side of the cell (ECF). This leads to the dissociation and release of sodium outside the cell, as well as the release of potassium inside the cell. Transport of $3Na^+$ ions to the outer side and $2K^+$ ions to the inner side occurs for every single ATP consumed. Each cell can have hundreds of such pumps.

Importance of Na-K-ATPase Pump

The primary purpose of this pump is to maintain the physiological concentration of potassium ions K^+ and sodium ions Na^+ inside and outside of the cell. This pump uses ATP-derived energy. This is necessary to maintain osmotic equilibrium. It is also helpful in the contraction of cardiac muscle, kidney function, and nerve signaling.

Sodium inside the cell, an osmotically active substance, attracts water, causing cell swelling. When extra Na^+ gets pumped outside the cell, its volume is regulated by creating low intracellular Na^+ and high intracellular K^+ concentrations relative to the external environment._Insulin enhances the pump, so K^+ enters the cell from the extracellular compartment (hypokalemia)._The beta-blockers inhibit the pump, producing higher potassium in the interstitial space (hyperkalemia). Each time, at the cost of one molecule of ATP, three Na^+ and two K^+ are transported across the plasma membrane as this pump has three binding sites for Na^+ and two for K. $^+$.

• A voltage-gated or Voltage-dependent Calcium Channel

Calcium ions are transported by the plasma membrane calcium ATPase (PMCA) pump and the sodium/calcium ion exchange (NCX). The pumps are located in the cell membrane, organelles (sarcoplasmic reticulum), and mitochondria. All calcium channels maintain calcium homeostasis and signaling in the cell. They are involved in a large number of different processes.

Voltage-gated calcium channels are closed at the resting state and open upon cell membrane depolarization. The channels regulate the concentration of calcium inside the cell and different cellular compartments through efflux and influx of calcium through the cell. Voltage-gated calcium channels transport calcium into the cytoplasm. PMCA and NCX channels transport calcium out of the cell to provide a deficient cytosolic concentration that can be used as a second messenger. Other calcium channels exist on the plasma membrane of specialized cells like neurons, sensory cells, endoplasmic reticulum (ER), and sarcoplasmic reticulum (SR), the main calcium storage sites inside the cell. Calcium channels found in the membrane of the ER/SR regulate the transfer of calcium between the cytoplasm and the ER/SR. Muscle Ca^{2+}-ATPase transports Ca2+ from the cytosol to the SR lumen to keep a low cytosolic calcium concentration level, which permits relaxation. Release of stored Ca^{2+} from SR to cytosol triggers muscle contraction.

Tissue hypoxia decreases ATP production, which affects the Ca^{2+}-ATPase pump and permits Ca^{2+} into the cell, where it activates various enzymes (e.g., phospholipases, proteases, endonucleases, caspases [pro-apoptotic enzymes]), leading to irreversible cell damage.

• Transport of Hydrogen Ions or Hydrogen Pump

The carrier protein for hydrogen ion transport is called a hydrogen pump. They are present in the stomach and kidneys. In the stomach, HCl is formed by the parietal cells using a hydrogen pump. The same pump at the kidney level secretes hydrogen ions from blood to urine at the distal convoluted tubule and collecting duct. They allow hydrogen protons but no other ions to cross the cell membrane.

B. Secondary Active Transport or Ion-Coupled Transport

In primary active transport, ATP-derived energy is used. In secondary active transport, the energy used is not derived from ATP but secondarily from the electrochemical gradient between the two sides of a membrane. This gradient is established together across a cell membrane by the chemical concentration gradient and the electric potential gradient. The difference in solute concentration across a membrane establishes a chemical gradient. The difference in the electrical charge (positive and negative) across a membrane creates the electrical gradient. Secondary active transport occurs after the occurrence of primary active transport. In mammals, the extracellular Na^+ is very high (145mM) against its low intracellular concentration (15mM). Secondary active transport involves the movement of an electrochemical ion (commonly Na^+ or H^+) down its electrochemical gradient (from higher to lower concentration) to the upward movement of another molecule or ion against a concentration/electrochemical gradient.

III. Bulk Transport

The movement of large molecules is possible through a plasma membrane. It involved the uptake and release of larger substances using energy. Both the processes, endocytosis and exocytosis, are active transport processes.

A. Endocytosis

This mechanism transports large molecular substances (macromolecules) inside the cell (entry), which otherwise cannot pass through the cell membrane through active or passive transport mechanisms.

Endocytosis includes: 1. Phagocytosis 2. Pinocytosis, and 3. Receptor-mediated endocytosis

1. Phagocytosis or Cell Eating (phago=eat, cyte=cell)

The process by which the larger solid particles are engulfed by the phagocytic cell. Cells like neutrophils, monocytes, and tissue macrophages (the most giant phagocytic cells) can engulf larger foreign bodies inside the cell.

Mechanism of Phagocytosis

The cell will send cytoplasmic extension (pseudopodium or false feet) around the foreign body (bacteria), which is in contact with the cell membrane. The pseudopodia then surround the particle, and when the plasma membrane of the projection meets the membrane, fusion occurs. This forms an intracellular endosome (like a vacuole), which is very large and usually known as a phagosome. During phagocytosis, the surface contact of the foreign body triggers the opening of calcium channels, increasing the internal Ca^{++} concentration. This situation triggers the formation of a phagosome or food vacuole. The primary lysosome fuses with the phagosome and forms the secondary lysosome. As a result, the hydrolytic enzymes in the secondary lysosome get activated, so digestion and degradation of the phagosomal contents occur.

Phagocytic cells remove foreign organisms (bacteria) and dead or damaged cells.

2. Pinocytosis (cell-drinking)

The extracellular fluid and its contents are taken into the cell by forming narrow channels through its membrane that pinch off into vesicles and fuse with lysosomes that hydrolyze or break down contents. This process requires energy in the form of ATP. Pinocytosis generates tiny vesicles. Pinocytosis is nonspecific in the substances that it transports.

3. Receptor Mediated Endocytosis

It is the transport of macromolecules mediated by a receptor protein found on the surface of the cell membrane. The cell membrane has some pits which contain a receptor protein called **clathrin**. Some are coated with another protein called **caveolin** and are concerned with transporting vitamins into the cell. Hormones or ligand substances bind to the receptors in the coated pits to form ligand-receptor complexes. The pit detaches from the cell membrane and becomes a coated vesicle or endosome. This mechanism is helpful in the transport of hormones, cholesterol, lipids, various growth factors, proteins, and antibodies.

B. Exocytosis

The process through which the substances from the cell are removed without passing through the cell membrane. Exocytosis is a process in which an intracellular vesicle (membrane-bounded sphere) moves to the plasma membrane, and subsequent fusion of the vesicular and plasma membranes ensues. Later, the content of the vesicles is released from the cell. Exocytosis is involved in the secretion of enzymes, peptide hormones, and antibodies from cells. Exocytosis is the opposite process compared to endocytosis.

Exocytosis is of Two Types in Multicellular Organisms

1. **Constitutive exocytosis**. The secretory macromolecule material release occurs outside the cell without any specific signals. Here, vesicles containing protein form continuously, and such vesicles containing proteins are transported from the Golgi to the cell surface. Example—release of ACTH pituitary gland or removal of a waste product or a toxin
2. **Regulated exocytosis**. The secretory vesicle material released outside the cell requires specific external signals for its release. Here, synthesized proteins are consolidated into vesicles and stored in the cell until secreted

in response to a specific signal. Examples: insulin secretion in response to blood glucose concentration. Release of zymogen granules from acinar cells of the pancreas. Release of neurotransmitter acetylcholine from synaptic vesicles.

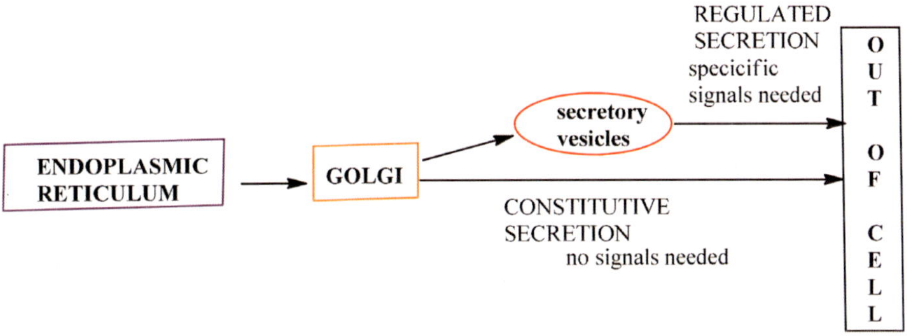

Fig. 12: Constitutive and regulated secretion

C. Transcytosis (vesicular transport, cytogenesis, cytopempsis)

It is a mechanism of transcellular transport of macromolecules. Endocytosis and exocytosis are also involved in this process. It may be either unidirectional or bidirectional. Macromolecules captured in vesicles on one side of the cell are ejected on the other side of the membrane. Blood capillaries are a well-known site for transcytosis. Examples of molecules include IgA, transferrin, and insulin. It is a strategy multicellular organisms use to selectively move material between two environments without altering the Macromolecules that enter the cell by endocytosis from one side and traverse the cell to the other. Macromolecules then leave the cell by exocytosis.

1.6 Mitochondria

These organelles produce ATP through oxidative metabolism using oxygen. They are considered the "powerhouses" of a cell. Each mitochondrion possesses two layers of membrane called the outer and inner membranes. The outer membrane is elastic. The inner membrane is repeatedly folded to provide ample surface area, extending into the matrix of the mitochondria, known as cristae. This surface area of the inner membrane contains many enzymes for oxidation/reduction, such as those of the electron transport chain. The inner membrane projecting into the matrix has cluster knobs called inner membrane particles or elementary particles. Mitochondria contain their DNA (mitochondrial DNA). They can duplicate themselves independently of other cellular genetic material.

The organic nutrients are oxidized via tricarboxylic (TCA) or Kreb citric acid cycle to a form that the cells can use. The ATP-producing system is present in the electron transfer particles of the electron transport system or respiratory chain. Their shape varies; even with altered conditions, mitochondria change their form and are pleomorphic. The number varies between the cell types but remains relatively constant within a kind of cell. The more active cell has more numbers. Absent in prokaryotic cells.

In length, it is 1 to 2 micrometers and between 1 to 0.5 micrometers in diameter. It has two layers of membrane, the outer and inner membranes. The fluid matrix found between the membranes separates the outer and inner. Different enzymes are located on the other membranes. The matrix provides communication between the two membranes and supplies coenzymes to the enzymes present in the membrane. The electron transport chain present in mitochondria aids in the formation of high-energy bonds.

Key Takeaways

- Carbohydrates contain C, H, and O elements. They are found in several simple and stored complex compounds (starch in plants and glycogen in animals). Carbohydrates are found as a storage structure for energy in emergency conditions.
- Carbohydrates are a chief energy source, provide structural components, have protective functions, and are involved in determining various blood groups. Their nonbiological use involves preparing adhesives of, papers, and fabric manufacturing.
- Uncontrolled consumption of carbohydrates is responsible for various pathophysiological conditions like obesity and type II diabetes.
- The modified carbohydrates make them specialized for commercial use in various industries like nutraceuticals, cosmetics, food, beverage, and pharmaceuticals. Their demand has increased in multiple sectors, e.g., pet animal nutrition, health management, sports nutrition, personal care products, and weight gain and loss products.
- Various bonds join the elements in carbohydrates.
- The cell membrane is made up of hydrophobic and hydrophilic layers. Hydrophilic layers are proteins. Proteins are of two types: integral (or intrinsic or transmembrane) proteins, and Peripheral membrane proteins. The lipid layer has hydrophilic globular (polar) heads that point outside, and the hydrophobic portion is a straight region (non-polar) that points inward.

- The complex lipoprotein structure of the cell membrane makes it selectively permeable in nature. Transporting of substances across the cell membrane uses passive and active transport mechanisms.
- The transport of ions/molecules occurs across the cell membrane and uses either passive or active transport. Passive mechanisms do not use energy, whereas active transport needs energy.
- The movement of a molecule/ion occurs with the help of transporter proteins, which help in uniport, symport, and antiport transport.
- Active transport types are Primary active transport and Secondary active transport.
- Glucose transporters (GLUTs) help transport glucose into cells either using sodium-glucose linked transporters (SGLTs) or glucose transporters (GLUTs), which do not depend on sodium.
- Bulk transport is an active transport mechanism used to transport large molecules. The transport may be exocytosis or endocytosis in nature.
- Endocytosis includes phagocytosis, pinocytosis, and receptor-mediated endocytosis
- Exocytosis may be constitutive exocytosis or regulated exocytosis. Constitutive exocytosis does not need signals, whereas regulated exocytosis is regulated under specific signals.
- Mitochondria are the powerhouse of a cell. The inner membrane is folded into cristae. The inner membrane contains enzymes forthe double-membrane oxidation/reduction at the electron transport chain.
- Mitochondria contain their DNA (mitochondrial DNA). They can duplicate themselves independently of other cellular genetic material.

Intext Questions

1. What is the importance of carbohydrates?
2. Why is it essential to modify carbohydrates?
3. What forces bind the elements to form a molecule?
4. What layers do one find in a cell membrane?
5. How does the diffusion process differ from the osmosis process?
6. Explain facilitated diffusion. What are its different types?
7. What types of endocytosis occur?
8. How is glucose transported in a cell?

9. How does primary active transport differ from secondary active transport?
10. What is bulk transport and transcytosis?
11. Describe the process of exocytosis.
12. What are the types of endocytosis?
13. Some types of endocytosis are non-specific processes. What does this mean?
14. Describe the process of receptor-mediated endocytosis.
15. What is the difference between endocytosis and exocytosis?
16. Why is pinocytosis a form of endocytosis?
17. Are vesicles involved in passive transport? Explain.
18. What is constitutive exocytosis?
19. Select the most appropriate answer for the following (The answer is given in **Bold** letters:

a. Bacteria form a glycocalyx and biofilm for

   i. **Protection** ii. Energy

   iii. Support

b. Carbohydrate storage structure(s) is/are

   i. Glycogen ii. Starch

   iii. **Both as in i and ii**

c. Carbohydrates provide

   i. Energy ii. Structural support

   iii. **Both i. and ii. are correct**

d. Biochemistry is the study of.

   i. **Chemistry of living things** ii. Chemistry of drugs

   iii. Chemistry of health

e. Protons and neutrons are found in

   i. **Nucleus** ii. Shell

   iii. Both i. and ii. are correct

f. Electrons have

i. Positive charge ii. **Negative charge**

iii. No charge

g. Bonds formed when participating atoms share an electron to create electron pairs between atoms are

i. Ionic bonds ii. Van der Waals

iii. **Covalent bonds**

h. Amphipathic molecules have

i. Hydrophobic parts ii. Hydrophilic parts

iii. **Both i. and ii. are correct**

i. The movement of molecules from an area of greater concentration to an area of lower concentration when separated by the plasma membrane is

i. Osmosis

ii. Diffusion iii. None of the above

j. Glucose transporters are

i. **GLUTs** ii. FLUTs

iii. JUT

k. Endocytosis includes

i. Phagocytosis ii. Pinocytosis

iii. **Both i. and ii.**

l. Exocytosis type when secretory vesicle material needs specific external signals to be released outside the cell.

i. Constitutive exocytosis ii. **Regulated exocytosis**

iii. None of the above

Further Readings

Active transport. https://study.com/academy/lesson/active-transport-in-cells-definition examples.html.

A textbook on animal physiology. A, K, Berry. 4th edition, Emkay publication. Delhi, India

Biochemistry. Naik, Pankaj.

Carbohydrates. James N. BeMiller. https://doi.org/10.1002/0471238961.0301180202051309.a01.pub2

Exploring the diverse applications of carbohydrates macromolecules in food, pharmaceutical, and environmental technologies.Ramaprabha K, Venkat Kumar S, Panchamoorthy Saravanan, R. Rajeshkannan, M.Rajasimman, Hesam Kamyab, Yasser Vasseghian. Environmental Research, volume 240, part 2, January 2024, 117521.

Fundamentals of biochemistry. J.L.Jain, Sunjay Jain and Nitin Jain.

https://microbiozindia.com/top-5-key-proteins-found-in-the-cell-membrane/

https://www.creative-biolabs.com/blog/index.php/membrane-protein-overview/

Life sciences. Fundamentals and practices I. sixth edition. Pranav Kumar and Usha Mina. (2017). Pathfinder publication. New Delhi

Molecular Biology of the cell. Alberts B, Johnson A, Lewis J, et al. Protein Processing III: secretin, lysosomal, and endocytic pathways. https://www.zoology.ubc.ca/

Objective Biochemistry. R.S.Senger and Reshu Chaudhary. (2014). New India publishing agency. New Delhi.

Pretest Physiology. James P. Ryan and Ronald F. Tuma. https://archive.org/details/physiologypretes00ryan_0

Textbook of Biochemistry for Medical Students. Vasudevan, D. M.; Sreekumar, S.; Vaidyanathan, Kanan.

Textbook of Veterinary Biochemistry. Dhanotiya, R.S. (2006) Second edition, reprint. Jaypee Brothers, medical publishers (P) lid. New Delhi. India

The Cell: A Molecular Approach. Cooper GM. Sunderland (MA). https://www.ncbi.nlm.nih.gov/books/NBK9839/

2

Chemistry of Carbohydrates

Carbohydrates are the most abundant class of organic compounds found in living organisms. Carbohydrates are also called sugars and are found in plants and animals. Plants synthesize them through photosynthesis (endothermic) by fixing carbon, hydrogen, and oxygen in the presence of solar energy and chlorophyll pigments.

$$nCO_2 + nH_2O + \text{energy} \rightarrow C_nH_{2n}O_n + nO_2$$

Carbohydrates are stored as starch in plants as a source of energy. They are also stored in animals as glycogen in muscles and the liver. They are also found in skeletal structures of cell walls in plants (cellulose), bacteria (peptidoglycans), insects, and crustaceans. This stored food is consumed by animals and man, and upon their oxidation in the body, energy is released for various metabolic activities (work, growth, and various productive functions). Composition-wise, plants are more prosperous in carbohydrates (30%) than animals (1%), with water at 60% and ash at 4%. Plants are poor in proteins (5%) and lipids (1%) as compared to proteins (20%) and lipids (15%) in animals.

2.1 Introduction to Carbohydrates

Carbohydrates or saccharides (sakcharon=sugar or sweetness) are a chemical term applied to a large group of substances, which are organic compounds containing elements Carbon, Hydrogen, and Oxygen in a ratio of 1:2:1. Hydrogen and Oxygen atoms are in the ratio of two to one as in water molecules. As the ratio of hydrogen to oxygen is like H_2O, they are also known as "hydrates of carbon" and are represented as **$(CH_2O)n$**. But there is an exception to this general formula. Deoxyribose ($C_5H_1OO_4$) and rhamnose ($C_6H_{12}O_5$) do not have the required ratio of 2:1 for hydrogen to oxygen. On the other side, some compounds possess this required ratio but are not carbohydrates, e.g., formaldehyde (CH_2O), acetic acid ($C_2H_4O_2$), and lactic acid ($C_3H_6O_3$). Further, some carbohydrates have nitrogen(glucosamine), phosphorus, and sulfur but do not satisfy the above general formula. The simplest sugars are called monosugars or monosaccharides. There are two main functional groups in carbohydrates: the carbonyl group (aldehyde or ketone) and several hydroxyl groups.

Carbohydrates, called sugars, are crystalline, sweet, and water-soluble substances. For example, glucose, fructose, and cane sugar. But some carbohydrates are tasteless, insoluble in water, and amorphous. For example. Starch, cellulose, etc.

Relative sweetness of various carbohydrates. The sweetness differs with various sugars or carbohydrates. The standard reference for measuring sweetness is sucrose (100% sweetness). Xylitol is as sweet as sugar (sucrose) but has fewer calories. Xylitol can maintain blood sugar levels more consistently than regular sugar.

Name of sugar	**Sweetness (%)**	**Name of sugar**	**Sweetness (%)**	**Name of sugar**	**Sweetness (%)**
sucrose	100%	tagatose	92%	maltose	32%
fructose	173%	glucose	74%	trehalose	45%
invert sugar (mixture of glucose and fructose)	120%	lactose	15%	galactose	32%
Dextrose	70 to 80%				

The most common sugar alcohols are also, to some extent, sweet as sugar. One gram of sugar provides four calories.

Sugar alcohol	**Calories/ gram**	**Percent as sweet as sugar**
Erythritol	0.2	60% to 80%
Isomalt	2	45% to 65%
Lactitol	2	30% to 40%
Maltitol	2.1	90%
Mannitol	1.6	50% to 70%
Sorbitol	2.6	50% to 70%
Xylitol	2.4	same sweetness as sugar

Artificial Sweeteners

Various types of artificial sweeteners are available. Sucralose is currently the sweetest commercially available compound and is 600 times sweeter than sucrose because of its differential binding properties to taste receptor proteins on the tongue. Another artificial sweetener, aspartame, is not a carbohydrate but a dipeptide derivative of aspartate and phenylalanine. Other artificial sweeteners are listed below:

- Saccharin. (first artificial sweetener,1879).
- Aspartame
- Alitame
- Sugar alcohols
- Stevia leaf extract.
- Acesulfame potassium
- Advantame.
- Sucralose
- Neotame

For healthy people, artificial sweeteners are safe in limited amounts. Sugar substitutes are low or zero in calories. It is best to avoid them or consume them in moderate quantities.

I. Definitions

A.Carbohydrates

Carbohydrates are derivatives containing polyhydroxy aldehydes (CHO group) or ketones (CO group). They can also be defined as substances that, upon hydrolysis, yield polyhydroxy aldehyde or polyhydroxy ketones. Glucose is a polyhydroxy aldehyde, while fructose is a polyhydroxy ketone. Polyhydroxy aldehyde is $CHO.(CHOH)n.CH_2OH$ and polyhydroxy ketone are $CH_2OH.CO.(CHOH)n.CH_2OH$.

B. Asymmetric or Chiral Carbon

It is known that the carbon atom has the shape of a tetrahedron in which the carbon nucleus resides in the center, and the four covalent bonds extend out to its corners. An asymmetric or chiral carbon has four different atoms or groups attached. In glucose, four different groups or atoms are attached (marked with an asterisk *) and are said to be asymmetric or chiral. Glucose has four asymmetric carbon-forming 16 isomers (8 D- and 8 L- derivatives).

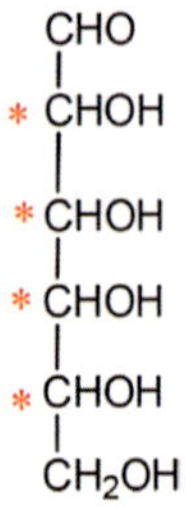

D- GLUCOSE

C. Tautomers and Tautomerism (desmotropism, kryptomerism, allotropism, merotropy)

Relocation of a hydrogen atom within a single chemical compound results in the formation of interconvertible structural isomers, and the formed two structures are tautomers. Tautomerization is a process of changing into a tautomeric form. It involves adding and removing an H-atom (proton) from one molecular site to another. It is possible by chemical reaction, interconverting the two tautomers in the presence of a catalyst.

Here, a single chemical compound exists in two or more interconvertible form. Tautomers are structural isomers (constitutional isomers) of chemical compounds. Tautomers will not have the same number of electrons and protons. Glucose and fructose, even though they have different structural arrangements of atoms, have the same molecular formula, $C_6H_{12}O_6$. Tautomers differ in chemical and physical properties. Keto form (more stable) and enol form (less stable) of chemical compounds are readily interconvertible. Tautomerism is a special kind of functional isomerism. These molecular changes occur in weak alkaline solutions of sugars. These enediols can easily break down at the double bond position and yield a complex mixture of aldehydes. The alkaline medium is more favorable for the reducing action of the sugars. Tautomers cannot be isolated into their components, whereas functional isomers can be.

D. Mutarotation

The change in the optical rotation of sugar is known as mutarotation due to the interconversion between two anomers. An aqueous monosaccharide solution shows a gradual change in optical rotation to reach a constant value. The alpha-D-glucose and beta-D-glucose are anomers as they differ at carbon one only. Except for a few ketoses, all reducing sugars undergo mutarotation. The open and ring forms of an unstable substance, show interconversion in such a way that the ring form predominates. This process needs some time.

The optical rotation of freshly prepared alpha-D glucose solution decreases and comes to a constant of +52.5. This is due to the formation of 2/3 of β-D glucose (optical rotation is +19.2) and 1/3 of α-D- Glucose (optical rotation is +112 in the open form, and when it passes back to the ring form, it gives rise to either anomer.

E. Anomers

Anomers are exceptional cases of epimers that differ in position **at the anomeric carbon**. For example, α-D-glucose and β-D-glucose. Exhibit mutarotation in aqueous solutions.

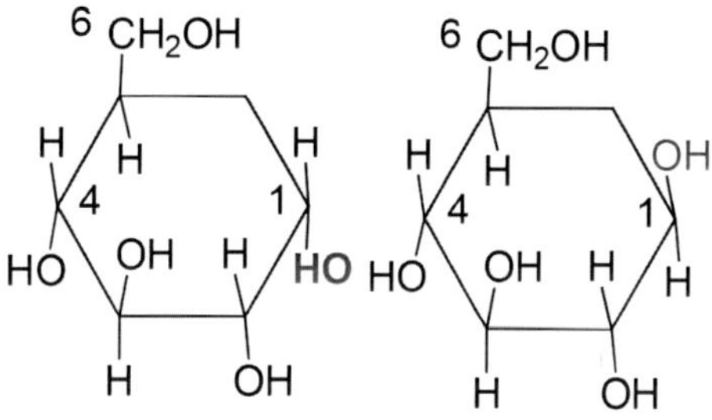

Alpha-D-Glucose Beta-D- Glucose

F. Isomers

Isomers are molecules with identical molecular formulas, having the same number of atoms of each element but different structural formulas.

Examples of fructose isomers are D-fructofuranose and D-fructpyranose.

G. Epimers

Epimers are isomers of carbohydrates with the same chemical formula but different structural arrangements. Examples are D-glucose and D-mannose. Epimers exhibit differences in spatial arrangement at only one chiral carbon.

	D-Glucose		D-Mannose
1	CHO	1	CHO
2	HO-C-H	2	**HO**-C-H
3	H-C-OH	3	HO-C-H
4	HO-C-H	4	H-C-OH
5	**OH**-C-OH	5	H-C-OH
6	CH2OH	6	CH2OH

D-Glucose D-Mannose

H. Diastereomers

They are stereoisomers that are not mirror images of each other. Diastereomers have the same molecular formula but differ in configuration at only one or multiple chiral centers. Examples of Diastereomer_1. Both L-erythrose(4C) and D-threose (4C), 2. α-D-glucose and β-D-glucose.

I. Enantiomers

Enantiomers are stereoisomers differing in configuration at every chiral center. They have the same formula, but the arrangement of atoms and groups is such that they are non-superimposable mirror images of each other. They have identical physical and chemical properties. D-glucose and L-glucose are enantiomers of glucose. When two enantiomers are present in equal proportions, the mixture is called a racemic mixture. The racemic mixture does not rotate polarized light due to the cancellation of the optical activity of one enantiomer by the other enantiomer.

II. The Aldose and Ketose Family tree

A. Aldehyde family tree

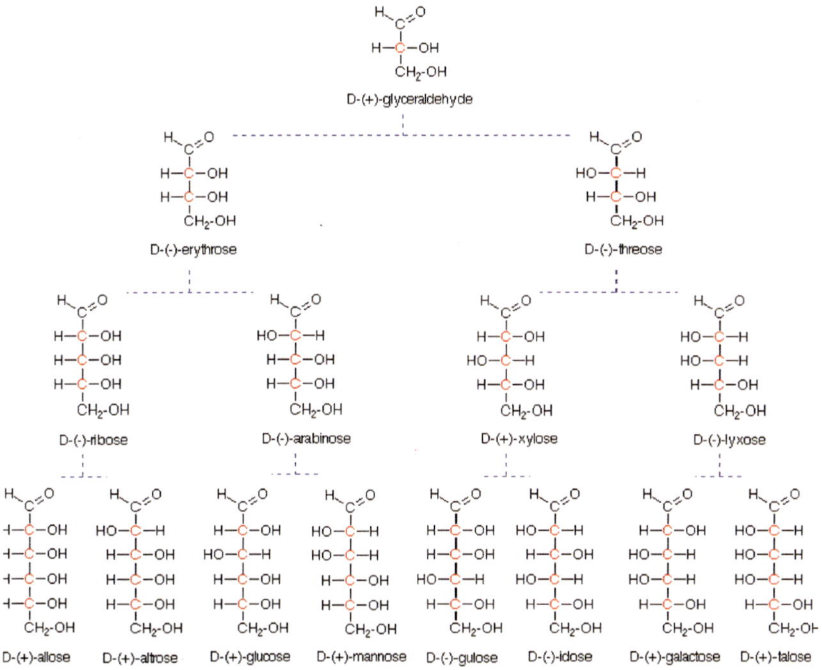

The L-family of aldoses are these structures' mirror images (enantiomers).

CHO	H-C-OH	CH_2OH — CHO	HO-C-H	CH_2OH D-glyceraldehyde L-glyceraldehy enantiomers	CHO	H-C-OH	H-C-OH	CH_2OH — CHO	HO-C-H	HO-C-H	CH_2OH D-erythrose L-erythrose enantiomers	CHO	HO-C-H	H-C-OH	CH_2OH — CHO	H-C-OH	HO-C-H	CH_2O D-threose L-ethreose enantiomers	CHO	H- C-OH	H-C-OH	H-C-OH	CH_2OH — CHO	HO- C-H	HO-C-H	HO-C-H	CH_2O D-Ribose L-Ribos enantiomers
CHO	H- C-OH	HO-C-H	H-C-OH	CH_2OH — CHO	HO- C-H	H-C-OH	HO-C-H	CH_2OH D-xylose L-xylose enantiomers	CHO	HO- C-H	H-C-OH	H-C-OH	CH_2OH — CHO	H- C-OH	HO-C-H	HO-C-H	CH_2OH D-Arabinose L-Arabino enantiomers	CHO	HO- C-H	HO-C-H	H-C-OH	CH_2OH — CHO	H- C-OH	H-C-OH	HO-C-H	CH_2O D-Lyxose L-Lyxos enantiomers	

Six Carbon Sugars

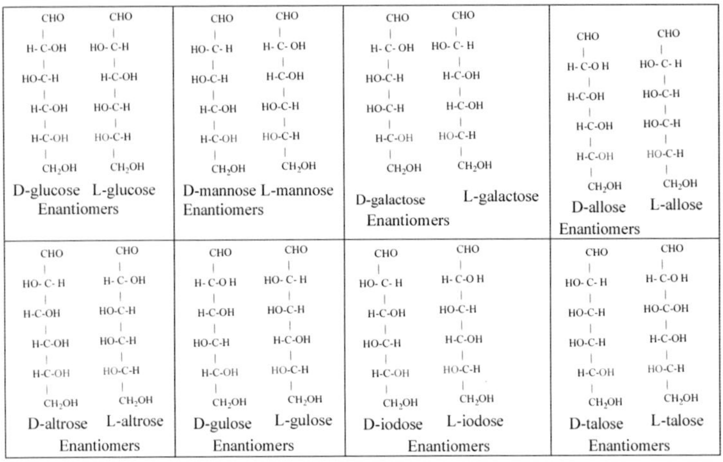

B. The Ketose Family tree

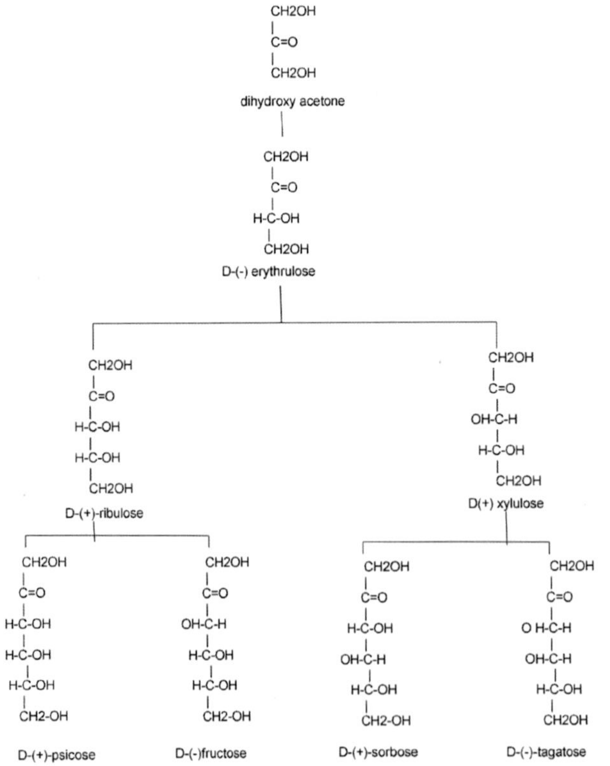

III. Important Functions of Carbohydrates

- They constitute a significant source of metabolic energy for different functions. Glucose is the primary sugar that provides energy.
- They are energy-reserved and stored as glycogen in the liver and muscles when in excess. They can then be broken down into glucose for future use. In plants, it is stored as starch.
- Sperms in seminal plasma use fructose as a source of energy.
- Acts as structural materials (e.g. chitin, xylans, cellulose, mannans).
- Cell recognition moieties. They are combined with proteins (glycoproteins) and lipids(glycolipids).
- The glycoproteins and glycolipids help in cell-to-cell recognition and signaling. Even through these structures, the cell interacts with its environment.
- Involved in blood groups (O, A, and B). The antigens, which determine blood types, belong to glycoproteins and glycolipids.
- All humans can synthesize the O antigen. Humans with A-type blood have N-acetylgalactosamine to the O antigen. Humans with B-type blood have galactose to the O antigen.
- Protective functions. Useful in the prevention of desiccation (bacterial). The membrane consists of peptidoglycan, a material made from 2 different polysaccharides - N-acetylmuramic acid (NAM) and N-acetylglucosamine (NAG).
- Chitin, comprised of many N-acetylglucosamines (NAG), is found in fungi cell walls.
- Cellulose, a complex sugar, is found in a plant cell wall. Soybean pectic polysaccharides induce protein synthesis that inhibits insect and microbial proteinases. Arabinoxylans have been found to inhibit intercellular ice formation and favor the winter survival of cereals.
- Useful in adherence of bacteria to surfaces. Involved in bacterial resistance to host defense.
- In cell-to-cell recognition.
- They are components of the energy transport compound ATP, recognition sites on cell surfaces,
- Carbohydrates are one of three essential components of DNA and RNA.

- Sugar derivatives are of biological significance, e.g., glucosamine (hyaluronic acid, heparin, and blood group substances) and galactosamine (in chondroproteins). They also create glycoproteins and proteoglycans.

2.2. Nomenclature and Classification of Carbohydrates

Carbohydrates can be considered from multiple angles:

I. Depending upon the complexity

A. **Simple Carbohydrates:** Monosaccharides

B. **Complex Carbohydrates:** Disaccharides, Oligosaccharides & Polysaccharides

II. Depending upon the number of sugar molecules

A. Monosaccharides

B. Oligosaccharides

C. Polysaccharides

III. Depending upon the reactivity of sugar

A. **Reducing sugars** – They reduce Tollen's reagents (or Benedict's or Fehling's). Sugars get oxidized.

B. **Nonreducing sugars**—They do not reduce Tollens or other reagents. Sugars not oxidized

IV. Depending upon the reactive groups present in the sugar

A. Aldehyde sugars- They have CHO Group.

B. Ketone Sugars-They have a C=O group.

2.3. Monosaccharides

Monosaccharides (mono=one, sakcharon=sugar) are the simplest form of carbohydrates that cannot be hydrolyzed further into simpler forms. They are, therefore, called simple sugars. They have two or more hydroxyl(-OH) groups with a free aldehyde group (-CHO) or ketone(=CO) group. All are reducing in nature. Their general formula is $C_n(H_2O)_n$ or $(CH_2O)_n$ or CnH_2nO_n.They are sweet in taste and soluble in water. Monosaccharides constitute the principal building blocks of complex carbohydrates.

They are sub-classified according to the following: 1. Number of carbon atoms they possess 2. Free ketone and aldehyde group they possess.3. Considering both criteria of the number of carbon and a reactive group

I. The monosaccharides Classification Depends Upon Carbon Atoms

Sugar Name	Carbon number	Formula	Examples
A.Dioses	2	$C_2H_4O_2$	Glycolaldehyde
B.Trioses	3	$C_3H_6O_3$	Glyceraldehyde
C.Tetrsoes	4	$C_4H_8O_4$	Erythrose, Threose
D.Pentoses	5	$C_4H_8O_4$	Arabinose, Ribose, Ribulose, Xylose, Xylulose, Lyxose
E.Hexoses	6	$C_6H_{12}O_6$	Allose, Altrose, Fructose, Galactose, Glucose, Gulose, Idose, Mannose, Sorbose, Talose, Tagatose
F..Heptoses	7	$C_7H_{14}O_7$	Sedoheptulose

II. The Monosaccharides Depending Upon Free Reactive Groups

A. **Aldoses.** They are sugars containing the aldehyde group(-CHO), e.g., glycerose, erythrose, ribose, and glucose. The suffix ***-ose*** is used for the aldoses.

B. **Ketoses** are those sugars containing the ketone group (-C=O), e.g., dihydroxyacetone, erythrulose, ribulose, fructose, and sedoheptulose. The suffix **- *'uloses'*** is used for ketoses.

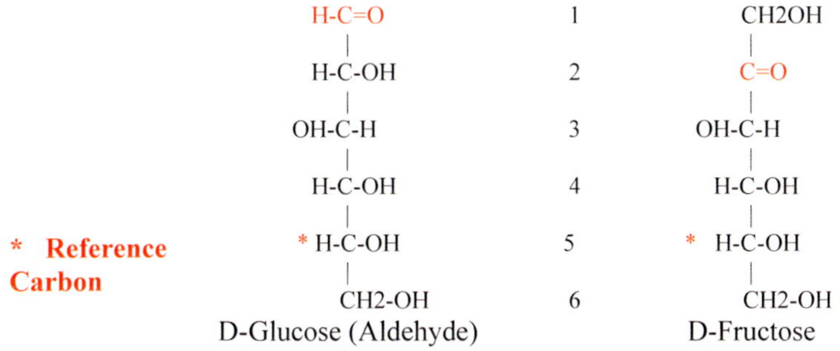

*The C5 atom in glucose molecules is called a reference carbon atom.

III. The Carbohydrate Classifications Considering the Number of Carbons and the Free Reactive Groups

Sugar and Carbon number	Aldehyde group sugars	Ketone group sugars
Dioses(2)	Glycoaldehyde	-
Trioses(3)	Aldotriose (glycerose or glyceraldehyde)	Ketotriose(dihydroxyacetone)
Tetroses(4)	Aldotetrose(erythrose, Threose)	Ketotetrose(erythrulose)

Pentoses(5)	Aldopentose(ribose, Arabinose,) Xylose, Lyxose)	Ketopentose(ribulose ,Xylulose)
Hexoses(6)	Aldohexose(glucose, Galactose,Mannose,Gulose, iodose,talose,altrose)	Ketohexose(fructose, Tagatose. Sorbose)
Heptoses(7)	Aldoheptose(glucoheptose, manoheptose,galactoheptose)	Ketoheptulose(sedoheptulose, perseulose, glucoheptulose and mannoheptulose)
Octoses(8)	Aldooctose(Glucooctose, mannoooctose,lactooctose)	
Nanoses(9)	Aldonanose(glucononose, mannononos)	
Decoses(10)	Aldodecoses(glucodecose)	

2.4. Oligosaccharides or Oligosacchroses or Low Molecular Weight Polysaccharides

The oligosaccharides (oligo=few in Greek) are carbohydrates containing two to ten monosaccharide molecules joined by glycosidic bonds. When the bond is broken down upon hydrolysis, monosaccharide molecules are released. The monosaccharide molecules can be the same or different. Depending upon the number of monosaccharide molecules, they are sub-grouped as follows:

I. **Disaccharides:** They are the carbohydrates (sugars) that contain two monosaccharide molecules. Examples are Maltose, Lactose, Sucrose, Gentiobiose, Melibiose, cellobiose, and trehalose. Their general formula is $C_n (H_2O)n-1$. If their reactive group(s) are free, they can react and are called ***reducing sugars***. If a reactive group(s) is not free, they cannot react and are called ***nonreducing sugars***. Sucrose is a nonreducing sugar.

II. **Trisaccharides:** They are carbohydrates (sugars) and contain three monosaccharide molecules. The examples are Rhamninose, Gentianose, Raffinose (=melitose), Rabinose, and Melezitose. Their general formula is $C_n(H_2O)_{n-2}$.

III. **Tetrasaccharides:** They are carbohydrates (sugars) that contain four monosaccharide molecules. Their general formula is $C_n(H_2O)_{n-3}$. Their examples are Stachyose and Scorodose.

IV. **Pentasaccharides:** They are carbohydrates(sugars) that contain five monosaccharide molecules. Verbascose. The general formula is $C_n(H_2O)_{n-4.}$

2.5. Polysaccharides or Polysacchrose (poly=many).

These carbohydrates contain monosaccharide units having more than 10 molecules, which can be recovered on their hydrolysis. Depending upon the same or different monosaccharide molecules, they are classified as;

I. **Homopolysaccharide.** They contain the same type of monosaccharide molecules, which get released due to the hydrolysis of polysaccharides. Starch (grain), Glycogen (liver, muscles), Inulin, Cellulose, Pectin, and Chitin

II. **Heteropolysaccharide.** Monosaccharide molecules released due to the hydrolysis of polysaccharides are of different types. Hyaluronic acid, Chondroitin

2.6. Isomers and Isomerism

Isomers are molecules with identical molecular formulas, having the same number of atoms of each element but different structural formulas (arrangements of atoms in space are different). This is possible in the case of many different carbohydrates, which will contain the same number of atoms and the same kind of groups. For example, formula $C_6H_{12}O_6$ or C_6 (H_2O) $_6$ represents 16 different simple sugars and possesses the structure $CH_2OH.(CHOH)_4CHOH$.

The phenomenon of the existence of isomers is known as isomerism. The isomers in broad groups are:

I. Structural (constitutional) isomers or constitutional

A. Chain isomers

B. Positional Isomers

C. Functional Isomers

D. Metamerism and metamers.

II. Stereoisomerism or spatial isomerism.

A. Enantiomerism by enantiomers (optical isomerism) and

B. Diastereomerism by Diastereomers. They are again of two types

1. conformational isomerism (conformers) and
2. cis-trans isomerism (geometric isomers).

I. Structural Isomers or Constitutional Isomers and Isomerism

They have the same molecular formula but differ in their **structures**. They have the same molecular formulae, but their physical and chemical properties differ. They can be split into three main subgroups as under:

A. Chain isomers, chain isomerism, nuclear isomerism, or skeletal isomerism: Chain isomers occur between two or more compounds having similar chemical formulas, but different arrangements of carbon atoms may be in straight or branched chains. This phenomenon is called chain isomerism. Examples

CH_3-CH-CH_3-CH_3 $\qquad$ CH_3-CH-CH_3

|

CH_3

Branched-chain $\qquad$ Straight chain

B. Positional Isomers: Compounds that have the same functional groups but are present on different positions of the chain. The basic carbon skeleton remains the same for example, butan-1-ol and butan-2-ol.

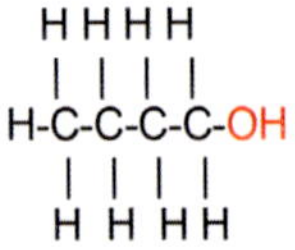

1-butanol

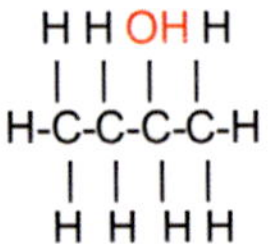

2-butanol

C. Functional Isomers: It is the type of isomer(structural) with the same molecular formula but different functional groups. For example, aldehydes (CH_3CH_2CHO OR C_3H_6O) and ketones (CH_3COCH_3 OR C_3H_6O) have the same molecular formula, but the functional group differs. The second example is ethanol and methoxy methane.

D. Metamerism and metamers: Metamers are isomeric compounds with the same molecular formula but different numbers of carbon atoms around the functional group. The alkyl groups are on two sides of the functional groups. For example, C_4H_{10} O is for diethyl ether and methyl propyl ether.

$C_2H_5O.C_2H_5$ $\qquad$ $CH_3O.CH_2CH_2CH_3$

Diethyl ether $\qquad$ Methyl propyl ether.

II. Stereoisomers or Spatial Isomers and Isomerism

When two or more compounds with the same molecular and structural formula but having different arrangements of their atoms in space (configuration) are stereoisomers, the phenomenon is known as stereoisomerism. The meaning of the word stereo is "three-dimensionality."

This includes

A. Enantiomerism by enantiomers (optical isomerism) and

B. Diastereomerism by Diastereomers. They are again of two types

1. conformational isomerism (conformers) and 2. cis-trans isomerism (geometric isomers).

A. **Enantiomers or optical isomerism.** This word has the Greek origin, enantiosis, meaning "opposite." The enantiomers are molecules that are mirror images of each other and are non-superimposable (non-identical). The epimers are enantiomers if there is only one stereocenter in the molecule. They rotate the plane-polarized light in different ways right and left sides). Enantiomers have similar physical properties but may exhibit different chemical properties.

A chiral carbon is that carbon which is characterized by the presence of four different groups attached to it. Only chiral molecules will have a non-superimposable mirror image. If two molecules are enantiomers, all chiral centers must have opposite configurations. The D and L forms are enantiomers or enantiomorphs. The majority of monosaccharides occur in the D-form in animals and the human body. However, sugars like arabinose and penta aldose occur in L-form in nature. The compound having asymmetric carbon atoms shows optical activity. A compound designated as d or (+) is a dextrorotatory compound. Such compounds rotate the plane of polarized light in the right or clockwise direction. However, the compound that rotates the plane of polarized light to the left or anticlockwise is called levorotatory and is designated as l or (-) type.

Do not confuse between D and L, d and l, and (+) and (-) letters used for sugars.

The number of asymmetric carbons present determines the number of stereoisomers. The relation between the number of stereoisomers and the number of asymmetric carbons is explained by the Le Bel-van't Hoff rule. According to this rule, the number of stereoisomers of a sugar is equal to 2^n, where n is the number of asymmetric carbons. Glucose has four asymmetric carbons, and the number of stereoisomers of glucose is 2x2x2x2=16. D-glucose is the enantiomer of L-glucose. Glyceraldehyde is the simplest aldose sugar, having one chiral carbon at the center, and so it has two enantiomeric forms, e.g., "D" and "L" glyceraldehyde.

H-C=O	1	H-C=O
|		|
H-C-OH	2	OH-C-H
|		|
CH_2OH	3	CH_2OH
D-glyceric aldehyde		L-glyceric aldehyde

German chemist Emil Fischer (1852-1919) identified the stereoisomers for aldohexoses in 1894. These sugars are Allose, Glucose, Mannose, Altrose, Glucose, Idose, Galactose and Talose. He received the Nobel Prize (1902) for chemistry for his work.

B. Diastereoisomers

Diastereomers are non-superimposable, non-mirror images of each other, whereas enantiomers are non-superimposable mirror images of each other. They are optical isomers that differ in configuration at more than one chiral carbon center. Any stereoisomer that is not an enantiomer is considered to be a diastereomer. They vary in their physical properties (melting points, boiling points, and densities). D-mannose and D-idose are diastereomers at positions 2 and 3.

The term epimer is used to designate diastereomers that differ in configuration at only one chiral center. Thus, ribose and arabinose are epimers at C-2, and D-glucose and D-galactose are diastereomers about position 4.

1. Conformational diastereomers: They are different conformers of a diastereomer molecule. They differ only in how groups are rotated about single bonds. The conformation of a given molecule will change constantly. There are an infinite number of possible rotations. The conformational isomer molecule does not have to be chiral or contain chiral centers. The confirmation may be eclipsed (less stable) or staggered (more stable). The term staggered means that the angle between a hydrogen on C1 and a hydrogen on C2 is 60 degrees, while eclipsed means this angle is 0.

i. Anomers

They are diastereomers (stereoisomers) where the molecule is cyclized, and the difference in configuration is only about the anomeric carbon. (In the case of aldoses, the anomeric carbon is C1; for ketoses, it is C2. e.g. sugar hemiacetal).

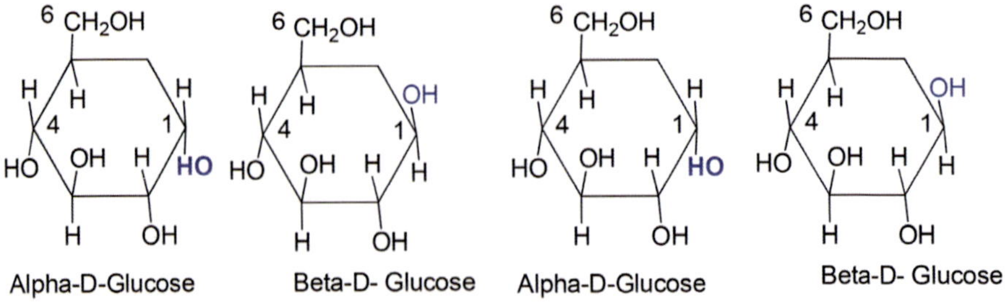

Glucose in open-chain form is not chiral at C1. But ring-form glucose has two optically active stereoisomers: ***alpha & beta*** glucose. These alpha and beta D sugars are known as anomers of one another, and the carbon is ***anomeric***. Alpha sugar has the presence of the OH group on the left side of the carbonyl carbon, while beta sugar has the presence of the OH group on the right side of the carbonyl carbon. Anomers are diastereomers (differ in their physical properties).

ii. Epimers

They are a particular case of diastereoisomerism. The epimers are enantiomers if there is only one stereocenter in the molecule. But if there are two or more stereocenters in the molecule, the epimers are **diastereoisomers**.

Enantiomers are always optically active, and their structural arrangement is a non-superimposable mirror image of each other. They are classified as dextrorotatory or levorotatory. Common examples of enantiomers are D-glucose and L-glucose.

Sugars differing from one another in the arrangement of groups on the carbon other than the **carbonyl carbon** are known as epimers of one another, e.g., Glucose and mannose are the epimers of one another at carbon 2. In contrast, glucose and galactose are the epimers of one another at carbon 4. ***D***-idose & ***L***-glucose (epimeric at C5) differ in configuration at only one position.

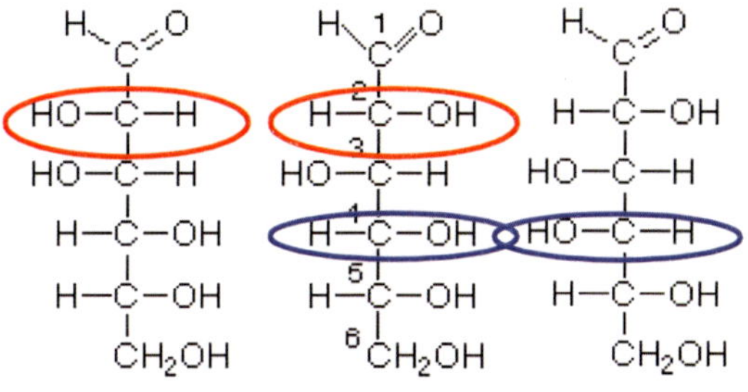

D-Mannose D-Glucose D-Galactose

Epimerization is important as it converts the primary sugar forms into their epimers, which are helpful in various metabolic and signaling pathways. It is also in the study of metabolism and drug development.

2. Geometric Isomers or cis-trans isomers

The Cis-trans (cis=same side: trans=opposite) isomers are also known as geometric isomers. Cis-trans isomerism is common in compounds with double bonds and in cyclic forms. The isomerism is due to different spatial arrangements of groups, which are isomeric about doubly bonded carbon atoms. In the cis isomer, similar groups lie on the same side, and in the trans isomer, the similar groups lie on opposite sides. It is also called cis-trans isomerism.

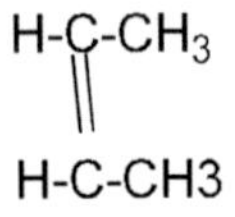

Cis 2 Butene

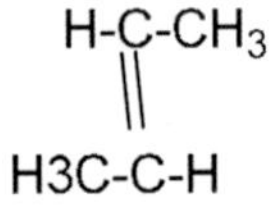

Trans 2 butene

These isomers differ in the physical and chemical properties of the geometrical isomers. The cis/trans nomenclature describes the geometric isomerism of alkenes.

2.7. Racemic Mixture or Racemic Modification or Dl-Mixture or (±) Conglomerate

It is known that the dextrorotatory and levorotatory enantiomers or substances show opposite optical rotation. When they are mixed in equal amounts, a racemic mixture is formed. This mixture is optically inactive as each one of them cancels the optical rotation of the other.

Key Takeaway

- The storage form of carbohydrates is starch in plants and glycogen (mainly in muscles and the liver) in animals. They are also found in skeletal structures of cell walls in plants (cellulose), bacteria (peptidoglycans), insects, and crustaceans.
- Carbohydrates containing elements Carbon, Hydrogen, and Oxygen in a ratio of 1:2:1, like water. They are represented as $(CH_2O)n$, except for a few exceptions.
- There are two main functional groups in carbohydrates: the carbonyl group (aldehyde or ketone) and several hydroxyl groups.
- One gram of sugar provides four calories. The sweetness differs with various carbohydrates.
- A carbon with four different atoms or groups attached is asymmetric or chiral. A Carbon with all 4 same atoms or groups attached is a symmetric carbon.
- Tautomers differ in chemical and physical properties. Keto form (more stable) and enol form (less stable) of chemical compounds are readily interconvertible.

- The alkaline medium is more favorable for the reducing action of the sugars.
- **Enantiomers** are **stereoisomers** differing in configuration at every chiral center. They have the same formula, but the arrangement of atoms and groups is such that they are **non-superimposable mirror images** of each other. They have identical physical and chemical properties.
- Involved in blood groups (O, A, and B). The antigens, which determine blood types, belong to glycoproteins and glycolipids.
- Cellulose, a complex sugar, is found in a plant cell wall. Soybean pectic polysaccharides induce protein synthesis that inhibits insect and microbial proteinases. Arabinoxylans have been found to inhibit intercellular ice formation and favor the winter survival of cereals.
- When two or more compounds with the same molecular and structural formula but having different arrangements of their atoms in the space (configuration) are stereoisomers, the phenomenon is known as stereoisomerism.
- The number of stereoisomers of sugar is equal to 2^n, where n is the number of asymmetric carbons.

Intext Questions

1. What is a chiral carbon? How many chiral carbons are present in glucose?
2. What are hydrates of carbon? Define the term "carbohydrates"?
3. What are tautomers and tautomerism?
4. What is a mutarotation.?
5. What are isomers and epimers?
6. Name the sugar with two carbon atoms.
7. Name the groups of the monosaccharides depending upon their free reactive groups.
8. How do dioses differ from disaccharides?
9. What are Homopolysaccharides and Heteropolysaccharides? Give an example of each.
10. Enlist different types of isomers.
11. How does an enantiomer differ from a diastereomer?
12. How do anomers and epimers differ from each other?

13. What difference is observed between the cis-isomer and the trans-isomer?

14. What happens to the optical property of a mixture containing equal amounts of dextrorotatory and levorotatory isomers? Why?

15. Select the most appropriate answer for the following (The answer is given in **Bold** letters:

a. The Sweetness of fructose relative to sucrose is

i. High ii. Low

iii. Same

b. Milk sugar sweetness relative to sucrose is

i. High **ii. Low**

iii. Same

c. Glucose and Fructose have

i. Different structural arrangements of atoms.

ii. Same molecular formulae.

iii. Both i. and ii. are correct

d. The pH of a medium that is more favorable for reducing the action of sugars

i. Acidic pH **ii. Alkaline pH**

iii. Neutral pH

e. The alpha D-glucose and beta D-glucose structurally differ at

i. **One carbon** ii. Two carbons

iii. All carbons

f. Xylulose and sorbose have the following carbon numbers:

i. **5 and 6 carbons, respectively** ii. 6 and 5 carbons, respectively

ii. 6 and 7 carbons, respectively

g. The number of carbon(s) in sedoheptulose is

i. Five ii. Six

iii. **Seven**

h. The example(s) of Ketohexose

   i. **Fructose** ii. Galactose

   iii. Mannose

i. The general formula $C_n(H_2O)_{n-2}$ stands for

   i. Disaccharides **ii. Trisaccharides**

   iii. Pentasaccharides

i. What is not valid for stereoisomers

   i. **Different molecular formula**

   ii. Same structural formula

   iii. Have different arrangements of their atoms in space

j. Which one is not correct for glyceraldehyde?

   i. Is aldose sugar **ii. Is ketose sugar**

   iii. It has two enantiomers.

k. Glucose and galactose are the epimers of one another at the carbon number

   i. Three **ii. Four**

   iii. Five

Further Readings

A Textbook of Biochemistry. Berry, A. K. (2009), sixth edition, Emkay Publishers, Delhi.

Biochemistry. Naik, Pankaj (2009). Second Edition, Jaypee Brothers, New Delhi.

Fundamentals of biochemistry. J.L.Jain, Sunjay Jain and Nitin Jain.

https://chemistry.stackexchange.com/questions/74853/are-glucose-and-galactose-cis-trans-isomers-of-each-other

https://www.chemistrysteps.com/aldoses-and-ketoses-classification-and-stereochemistry/

Life sciences. Fundamentals and practices I. sixth edition. Pranav Kumar and Usha Mina. (2017). Pathfinder publication. New Delhi

Objective Biochemistry. R.S.Senger and Reshu Chaudhary. (2014). New India publishing agency. New Delhi.

Textbook of Biochemistry for Medical Students. Vasudevan, D. M.; Sreekumar

Textbook of Veterinary Biochemistry. Dhanotiya, R.S. (2006) Second edition, reprint. Jaypee Brothers, medical publishers (P) lid. New Delhi. India

3

Monosaccharides

Hydrates carbons (saccharides) found in nature (animals and plants) are classified into three groups based on the number of monosaccharides found or released upon acid hydrolysis. Monosaccharides the groups of carbohydrates.

3.1 Formulation of Monosaccharides

The molecular formula of any monosaccharide (mono=one and saccharide=sugar) is $C_n(H_2O)_n$ or $(CH_2O)_n$. Rhamnose and Deoxyribose are exceptions to the general formula of carbohydrates. Acetic acid ($C_2H_4O_2$) and formaldehyde (CH_2O) are not carbohydrates but have the formula $C_n(H2O)_n$ or $(CH2O)_n$. **Fitting and Baeyer** proposed the formula $CHO\text{-}(CHOH)_4\text{-}CH2OH$, which indicates the presence of aldehyde and five hydroxyl groups.

More than one monosaccharide molecule unites to form disaccharides, oligosaccharides, and polysaccharides. Monosaccharides, being simpler units, cannot be broken down further. Generally, they have two to ten carbon atoms per molecule (dioses to decoses). The monosaccharides may have either aldehyde or ketone groups. The one possible aldodiose ($CH_2OH.CHO$) is glycolaldehyde (2-hydroxyethanal), but ketodiose is not possible since there are only two carbons. The glucose molecular formula $C_6H_{12}O_6$ represents the number of C, H, and O atoms but not groups. However, the proposed formula, $CHO\text{-}(CHOH)_4\text{-}CH_2OH$, indicates the presence of an aldehyde and five hydroxyl groups. These simple sugars or monosaccharides have carbon backbones linked by single bonds with hydrogen and hydroxyl groups attached in an unbranched manner.

Commonly occurring sugars can exist in linear (straight) or ring forms. About a small proportion (10%) of the total sugars occur in straight chains, whereas a large proportion of sugars occur in closed ring form. The 3D arrangement of atoms in carbohydrates is illustrated in two ways: Fischer projection open-chain (acyclic) form and Haworth closed-chain (cyclic) form.

I. Linear form or Fischer (Hermann Emil Fischer) projection formula

This kind of linear formula gives more information, such as the presence of asymmetric carbon atoms (4 for glucose) in the molecule. It indicates the existence of 2^n number of stereoisomers ($2^4 = 16$ for glucose). The structures that Fischer proposed represent a very minor percentage of the forms that hexose sugars adopt in solution. The orientation of the carbon chain is so that the C1-carbon is at the top. The Fischer projection formula of a few monosaccharides is given below:

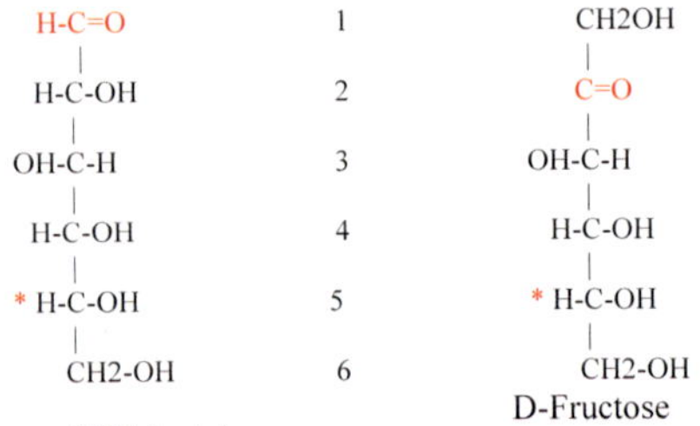

D-Fructose

D-Glucose (Aldehyde)

* **Reference Carbon**

An open ring form of glucose was developed to present alpha-D-glucose and beta-D-glucose. He was awarded the Nobel Prize (1902) in Chemistry for his work in determining the structure of the D-aldohexoses.

D and L Sugars

Fischer also developed the D/L system to specify the structures of sugars. Glucose has one primary alcoholic group, which is at the sixth carbon atom. The sugars are designated as D or L sugar depending upon the distribution of H and OH around the carbon atom preceding the carbon atom containing the primary alcohol group of the sugar molecule. The D sugar will have H on the left side of the plane and OH on the right side. The L sugar has H on the right side of the plane and OH on the left side. The L sugar is the mirror image of the D sugar.

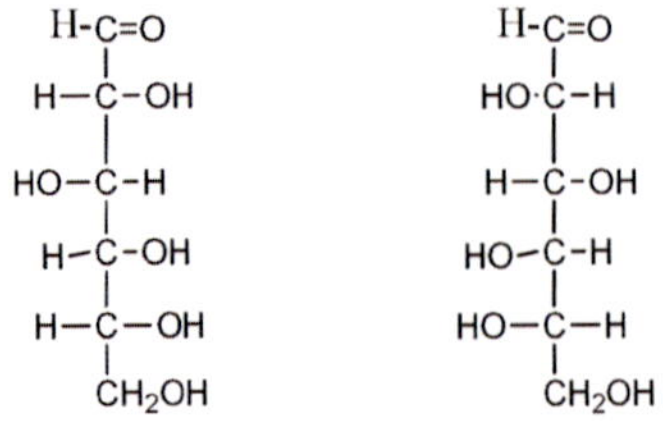

D-GLUCOSE L-GLUCOSE

The proposed sugar ring structure can result in the formation of a structure known as a pyranose ring, containing five carbons, or a furanose ring, containing four carbons.

PYRANOSE

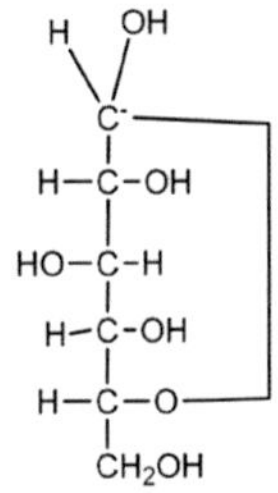

ALPHA- D- GLUCOSE PYRANOSE
(Fischer projection)

FURANOSE

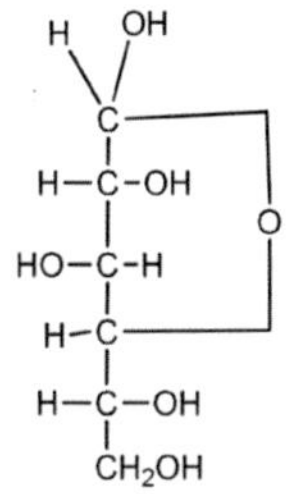

ALPHA- D- GLUCOSEFURANOSE
(Fischer projection)

Thermodynamically, the pyranose form is more stable than the Furanose form. Furanose rings are more flexible than pyranose rings and may be a component of DNA and RNA. The glucose ring form will produce two isomeric glucose forms, alpha-D-glucose and beta-D-glucose. These isomeric forms in solution change to the same equilibrium mixture on standing when sugar is dissolved in water to make its solution. Due to ring formation, the C_1, instead of being CHO, becomes HCOH or HOCH. They are called 'anomers'. As the Fischer formula failed to explain the other properties of Glucose (fails to react with Schiff reagent, mutarotation, etc.), he proposed a ring structure that can explain other properties satisfactorily. In many reactions, the aldehyde sugars do not behave as aldehydes but as hemiacetals.

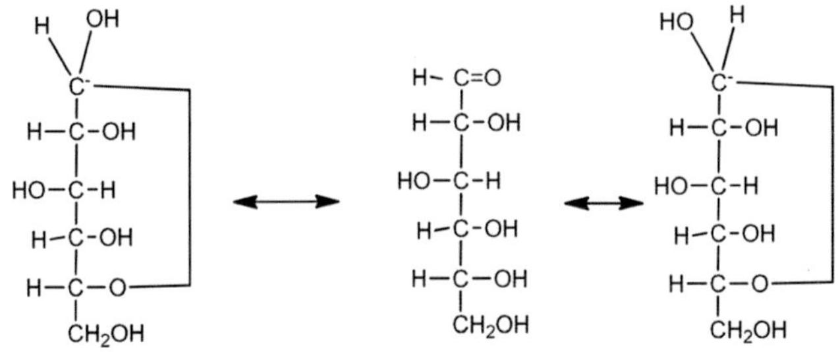

ALPHA- D- GLUCOSE
(Fischer projection)

ALDEHYDE FORM
CHAIN FORM OF
D- GLUCOSE

BETA-D-GLUCOSE
(Fischer projection)

The **aldehyde group** of a sugar can form hemiacetals or furanose. The Aldehyde group at carbon one can react with the hydroxyl group of alcohols of carbon 5 of sugar or 4, forming hemiacetals. A furanose is formed if the reaction is between the C-4 hydroxyl and the aldehyde.

The **ketone group** of a sugar forms hemiketals when the ketone group reacts with the alcohol group present at carbon 5 or 4. When the ketone group (carbonyl group) of fructose reacts with the hydroxyl group present at carbon 5, a five-membered ring resembling furan is formed and is called D-fructofuranose. Furanoses can exist in both alpha and beta forms. Here, carbon C-2 is an anomeric carbon atom.

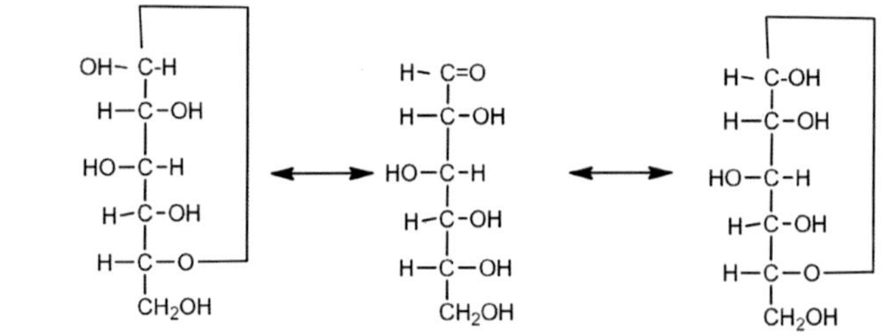

BETA D-GLUCOSPYRANOSE (cyclic Fischer projection)

ALDEHYDE FORM CHAIN FORM OF D- GLUCOSE

ALPHA D- GLUCOPYRANOSE (Cyclic Fischer projection)

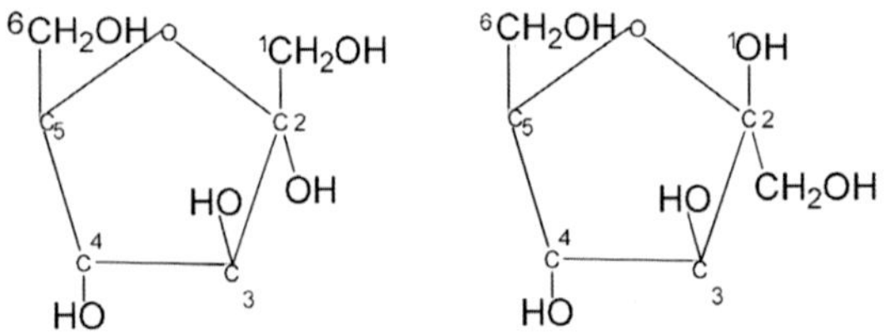

Alpha-D-fructose(furanose) Beta-D-fructose(furanose)

II Haworth Perspective Formula

Walter Norman Haworth (awarded the Nobel Prize) proposed a structure resembling a hexagon (pyranose form) or a pentagon (Furanose form). A Haworth projection represents a simple three-dimensional cyclic structure of monosaccharides.

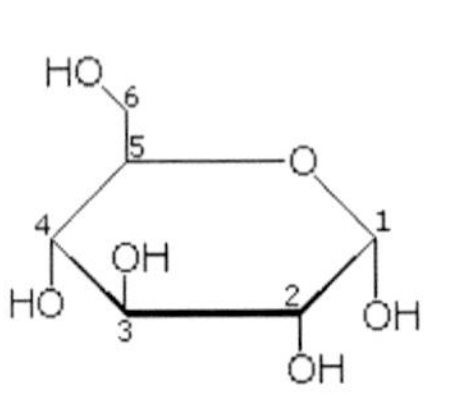

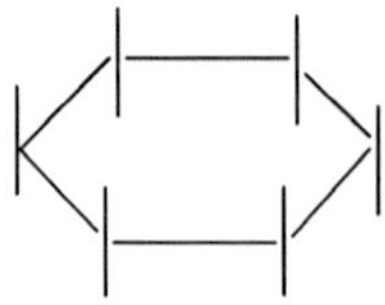

Haworth projection lines

In the above figure, a thicker line indicates atoms closer to the observer (atoms 2 and 3 -and their corresponding OH groups), atoms further from the observer are 1 and 4, compared to the remaining atoms, 5, etc, which are the furthest. Haworth drew the ring as a flat hexagon with groups above and below the ring's plane. Haworth projection lines are shown in the ring structure.

The Haworth developed a hexagonal ring resembling the pyran, containing five carbon atoms and one oxygen in the ring. From this, names like alpha-D-glucopyranose were given for the hexagonal structure of alpha-D-glucose, and beta-D-glucopyranose given for the hexagonal structure of beta-D-glucose.

Walter Haworth conclusively determined that the hexose sugars preferentially form a ***pyranose***, or six-membered, ring. The ring form is favored in aqueous solutions, and the mechanism of ring formation is similar for most sugars. The pyranose form, compared to furanose form, is thermodynamically more stable.

The cyclic structure of saccharides can also be represented as chair and boat conformations**.** The carbon 1 is always anomeric or hemiacetal. The chair conformation has the lowest energy. On the other hand, the boat conformation has higher energy than the chair form. The chair form predominates. The stable chair form of Beta d-glucose is given below:

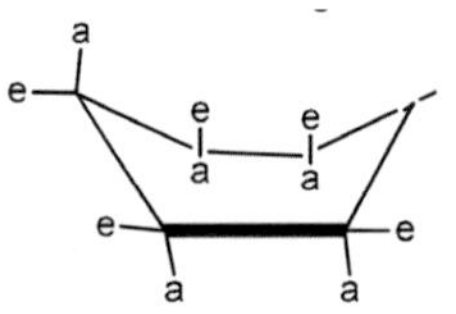

Boat form

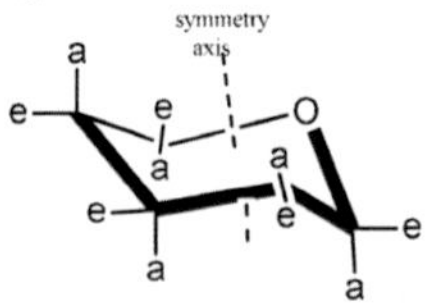

Chair form

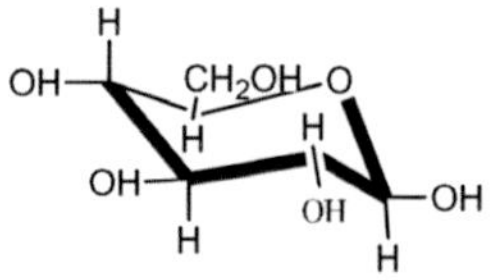

Chair form
of B D-glucose

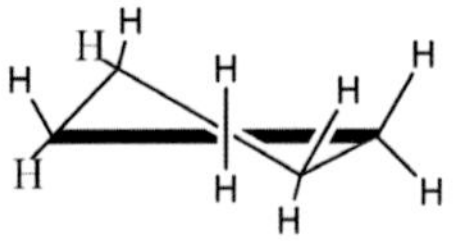

Half chair conformation

The pyranose ring is puckered (a flexible ring can adopt different conformations), leading to 38 distinct basic pyranose conformation structures: two chair forms, six boat forms, six skew boat forms, 12 half-chair forms, and 12 envelope forms. Furanose rings may be a component of DNA and RNA as they are more flexible than pyranose rings.

Haworth projection presents the cyclic structure with α- or β- forms. The isomer's structure differs in terms of the OH-group arrangement attached to the ring structure's anomeric carbon. In the case of α- form, the OH- group is under the ring structure, and in the case of β-isomer, it is on top of the ring structure.

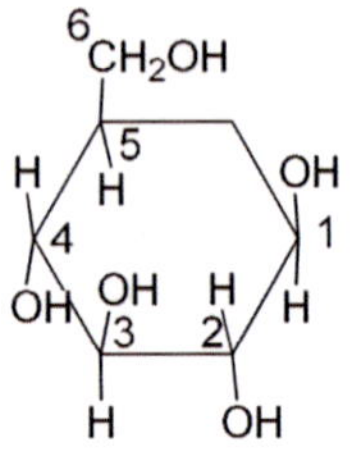

Alpha-D-Glucopyranose
Haworth projection

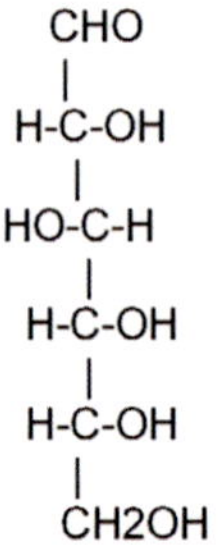

Glucose

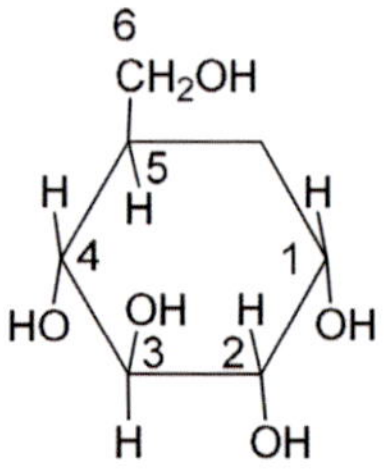

Beta-D- GlucoPyranose
Haworth projection

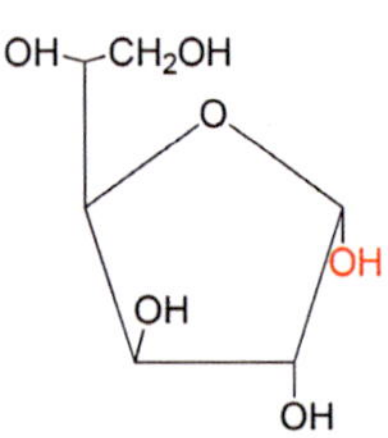

Alpha-D-glucofuranose

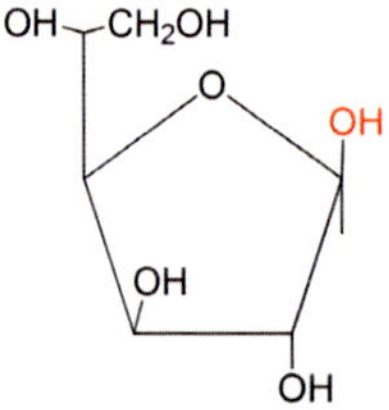

Beta-D-glucofuranose

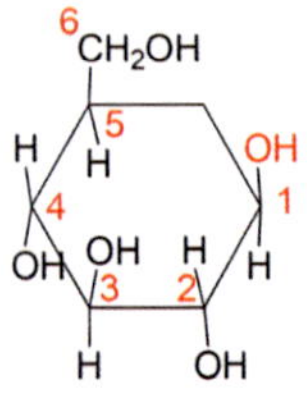

Alpha-D-Glucopyranose
Haworth projection

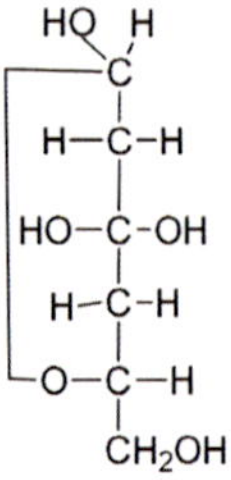

Alpha-L-glucopyranose

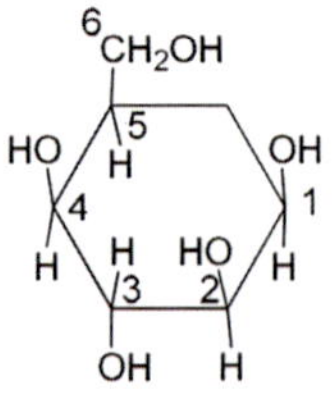

Alpha-L-Glucopyranose
Haworth projection

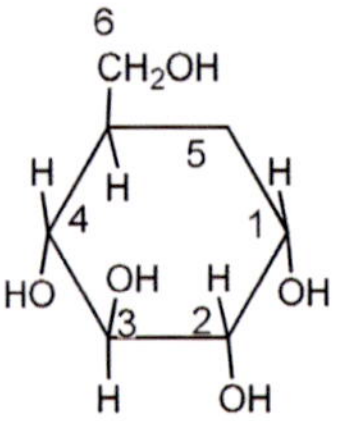

Beta-D- GlucoPyranose
Haworth projection

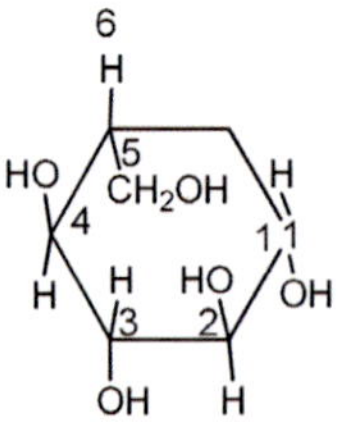

Beta-L- GlucoPyranose
Haworth projection

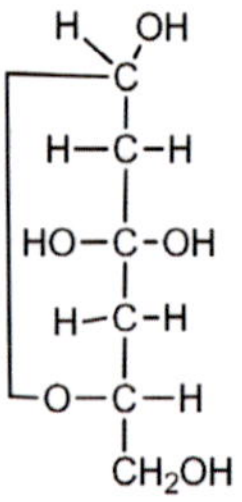

Beta-L-glucopyranose

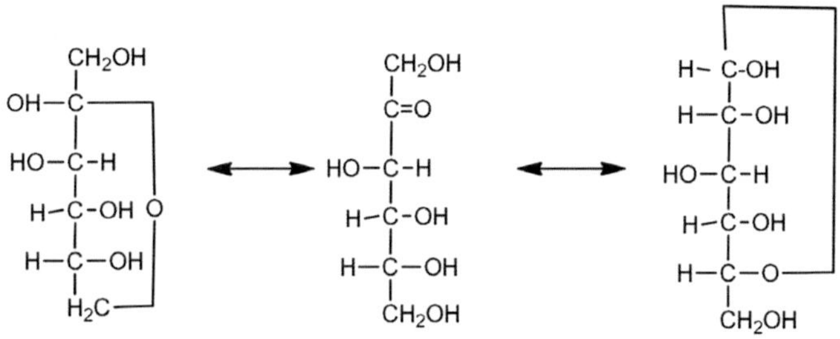

Beta D-fructose or Beta D-fructopyranose six membered ring

D-fructose ketoform

5 membered ring Beta-D-fructofuranose

3.2 General Properties/Reactions of Monosaccharides

The 2C-containing diose is an aldehyde, e.g., glycolaldehyde (CH_2OHCHO or 2-hydroxyethanol)). A ketodiose is not possible due to two carbons. It does not meet the definition of monosaccharides as the chemical formula of an unmodified monosaccharide is $(C_nH_2O)_n$, where n is a value of three or greater. It is (glycolaldehyde) sometimes considered the most basic sugar.

The smallest monosaccharide is **glyceraldehyde**, a triose sugar having one chiral center. Monosaccharides of five, six, or seven carbons are often more stable in aqueous solution as cyclic structures than open chain. Cyclic monosaccharides form spontaneously through a covalent linkage of the carbonyl carbon with a hydroxyl group in the carbon backbone. This bond results from a reaction between an alcohol group and an aldehyde group of an aldose sugar to form a **hemiacetal** or between an alcohol group and a ketone group of a ketose sugar to create a **hemiketal**.

I Physical Properties of Monosaccharides

Monosaccharides are colorless solids with a sweet taste. Their solubility depends on the type of solvents. It is higher in water, sparingly in alcohol, and is not soluble in ether. Their asymmetric carbon atoms create different isomeric forms. Monosaccharides are reducing in nature.

II Important Chemical Reactions of Monosaccharides Due to Aldehyde, Hydroxyl, and Ketone Groups/ Derived Sugars/Derivatives of Sugars

Monosaccharides exhibit various reactions due to the presence of -CHO, =CO, and OH groups. Reduction or oxidation reactions are due to the carbonyl groups, and esterification reactions are due to their hydroxyl groups. The important reactions of monosaccharides are as follows.

A. Formation of Iodo Compounds

After heating with concentrated HI (hydroiodic acid), an aldose sugar loses its oxygen and is converted into an iodo compound, e.g., Glucose to iodohexane($C_6H_{13}I$).

B. Ester Formation

Alcoholic groups form esters with acids. Important esters with phosphoric acid are glucose-1-phosphate and glucose-6-phosphate, which are formed as important intermediates during metabolism in the body. Pentose phosphates are involved in the formation of nucleic acids.

```
                        OH                 OH
  HCOH                  |                  |
   |          +  HO-P=O     →     HCO-P=O
   R                    |                  |    |
                        OH                 R   OH
D-glucose     Phosphoric          D-glucose-
              acid                1-Phosphate
```

The acetate esters from cellulose can be spun into rayon-like fibers (fire resistant).

C. Dehydration by Strong Acid or Furfural Formation

Concentrated sulphuric acid / strong hydrochloric acid, removes water molecules (dehydration), and forms furfurals. Pentoses form furfural, while the hexoses form hydroxymethyl furfural (HMF). Furfural formation from glucose must undergo three processes, e.g., isomerization, decarboxylation, and dehydration. Therefore, furfural yield from substrate glucose is generally lower than that of HMF yield.

This reaction forms the basis for a general test known as Molisch's test for identifying carbohydrates. Similarly, the Anthrone reagent (prepared in sulphuric acid) is treated with glucose in the Anthrone reaction. It produces a green color due to the furfural derivative.

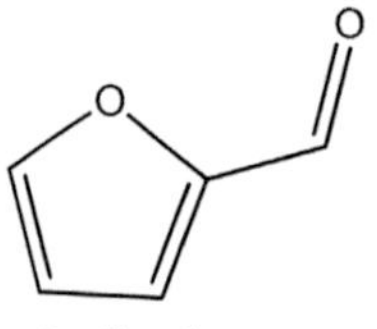

furfural

D. Acetylation

The reaction with acetylcholine indicates the presence of the -OH group in the sugar. Glucose has 5 OH groups, which get replaced by the acetate group and form Penta acetate.

$$\begin{array}{l} CHO \\ | \\ (CH_2OH)4 \\ | \\ CH_2OH \\ \text{Glucose} \end{array} + \begin{array}{l} 5CH_3COCL \\ \text{Acetic acid} \end{array} \xrightarrow{ZnCl_2} \begin{array}{l} CHO \\ | \\ (COOCCH_3)_4 \\ | \\ CH3OOCCH_3 \\ \text{Glucose penta acetate} \end{array} + HCL$$

It is insoluble in water. It is soluble in chloroform and methanol. Its chemical stability is very high, and it has a long shelf life.

E. Methylation or Ether Formation

Methylating agents, when they react with monosaccharides, yield methyl glycosides.

C6H12O5OH + CH3OH ⟶ C6H11OCH3 + H2O
Methanol Dry HCl Methyl glycoside

F. Fermentation

Monosaccharides like glucose and fructose can be fermented by microorganisms (yeast) to produce ethanol and CO_2.

G. Oxidation

The Aldehyde group, upon oxidation, forms **'aldonic acids'** where carbon is first converted from CHO to COOH. When the last carbon (C6) CHOH is converted to COOH, then ***'uronic acids'*** are formed (with C1 containing CHO group). A **'saccharic acid'** is formed when C1 (containing CHO) and C6 (containing CHOH) carbons are converted to COOH.

1. Reaction with Mild Oxidizing Agents

When monosaccharides are treated with ***mild oxidizing agents*** like 1. Tollen's reagent (ammoniacal Ag_2O)/hypobromous acid or 2. Fehling solution (alkaline $CuSO_4$) oxidizes at carbon number one, forming ***aldonic acids*** (CHO group to COOH GROUP). Glucose forms gluconic acid. The Tollens' reagent will only act on the open-chain form in mannose and galactose. Mannose forms mannoic acid, and galactose forms galactonic acid.

The Reduction of *Tollen's reagent* yields silver, as Polishing on the Surface of the Tube

CHO COOH
| |
(CHOH)4 $+Ag_2O$ (CHOH)4 + 2Ag
| Tollen's reagent | silver mirror
CH2OH CH2OH

Glucose Gluconic acid

Tollen's reagent, an alkaline solution of ammoniacal silver nitrate, detects the presence of aldehyde or α–hydroxy ketone. Either of these groups and the silver ions are reduced to metallic forms, which get deposited on the sides of the test tube._Both glucose and fructose are **reducing sugars**, and they reduce Tollen's reagent to metallic silver.

Alpha hydroxy ketone gets converted into an enol by tautomerism and then eventually turns into an aldehyde group because of tautomerism. And that aldehyde group gives a positive test with Fehling's solution.

Normal Ketones do not react with Tollens' reagent as they lack reducible hydrogen, or a ketone does not have an H atom directly bonded to a carbonyl group.

CHO COOH
| |
(CHOH)4 +2Cu(OH)2 $(CHOH)_4$ $+Cu_2O+H_2O$
| Fehling solution |
CH2OH CH2OH

Glucose Gluconic acid

The reaction with mild oxidants, such as HOBr, produces gluconic acid. As ketoses do not show such a reaction, this reaction can differentiate aldoses from ketoses.

2. Slow Oxidation Reaction

When the last carbon (CHOH) of monosaccharides gets oxidized to form COOH under special conditions, uronic acids are formed. In the animal body, the primary alcohol group on carbon six (CHOH) gets oxidized to a carboxylic group (COOH) to yield ***glucuronic acid.*** Mannose yields mannuronic acid. Glucuronic acid in the body combines with some toxic substances, and the conjugation products are excreted. It is also found as a component in hyaluronic acid, heparin, chondroitin sulfate, and mucoitin sulfate, which are found in blood, skin, and cartilage.

CHO → CHO

| |

(CHOH)4 + oxidation ⟶ (CHOH)4

| |

CH2OH COOH

Glucose Glucuronic acid

D-galacturonic acid is a constituent of pectin and many plant gums and mucilages.

3. Strong Oxidation Reaction

Under ***strong oxidation*** conditions using conc. HNO3, the first (CHO), and the sixth (CHOH) OH are oxidized to COOH, and ***saccharic acid*** is formed. Glucose forms glucosaccharic acid, mannose to mannaric acid, and galactose to mucic acid. Mucic acid forms crystals, which are used for the identification of galactose.

CHO COOH

| |

(CHOH)4 +Conc.HNO3 (CHOH)4

| |

CH2OH COOH

Glucose Glucaric acid

(Saccharic acid)

H Reduction Reactions

1. Reaction with Sodium/Mercury Amalgam and Water to form *sugar alcohols*

The monosaccharides are converted to corresponding alcohols by reducing agents like sodium amalgam. This reaction helps to form corresponding sugar alcohols. Both aldoses and ketoses may be reduced to their polyhydroxy alcohols. A few examples of sugar alcohols are given below.

Sugar	**Alcohol**	**Sugar**	**Alcohol**
Glucose	Sorbitol	Fructose	Sorbitol and Mannitol
Mannose	Mannitol	Ribose	Ribitol
Glyceraldehyde	Glycerol	Galactose	Dulcitol or galactitol
Erythrose	Erythritol		

Ketoses form two alcohols because of the asymmetric carbon formed at the second carbon. Mannitol is used to reduce intracranial pressure by forced diuresis. In the cataract of the lens, the dulcitol and sorbitol accumulate in excess amounts.

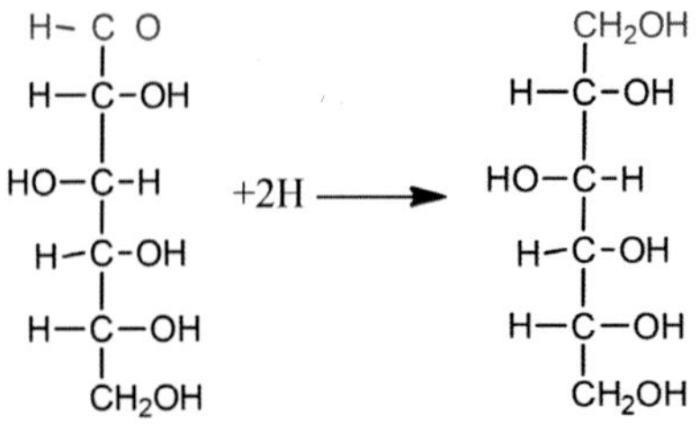

D- Glucose Sorbitol

Applied aspects

- Frequently used low-calorie or "sugar-free" products generally contain sugar alcohols, as they have half the calories of sugar are
- Used in confectionery products, bakery products, desserts, jams,
- Sorbitol is used for the preparation of elixirs and syrups.

2. Reaction with *Strong mineral acids* to Form Furfural and *furfural derivatives.*

Monosaccharides are stable to the action of dilute mineral acids, but when treated with concentrated sulphuric acid, they form furfural and furfural derivatives. This is due to the dehydration of sugars and the removal of three water molecules by concentrated sulfuric acid. Moreover, the sugar gets converted to black in color due to the formation of carbon.

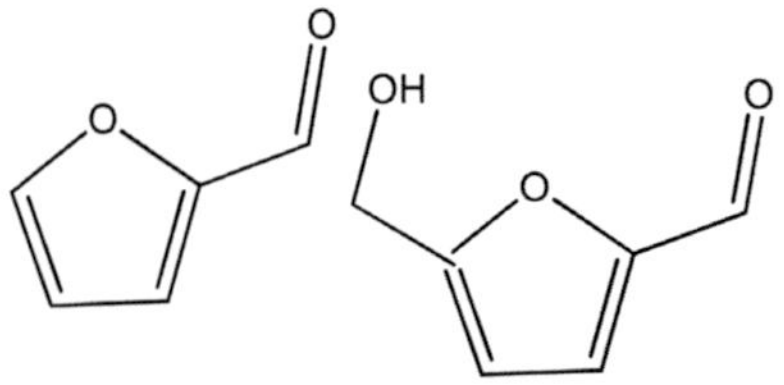

Furfural HMF

Furfural is formed by dehydrating five-carbon sugars(pentoses) such as xylose and arabinose. The hexoses form hydroxy methyl furfural (HMF) in a two-step reaction pathway comprising (i) glucose isomerization to fructose under basic conditions and (ii) fructose dehydration to HMF under acidic conditions. This is the principle of Molisch's test in which Molisch reagent reacts with glucose, forming a furfural derivative. Similarly, when the anthrone reagent (prepared in sulphuric acid) reacts with glucose in the anthrone reaction, it produces a green color due to the furfural derivative formed with a dilute and strong alkali.

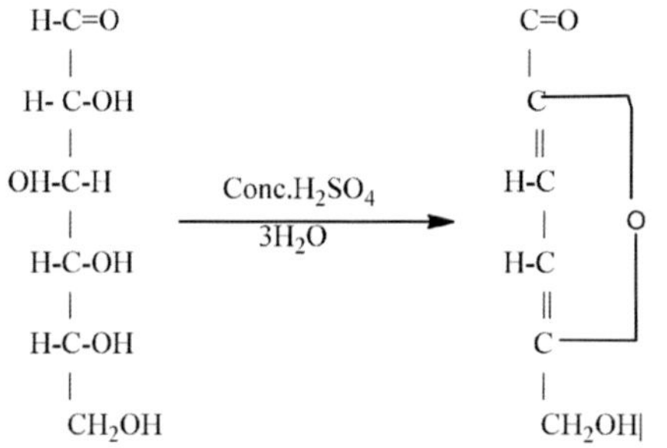

Hydroxy methyl furfural

D-Glucose-

3. Reaction with Phenylhydrazine *hydrochloride* and *Osazone* Formation

When monosaccharides are treated with phenylhydrazine hydrochloride under hot conditions, they first form sugar hydrazone; this further reacts with one more molecule of phenyl hydrazine to form sugar osazone. The glucosamine crystals are seen in hot solution. The osazone crystals of glucose, fructose, and mannose molecules are identical (needle-shaped) even though both the sugar molecules are different. The reason is that the glucose, fructose, and mannose molecules are similar in the lower four carbon atoms. The reaction takes place at carbon numbers one and two. Therefore, the sugar molecules become identical for all six carbon arrangements after the osazone formation reaction. Hence, the shape of their osazone crystals is the same (needle-shaped) as that of glucose, fructose, and mannose. Lactose gives hedgehog-shaped crystals, while maltose gives sunflower-shaped crystals. The crystals of lactosazone and maltosazone are soluble in hot water and appear upon cooling the solution. Osazone is of prime importance in the identification of sugars as well as in determining the structural configurations of sugars.

i. Glucose

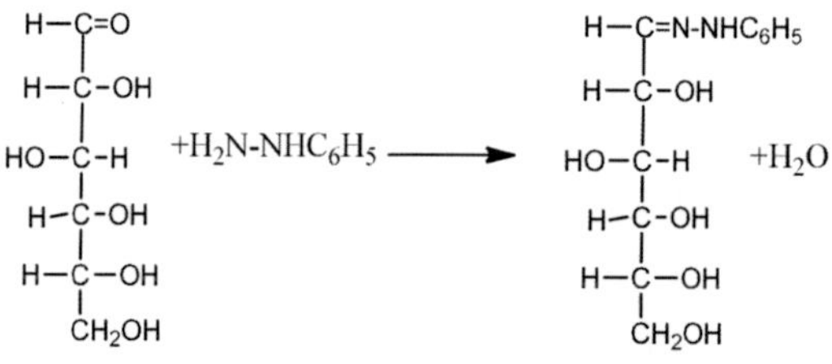

D-Glucose

D-Glucophenyl hydrazone

The crystals of sugar phenyl hydrazone are soluble in water and difficult to isolate.

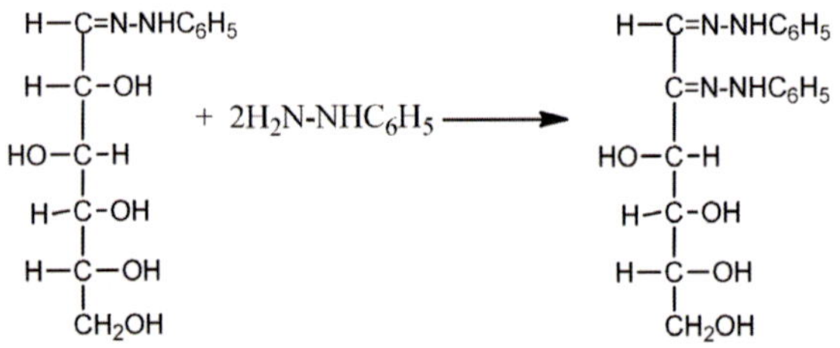

D-Glucophenyl hydrazone — D-Glucose phenyl osazone

The sugar osazones are insoluble in water and can be observed under a microscope.

ii Fructose

Fructose and other ketoses react similarly to form hydrazones and osazones.

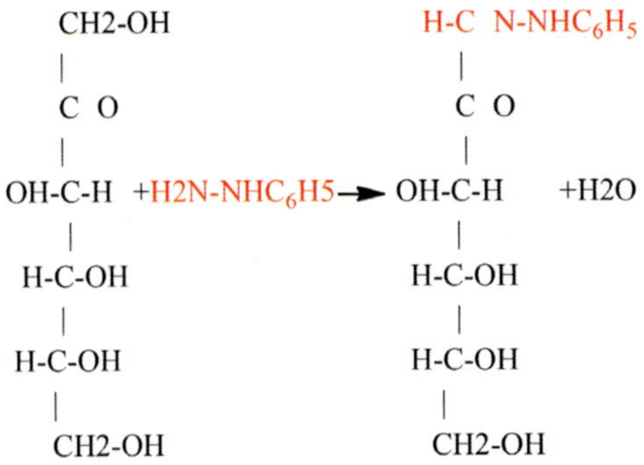

D-Fructose — D-Fructose phenyl hydrazone

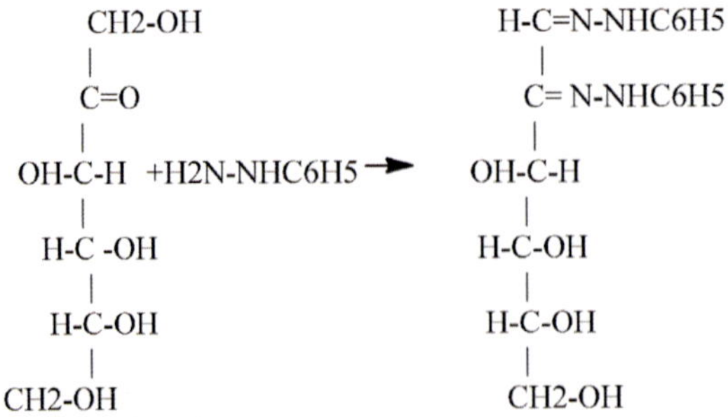

D-Fructose phenyl hydrazone — D-Fructose phenyl-osazone

iii Galactose phenyl-osazone and Mannose phenyl-osazone

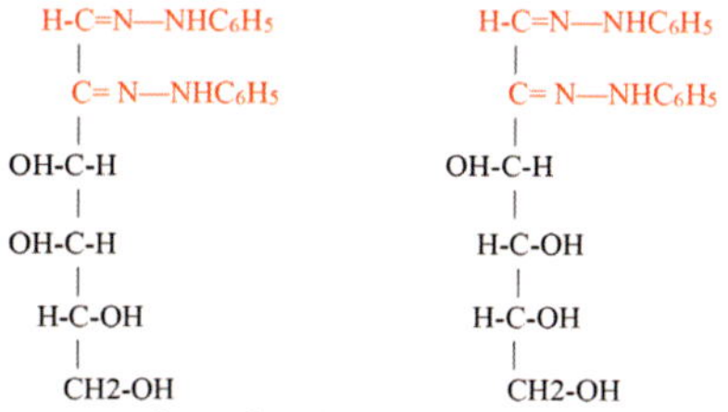

D-Galactose phenyl-osazone   D-Mannose phenyl-osazone

Three different sugars, after osazone formation, look identical in structural formula. Their osazone crystals, therefore, are also similar.

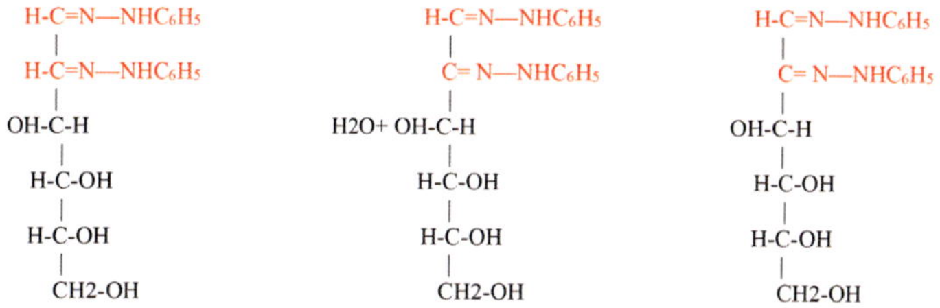

D-Glucose phenyl osazone   D-Fructose phenyl-osazone   D-Mannose phenyl-osazone

I. Formation of the corresponding ketose from osazone

Osazone, when treated with strong HCl, are hydrolyzed to corresponding osone.

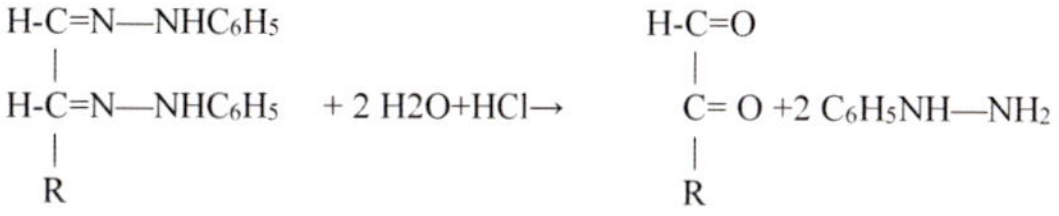

Osazone   Osone

When osones are treated with zinc and acetic acid, the aldehyde group is preferentially reduced to form the corresponding ketose:

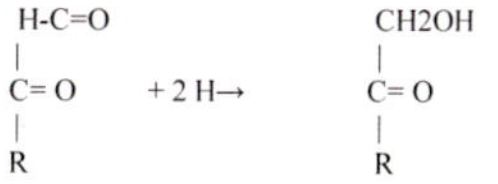

Osone   Ketose

These reactions help convert an aldose into a ketose, i.e., glucose, and then fructose.

J. Formation of enediols- Action of Alkali Upon Sugars or Tautomerization

In mild alkaline solutions, carbohydrates containing free sugar groups -aldehyde or ketone, will tautomerize to form enediols. During tautomerization, the H atom (proton) removed from one molecular site is added to another site to form an enolic compound. An enediol has a double-bonded carbon having two OH groups attached to this carbon. When glucose (aldose) is treated with a mild alkali, it forms enediols (1,2-enediol and 2,3-enediol), which are unstable and result in the formation of mannose or fructose.

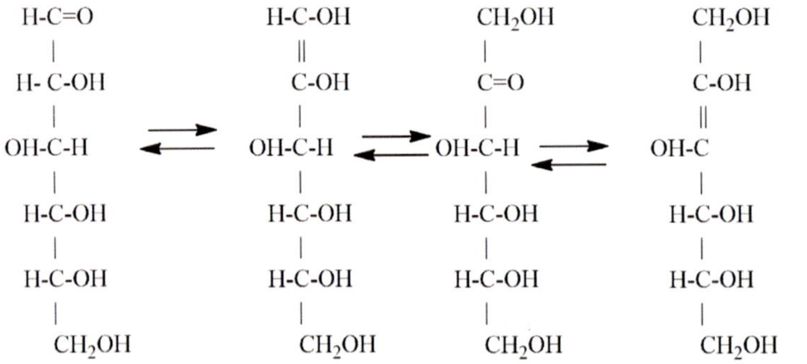

D-Glucose- 1,2-Enediol Fructose 2.3-Enediol

Enediols from glucose and fructose

In the equilibrium, the final product is a mixture of D-glucose, D-fructose, and D-mannose. The reaction is known as Lobry-de Bruyn-van Ekenstein's transformation reaction. An alkaline medium is favorable for the reducing action of sugars.

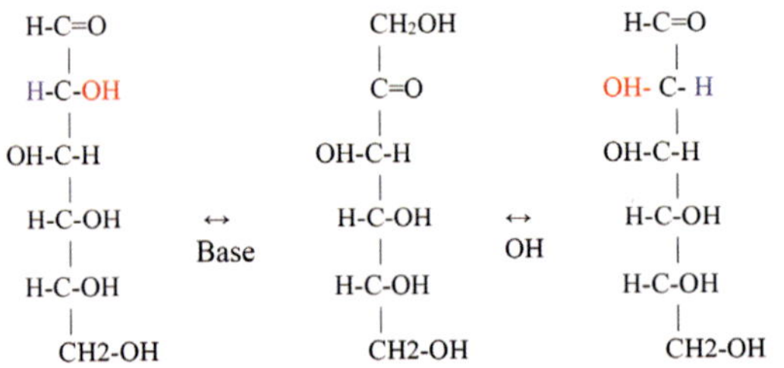

D-Glucose D-fructose D-Mannose

In concentrated alkali, sugar develops the characteristic brown color and nutty flavor of caramel (caramelization). Yellow and brown pigments develop, and many double bonds between the carbon atoms are formed, and carbon-to-carbon bonds may rupture.

K. Reduction Action of Sugars

The aldehyde and ketone are reactive groups of the monosaccharides. They express their reducing power more in an alkaline medium. They act as reducing sugars, forming metallic silver from silver and forming hydroxide or oxide in the case of bismuth, mercuric, and cupric salts. Ferricyanide is reduced to ferrocyanide.

The presence of reducing sugars can be identified by the following tests:

Benedict's qualitative reagent test. It is a test reagent for the presence of sugar (urine) and contains copper sulfate (cupric sulfate), sodium carbonate (provides an alkaline medium), and sodium citrate (keeps cupric hydroxide in solution). When glucose is made to react with Benedict's reagent, it first forms enediols (due to reaction with sodium carbonate), which react with CuSO4 and convert the cupric ions to cuprous ions and then form cuprous oxide. The red and yellow color obtained are due to the formation of cuprous oxide.

Fehling's test. It uses Fehling's A solution (prepared using copper II sulfate) and Fehling's B solution (prepared using potassium sodium tartrate-Rochelle salt and sodium hydroxide, a strong alkali). Fehling's A is blue, and Fehling's B is a colorless, alkaline, transparent liquid. Both solutions are stored separately (mixing two of them will result in the formation of an unstable copper tartrate complex). Fehling's reagent contains cupric sulfate in a solution containing NaOH (provide alkaline pH) and sodium potassium tartrate (maintain the solubility of cupric hydroxide).

Testing a solution for the presence of reducing sugar is done by taking a sample of about 5 ml in a test tube, adding 2 to 3 drops of Fehling's reagent, mixing well, and heating for 1 to 2 minutes. A reddish-brown precipitate formed indicates the presence of reducing sugar. Sugars with an aldehyde group (as a carbonyl group) give a positive test.

Barfoed's reagent test. Barfoed's reagent is prepared using copper acetate in acetic acid to provide an acidic medium. It is used to distinguish monosaccharides from reducing disaccharides. Monosaccharides, being powerful reducing substances, bring about a light reduction and produce a slight reddish precipitate. Monosaccharides, which are strong reducing agents, red precipitates are formed in about 1-2 min, and if it takes 7 to 8 minutes, it indicates the presence of reducing disaccharides. Disaccharides take longer due to the longer time spent in their hydrolysis and followed by the reduction reaction with the reagent.

L. Reaction with Hydrogen Cyanide (HCN) or Killiani Fischer's Synthesis

In this reaction, monosaccharides like glyceraldehyde react with HCN, forming two cyanohydrins due to the creation of a new chiral center. This reaction adds one more C to the existing number of C atoms of the sugar. Thus, glyceraldehyde, having 3 C, will be converted to four carbons containing sugars, namely D-erythrose and D-threose. Similarly, tetroses are converted to pentoses and pentoses to hexoses.

When glyceraldehyde and HCN react, the H and the CN groups of HCN participate in this reaction. The H is added to the O- present in CHO (aldehyde group) to form an OH or C-OH bond. Second, CN itself reacts with the C of CHO, forming a C-CN bond, so the cyanohydrins of glyceraldehyde formed are two in number, with four carbons in place of three carbons. This takes up many steps.

This glucose reaction with HCN confirms the presence of the carbonyl group in glucose.

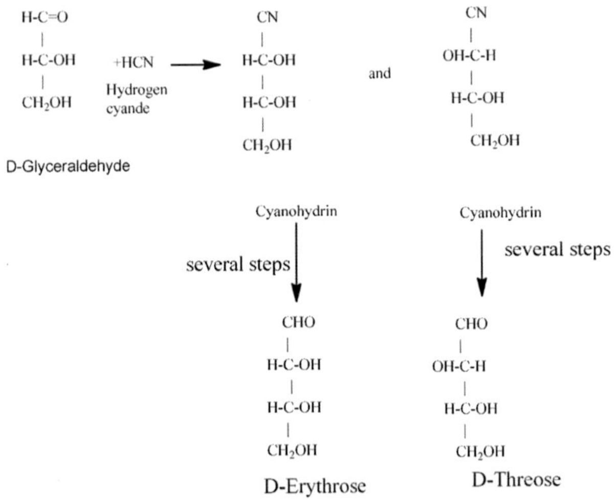

When these cyanohydrins are hydrolyzed, the cyanide group will convert to the carboxylic group, forming carboxylic acid.

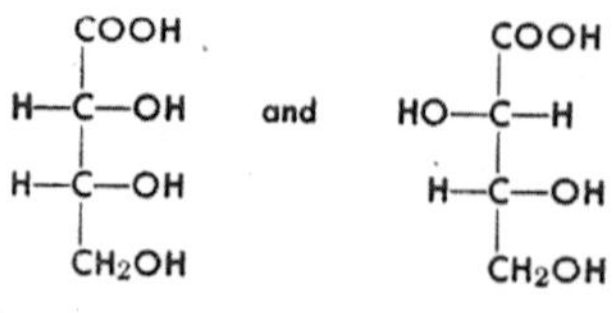

Similarly, L-aldehydes will have a mirror image of D-forms. The following aldehydes can be derived from D-glyceraldehyde.

D-Glyceraldehyde							
D-erythrose				D-threose			
D-ribose		D-arabinose		D-xylulose		D-lyxose	
D-allose	D-altrose	D-glucose	D-mannose	D-gulose	D-idose	D-galactos	D-talose

M. Condensation Reaction with hydroxylamine to form oximes, or a reaction for shortening the sugar chain by one carbon atom at a time

When aldehyde glucose reacts with hydroxylamine, it forms oximes by a condensation reaction.

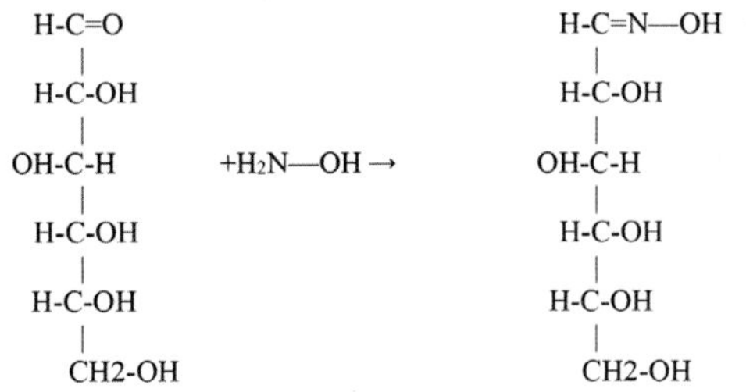

D-Glucose D-Glucose oxime

When glucose oxime is treated with acetic anhydride, one molecule of H_2O is removed forming cyanohydrin or nitrile, and acetylates the hydroxyl groups. The cyanohydrin or nitrile is treated with ammoniacal silver nitrate solution, HCN is removed, and forms an acetylated sugar, which, upon decomposition, forms a sugar with one less carbon atom than the original (D-glucose to D-arabinose). The process cannot be applied to ketoses.

N. Formation of Deoxy Sugars

Replacing the hydroxyl group of sugar with hydrogen yields deoxy sugar (without oxygen). The most critical example is 2-deoxy-ribose (DNA) from ribose. Fucose is found in glycoproteins and glycolipids of animal cells.

O. Formation of Amino Sugars or Aminosachharides

When an amino group (-NH_2) replaces the hydroxyl group (at C-2) of sugar, it yields respective amino sugars. Glucosamine from glucose, mannosamine from mannose, and galactosamine from galactose. These sugars with acetic acid form N-acetyl amino sugars. Glucosamine is an important amino sugar used to treat cartilage damage and reduce the pain and progression of arthritis.

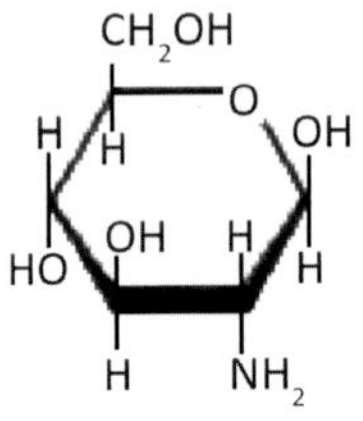

Glucosamine
(an amino sugar)

P. Formation of Glycosides or Reaction Due to Alcoholic Group

This reaction is due to the sugar's hydroxyl group(C-1). Simple sugars in a ring form react with alcohols in the presence of hydrogen chloride as a catalyst and form the Alpha and Beta glycosides. The pyranose glycosides are known as "pyranosides," whereas the furanose glycosides are "furanosides."

Some natural glycosides are phlorizin (glucose+phloretin), Digitonin (4-galactose+xylose+digitogenin), saponin (sugar+sapogenin), and indican (glucose+indoxyl). The group attached to the sugar in a glycoside is known as an aglucone.

The above reactions are due to aldehyde and hydroxyl groups. The reactions due to the aldehyde group of sugars are the formation of ----osazone, cyanohydrins, oximes, sugar alcohols, sugar acids, and furfurals. The reactions due to hydroxyl groups of sugars are the formation of glycosides, ethers (methylation), deoxy sugars, and amino sugars.

3.3 Monosaccharides of Biological Importance

I. Glucose (grape sugar)

It is a six-carbon structure monosaccharide ($C_6H_{12}O_6$). It is an aldose sugar with a reducing property. It has an anomeric carbon (C1), which forms alpha and beta anomers. Its enantiomers exist in L and D forms.

Cyclization reaction occurs when the C-5 hydroxyl group of D-glucose reacts with the oxygen atom of the C-1 aldehyde group. The product formed is a cyclic hemiacetal. In this conformation, the C-1 carbon of D-glucose becomes the new chiral center. As such, cyclic forms of glucose exist as either -D-glucose with the hydroxyl group at C-1 above the plane of the ring or as -D-glucose with the hydroxyl group below the plane of the ring.

Crystalline glucose exists in the cyclic --D-glucose form, but once it is dissolved in an aqueous solution, a very small amount of the monosaccharide is found in the open chain conformation (<0.05%). When glucose is dissolved in water, heat energy from water is absorbed, and an endothermic reaction occurs. This is why water becomes cold. Compared to the four isomers, the beta-form is the most stable isomer of glucose due to the minimum steric hindrance and equatorial position of the hydroxyl groups. The beta-D-glucose is more stable than alpha-D-glucose, which is the opposite of the case of mannose.

Its pyranose form is known as glucopyranose and furanose form as glucofuranose. The Structure of glucose can be written in two forms, namely in linear form or by **Fischer projection,** and in ring form or **Haworth projection**.

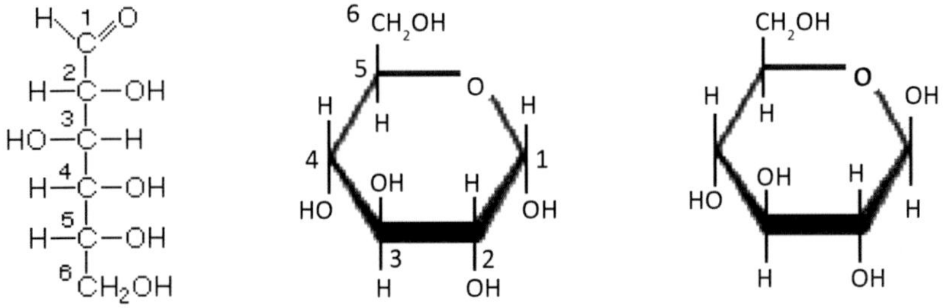

A. Tautomeric Forms of Glucose

In solution, **Glucose** can be present in at least five different tautomeric forms:

1. α-D-Glucopyranose62%); 2. β-D-Glucopyranose (38%);3. α-D-Glucofuranose (trace); 4. β-D-Glucofuranose (trace); 5. Linear D-Glucose (0.01%).

B. Evidence for Glucose as Containing Six Carbons

1. Hot hydriodic (HI) treatment, when given to glucose for a long time, causes n-hexane to develop (in low yield). From this, it was concluded that six glucose carbons are in an unbranched or straight chain.
2. The presence of an aldehyde carbonyl group was deduced from Cyanohydrin formation, its reduction to sorbitol, and mild oxidation to glucuronic acid. Oxidation by dilute nitric acid gave glucaric acid, supporting a six-carbon chain.

C.Importance of Glucose

Glucose or grape sugar is an essential sugar. It is the chief source of energy for the body. The brain also uses it to get energy. Glucose is the end product of the digestion of most of the carbohydrates. For body requirements, the digested glucose is absorbed into the blood for transport to the different tissues in the body. It forms many biologically important carbohydrates, from disaccharides to polysaccharides. It is present in ripe fruits. As a structural component, it occurs in many biological materials, e.g., woody plants, exoskeletons of insects and other arthropods, cell walls of bacteria, amylose, and Amylopectin (in plants), glycogen (in animals), glucans, and glycoproteins. It is stored as starch in plants and as glycogen in animals.

Blood sugar levels higher than normal lead to Diabetes Mellitus. The kidney completely absorbs filtered glucose in the case of a normal person; in a diabetic person, glucose appears in the urine. Glucose can be available as polymers of α (1-4) linkages for the energy source, but polymers with β(1-4) linkages are found as structural components and are not digestible by humans.

D.Glucose General Data

Chemical name: 6-(hydroxymethyl)oxane-2,3,4,5-tetrol

Molecular mass: 180.16 g/mol

Density: 1.54 g/cm3

Melting point: α-D-Glucose: 146°C

β-D-Glucose: 150°C

E Important Reaction of Glucose

Glucose shows its reactivity due to its reducing nature by the presence of aldehyde, hydroxyl, and carbonyl groups in its structure. Reaction with Tollen's, Benedict's, or Fehling's reagents

- Formation of sugar acids
- Formation of alcohol--sorbitol
- Furfural derivatives
- Ester formation and acetylation
- Glucose Osazone formation
- Chain shortening (degradation) and lengthening (Killiani-Fischer synthesis).
- Mutarotation
- Optical activity and optical isomers
- Tautomerization and enediol formation

When exceeded, its average value in blood causes the pathological condition known as Diabetes Mellitus. It provides energy and is oxidized by every cell. It is usually reabsorbed by the renal tubules so that it is not excreted in normal urine.

II Fructose (levulose, fruit sugar)

Fructose. It is a six-carbon polyhydroxy ketone sugar. It is an isomer of glucose with a $C_6H_{12}O_6$ molecular formula. It forms the disaccharide sucrose with glucose. It is a reducing sugar. It occurs in plant sources (fruits, root vegetables) and animal sources (honey, seminal plasma). Inulin and table sugar contains fructose. It can form cyclic hemiketals, furanose (5-membered rings), and pyranose (6-membered rings) ring structures.

A. Importance of Fructose

Fructose is a polyhydroxy ketone sugar with six carbons. It can be converted to glucose to become an energy source. It is also helpful for fat synthesis. It is a natural sweetener, more so due to its sweetness than sucrose. Its presence in semen is a source of energy for sperm.

Formula: $C_6H_{12}O_6$

Molar mass: 180.16 g/mol

Melting point: 103 °C

Boiling point: 440 °

III Galactose

Galactose (Gal) is a monosaccharide known as brain sugar. It helps in the brain development of infants. Its molecular formula is $C_6H_{12}O_6$. It is less sweet than glucose. It can form a polymer known as Galactans.

Its molar mass is 180.156 g/mol. Galactose solubility in water is 68.30 grams per 100 grams at 20–25°C.

Importance of Galactose

- It combines with glucose to form milk lactose.
- It is the precursor for synthesizing glycolipids, sphingolipids, membrane glycoproteins, and mucopolysaccharides.
- It is a fundamental and structural substance for cells, cell walls, and intracellular matrix.

- When galactose is not metabolized in the body, the condition known as galactosemia occurs when there is a deficiency/nonfunctional state of the liver enzyme **galactose-1-phosphate uridyltransferase (GALT).** A galactose tolerance test helps to detect this condition.

IV Mannose

It is present in some fruits and plant products. It is an aldehyde sugar and is a stereoisomer of glucose. It is a constituent of several glycoproteins. Mannose is not well metabolized in humans. The oxidation of mannitol can form mannose. Alpha-D-mannose is more stable than beta-D-mannose because of the equatorial position of the bulky-OH group in alpha-D-mannose at the anomeric carbon. The bulky -OH group is arranged in the axial position in the case of beta-D-mannose.

D-mannose is absorbed more slower than glucose. Mannose, when ingested, is not converted to glycogen or stored in the liver. Most D-mannose from the blood is filtered through the kidneys and excreted into the urine. High doses of D-mannose may damage the kidneys. D-mannose may play a helpful role as a "prebiotic." Mannose has immune regulatory functions.

V Ribose and Deoxyribose

They are pentose monosaccharide polyhydroxylated aldehyde sugars with the open chain structure:

CHO	CHO
|	|
H-C-OH	H-C-H
|	|
H-C-OH	H-C-OH
|	|
H-C-OH	H-C-OH
|	|
CH_2OH	CH2OH
Ribose	Deoxyribose

Deoxyribose is the ribose molecule used by nature for constructing DNA and RNA. The enzymes called ribonucleotide reductases form Deoxyribose from ribose 5-phosphate.

VI Amino sugars

Amino sugars are formed when one or two hydroxyl groups of sugar are substituted with an amino group (−NH2). These sugars help to form collagen in the body to form cartilage.

A. Glucosamine ($C_6H_{13}NO_5$)

The biosynthesis of glycosylated proteins and lipids is possible using an amino sugar. Different types include N-acetyl glucosamine, glucosamine hydrochloride, and glucosamine sulfate. it is found in chitosan chitin. Glucosamine builds cartilage, ligaments, tendons, and the thick fluid surrounding joints. Glucosamine sulfate and glucosamine hydrochloride are preferred to alleviate osteoarthritis (to rebuild cartilage).

It is an amino sugar. It is helpful in the biosynthesis of glycosylated proteins and lipids. It is found in chitosan chitin. Glucosamine helps build tendons, ligaments, cartilage, and the thick fluid surrounding joints. Oral glucosamine, a precursor for glycosaminoglycans, is used to treat osteoarthritis (to rebuild cartilage). Glucosamine sulfate and glucosamine hydrochloride are preferred for osteoarthritis.

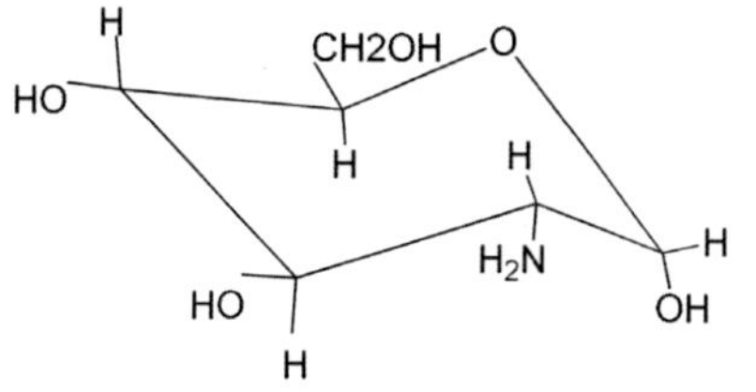

Alpha D-glucosamine(Alpha pyranose ring structure)

B. Galactosamine

It is an amino derivative of galactose where the NH2 group is inserted after replacing the -OH group. It is a significant component of glycoproteins. Its molecular formula is $C_6H_{13}NO_5$.

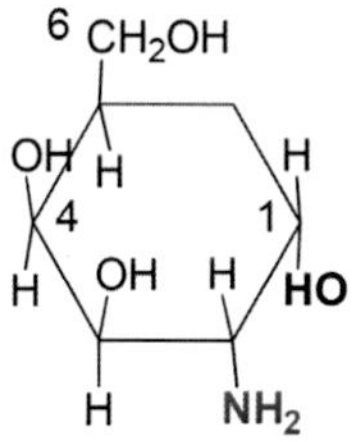

Galactosamine

Some glycoprotein hormones (FSH and LH) do possess galactosamine. It is used to study hepatotoxic or liver-damaging effects for research purposes. A N-acetylgalactosamine is found in the A blood group antigen (D-galactose for the B antigen).

It is a hexosamine derived from galactose, having a molecular formula of $C_6H_{13}NO_5$. It occurs in some glycoprotein hormones like FSH(follicle-stimulating hormone) and LH (luteinizing hormone). These hormones also contain glucosamine, galactose, and glucose.

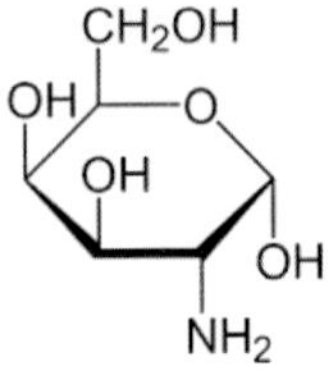

Galactosamine

Galactosamine is a hepatotoxic or liver-damaging agent. It is the terminal carbohydrate forming the antigen of human blood group A.N-acetylgalactosamine, which is necessary for intercellular communication. N-acetylgalactosamine is used as a starting material in carbohydrate chemistry and the development of glycotherapeutic compounds.

Formula: $C_8H_{15}NO_6$ **Molecular Weight**: 221.21

c. Sialic Acid or Neuraminic Acid

Sialic acids exist in bodily fluids. The goblet cells have an acidic glycoprotein due to the presence of sialic sulfate groups. Sialic acids play essential roles in immune modulation and cell adhesion. The highest level of sialic acids in the body is found in the brain. Sialuria is the condition of finding sialic acid in urine. It is the derivative of neuraminic acid (a monosaccharide with a nine-carbon structure). It is found widely in animal tissue. The hydroxyl group substituents include acetyl, methyl, sulfate, and phosphate. NANA is N-acetyl neuraminic acid, a sialic acid.

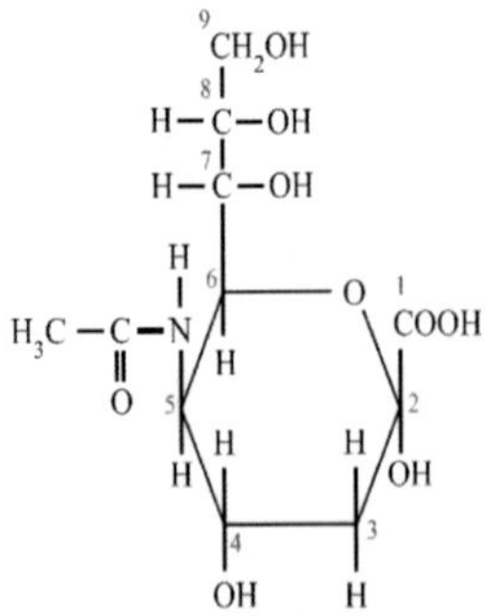

Sialic acid

Muramic acid is a nine-carbon sugar, a component of bacterial cell walls.

Key Takeaway

- The molecular formula of any monosaccharides (mono=one and saccharides=sugars) is $C_n(H_2O)_n$ or $(CH_2O)_n$.
- Monosaccharides cannot be broken down further into simpler units.
- Commonly occurring sugars can exist in either linear (straight) or ring forms
- Glucose has one primary alcoholic group, which is at the sixth carbon atom.
- The D sugar will have H on the left side of the plane and OH on the right side. The L sugar has H on the right side of the plane and OH on the left side. The L sugar is the mirror image of the D sugar.
- The sugar ring structure containing five carbons is a pyranose ring, and a ring structure containing four carbons is furanose.
- Thermodynamically, the pyranose form is more stable than the Furanose form. Furanose rings are more flexible than pyranose rings
- Monosaccharides exhibit various reactions due to the presence of -CHO, =CO, and OH groups. The Carbonyl group undergoes reduction or oxidation reactions, and the hydroxyl groups form esters
- Pentose phosphates are involved in the formation of nucleic acids.
- The Aldehyde group, upon oxidation, forms 'aldonic acids' where carbon is first converted from CHO to COOH. When the last carbon (C6) CHOH is converted to COOH, then ***'uronic acids'*** are formed (with C1 containing a CHO group). A 'saccharic acid' is formed when C1 (containing CHO) and C6 (containing CHOH) carbons are converted to COOH.
- Ketoses form two alcohols because of the asymmetric carbon formed at the second carbon.
- Three different sugars, after osazone formation, look identical in structural formula. Their osazone crystals, therefore, are also similar.
- The reactions due to the aldehyde group of sugars are the formation of ----osazone, cyanohydrins, oximes, sugar alcohols, sugar acids, and furfurals. The reactions due to hydroxyl groups of sugars are the formation of glycosides, ethers (methylation), deoxy sugars, and amino sugars.

Intext Questions

1. Write the general molecular formula for monosaccharides.
2. How does the position of the bulky-OH group decide the stability of the mannose?
3. Name the most stable isomer of glucose.
4. Why does the water become cooler when glucose is dissolved in it? Why?
5. Write the structures of alpha-D and beta-D fructose furanose.
6. How do the structures of alpha-d differ from alpha-L glucose?
7. How are hemiacetal and hemiketal formed in the case of monosaccharides?
8. How are iodo compounds of glucose obtained?
9. Name the type of reaction occurring between a. glucose and acetic acid, and b. glucose and methanol?
10. What oxidation products are obtained in a chemical reaction occurring with monosaccharides? How do these products differ from each other?
11. How are furfural derivatives obtained from monosaccharides?
12. Which two sugars have similar osazone crystals? Why?
13. What are enediols? How are they obtained?
14. What is Killiani Fischer's synthesis?
15. How do amino sugars form? Write their medical importance.
16. Write the names of four carbohydrates with glucose as one component.
17. Write about the biochemical importance of galactose.
18. Which sugars are components of nucleic acids?
19. What is sialic acid?
20. What happens to D-glucose treated with the following reagents?

    a. HCN   b. HI
21. What happens when any aldehyde sugar reacts with HCN?
22. MCQ choose one correct option n from the given below (Answer is given in bold letters):

a Commonly occurring sugars can exist in

i. Straight chains ii. Ring forms.

iii. **Both i. and ii.**

b Which of the following is correct for D and L sugars?

i. **The D sugar will have H on the left side of the plane and OH on the right side of the bottom-most chiral center of the Fischer projection.**

ii. The D sugar will have H on the right side of the plane and OH on the left of the bottom -most chiral center of the Ficher projection.

iii. The D sugar will have two H on the left side of the plane and one OH on the right of the bottom -most chiral center of the Ficher projection.

c Identify the below-given L-sugar.

```
  H-C=O
     |
HO-C-H
     |
   H-C-OH
     |
  HO-C-H
     |
  HO-C-H
     |
   CH2OH
```

i. Fructose ii. Galactose

iii. Glucose

d The furanose ring contains the following number of carbons:

i. **Four** ii. Six

iii. Three

e Identify the correct statement from the following.

i. **Thermodynamically, the pyranose form is more stable than the Furanose form. Furanose rings are more flexible than pyranose rings**

ii. Thermodynamically, the pyranose form is less stable than the Furanose form. Furanose rings are more flexible than pyranose rings

iii. Thermodynamically, the pyranose form is more stable than the Furanose form. Furanose rings are less flexible than pyranose rings

f The carbonyl group in aldoses is found on the carbon number

i. **On the terminal carbon atom**

ii. On the fourth carbon atom

iii. On the second carbon atom

g The carbonyl group in ketoses is typically found

i. On the terminal carbon atom ii. On the fourth carbon atom

iii. **On the second carbon atom**

h Which number of glucose carbon is converted to COOH to form uronic acid derivative?

i. Carbon one ii. **Carbon six**

iii. Carbon four

i A 'saccharic acid ' is formed when the following carbon number of glucose is converted to COOH.

i. One ii. Six

iii. **Both one and six.**

j The reduction of which of the following reagents yields silver as polishing on the surface of the tube.

i. **Tollen's reagent** ii. Barfoed's reagent

iii. Fehling's reagent

k The alcohol product from glucose is

i. **Sorbitol** ii. Ribitol

iii. Glycerol

l Osazone formation reactions occur at two carbons of Glucose, fructose, and mannose, so the osazone formation looks identical. The two carbon atom numbers that react are

i. At carbon 1 and 6 ii. **At carbon 1 and 2**

iii. At carbon 2 and 4

Further Readings

Biomolecules: (introduction to structure and function) Carbohydrates Suman Khowala, Deepak Verma, Samudra P. Banik. https://www.studocu.com/in/document/panjab-university/biotechnology-hons/carbohydrates functions/12432089

European Journal of Clinical Nutrition. The role and potential of sialic acid in human nutrition.

https://www.verywellhealth.com/the-benefits-of-glucosamine-88809

Jain J.L., Sunjay Jain, Nitin Jain (2005). Fundamentals of Biochemistry. S. Chand & Company Ltd. New Delhi..

Life sciences. Fundamentals and Practice I. Pranav Kumar and Usha Mina (2006). Sixth Edition. Pathfinder publication. New Delhi.

Medical Biochemistry (2004). Ethiopia public health initiative.

Objective Biochemistry. Sengar, R.S. and Chaudhary Rishu (2014). New India Publishing Agency. New Delhi.

Textbook of Biochemistry for Medical Students. Vasudevan, D. M.; Sreekumar, S.; Vaidyanathan, Kanan.

4

Oligosaccharides

Oligosaccharides comprise 2 to 10 units of monosaccharide molecules, which are released upon hydrolysis. The oligosaccharides are classified depending upon the number of monosaccharide units, e.g., Disaccharides (two monosaccharide units), Trisaccharides (three monosaccharide units), Tetrasaccharides (four monosaccharide units). Disaccharides are the most common oligosaccharides found in nature.

4.1 Disaccharides

Disaccharides are formed when glycosidic linkages covalently bond two monosaccharides together. A glycosidic bond is formed when the hydroxyl group on one of the sugars reacts with the monomeric carbon on the second sugar. A glycosidic bond can be alpha or beta, depending on the steric configuration at carbon one on the monosaccharide unit. During the union of two monosaccharide units, water molecules are eliminated (dehydration reaction or condensation reaction, or dehydration synthesis), and the f***ormation of a glycosidic bond*** occurs, through which the units are linked via an oxygen bridge.

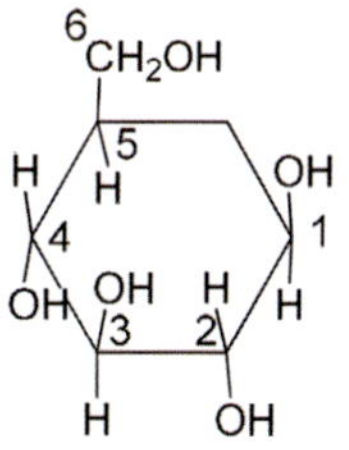

Alpha-D-Glucopyranose
Haworth projection

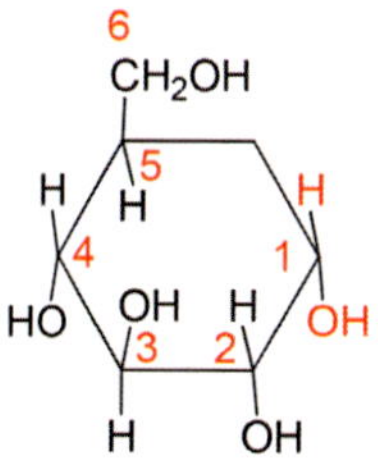

Beta-D- GlucoPyranose
Haworth projection

If the bond is below the plane of the ring, it is said to be an alpha glycosidic (α-glycosidic) bond; if it is above the plane of the ring, it is known as a beta glycosidic **(β-glycosidic)** bond. The numbers 1,4,6 identify the carbon atoms at each end of the glycosidic bond. Numbering starts clockwise next to the ring oxygen at the anomeric carbon. Alpha linkages are digestible, and beta linkages are not. Substances joined with glucuronic acid via glycosidic bonds increase their water solubility (**glucuronidation**).

Acetals (two -OR groups) and hemiacetals (one -OR and one -OH) are recognized functional groups. Hemiacetal is formed before (intermediate) the formation of acetal.

List of Various Disaccharides

The disaccharide is named by listing the non-reducing monosaccharide on the left, then the glycosidic linkage between the two monosaccharides, and then the reducing Monosaccharide on the right.

Common Disaccharides

Disaccharide	**Component monosaccharides and linkage**
Sucrose (Table sugarcane sugar or beet sugar)	glucose 1α→2 fructose
Maltose (product of starch hydrolysis)	glucose 1α→4 glucose
Lactose (milk sugar)	galactose 1β→4 glucose
Cellobiose	glucose 1β→4glucose
Trehalose (in fungi)	glucose 1α→1 glucose

Less Common Disaccharides

Disaccharide	**Units**	**Bond**
Kojibiose	two glucose monomers	α (1→2)
Nigerose	two glucose monomers	α (1→3)
Isomaltose	two glucose monomers	α (1→6)
ß, ß,-Trehalose	two glucose monomers	β (1→1)
Sophorose	two glucose monomers	β (1→2)
Laminaribise	two glucose monomers	β (1→3)
Gentiobiose	two glucose monomers	β (1→6)
Turanose	a glucose monomer and a fructose monomer	α (1→3)
Maltulose	a glucose monomer and a fructose monomer	α (1→4)
Palatinose	a glucose monomer and a fructose monomer	α (1→6)
Gentiobiulose	a glucose monomer and a fructose monomer	β (1→6)
Manobiose	two mannose monomers	It can be either α (1→2), α (1→3), α (1→4), or α (1→6)

Disaccharide	Units	Bond
Melibiose	a galactose monomer and a glucose monomer	α (1→6)
Melibiulose	a galactose monomer and a fructose monomer	α (1→6)
Rurinose	a rhamnose and a glucose	α (1→6)
Rutinulose	a rhamnose monomer and a fructose monomer	β (1→6)
Xylobiose	two xylopyranose monomers	β (1→4)

Disaccharides are sometimes crystalline, sometimes water-soluble, and sometimes sweet-tasting and sticky-feeling. The most common disaccharides are grouped as

I. Reducing disaccharides and

II. Non-reducing disaccharides.

I. Reducing Disaccharides

In reducing disaccharides, monosaccharide components are bonded by hydroxyl groups. Here, monosaccharide, the reducing sugar of the pair, still has a free hemiacetal unit that can perform as a reducing aldehyde group; **lactose, maltose, and cellobiose** are examples of reducing disaccharides, each with one hemiacetal unit, the other occupied by the glycosidic bond. Reducing sugar acts as a reducing agent. Reducing sugars form some aldehyde or ketone in an alkaline solution, which allows it to act as a reducing agent (Benedict's reagent). In such a reaction, the sugar becomes a carboxylic acid.

A. Lactose (Milk Sugar)

It is reducing sugar as it has a free carbonyl group on the carbon atom 1of the glucose unit. It exhibits mutarotation ($+55.2^0$) and forms osazone and oxime. Its crystals are rhombic prisms in shape.

It comprises β-D glucose and β-D galactose units joined together through carbon 1 of D-galactose and carbon 4 of D-glucose. A β-1-4 glycosidic bond binds them together. It is a reducing sugar due to a free carbonyl group on the glucose. Hydrolysis gives glucose and galactose units.

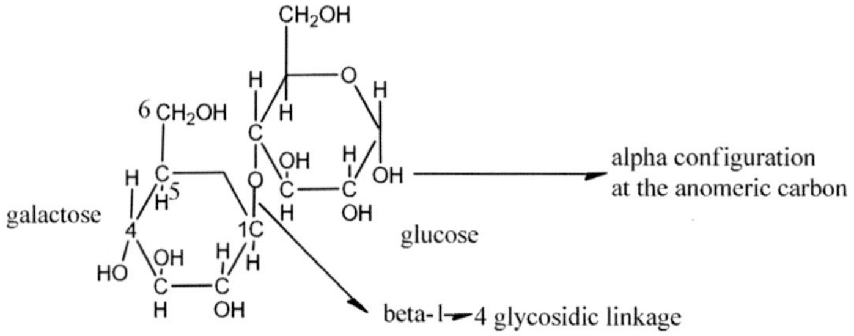

ALPHA- D-LACTOSE

Its systematic name is β-D-galactopyranosyl-(1-4)-D-glucose. The glucose can be in either the α-pyranose form or the β-pyranose form, whereas galactose is always in the β-pyranose form. As a result, α-lactose and β-lactose correspond to the anomeric form of the glucopyranose ring alone.

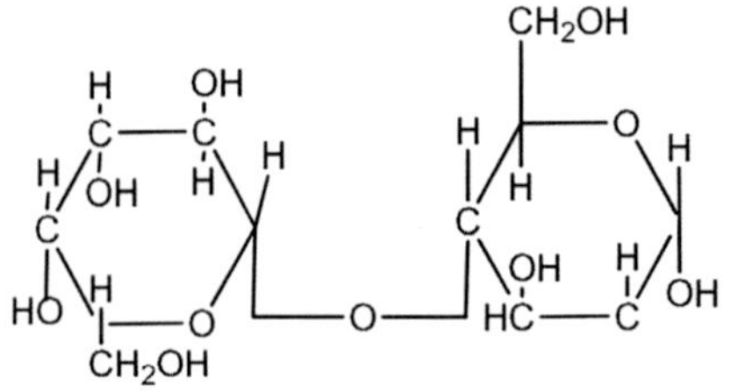

Alpha lactose

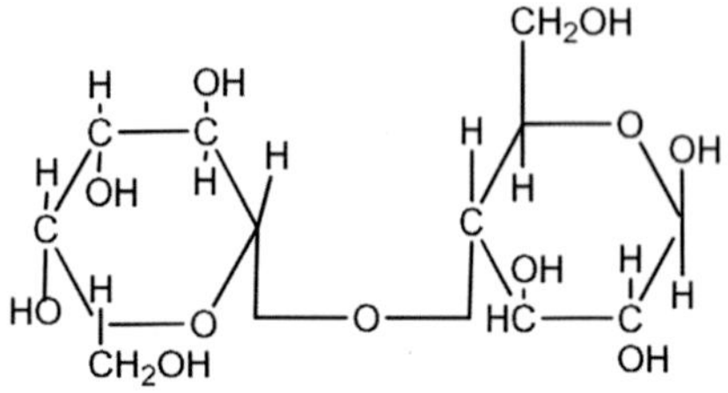

Beta a lactose

Lactose is formed in the lactating mammary gland and is found exclusively in mammals' milk (milk sugar). Human milk contains about 6-8%, and cow milk about 4.8%. The only body part that produces it is the mammary gland in the animal's body. There is no lactose in the milk of seals and other close relatives. It is a white, crystalline, and dextrorotatory. It is less soluble in water and much less sweet than sucrose. In terms of sweetness, this sugar is about one-sixth as sweet as sucrose.

Lactose is digested in the intestine by the enzyme lactase. A deficiency of lactase in the body leads to the condition known as lactose intolerance. When not digested, lactose ferments in the colon and causes bloating, gas, abdominal pain, and diarrhea. These problems are not found with Yogurt, as lactose is consumed by the bacteria that transform milk into yogurt.

B.Maltose (Malt sugar)

Its two monosaccharide units are D- glucose, in pyranose form, joined together by an α-1, 4-glycosidic linkage (through their 1 and 4 carbon atoms). In other words, the glycosidic linkage is found between OH at the first carbon atom of the first glucose residue and OH at the fourth carbon atom of the second glucose.

Maltose with formula $C_{12}H_{22}O_{11}$ is formed, releasing a H_2O molecule due to the linkage formed between two glucose units. Maltose is a reducing sugar, as one of its units is typically an open chain with a free aldehyde group. It can form osazone and exhibits mutarotation (+130^0). Maltose is usually found in beta form, and it is also found in alpha and aldehyde forms. It is produced by the hydrolysis of starch (acid or enzymatic-- amylase or diastase). Commercial corn syrup is a source of maltose.

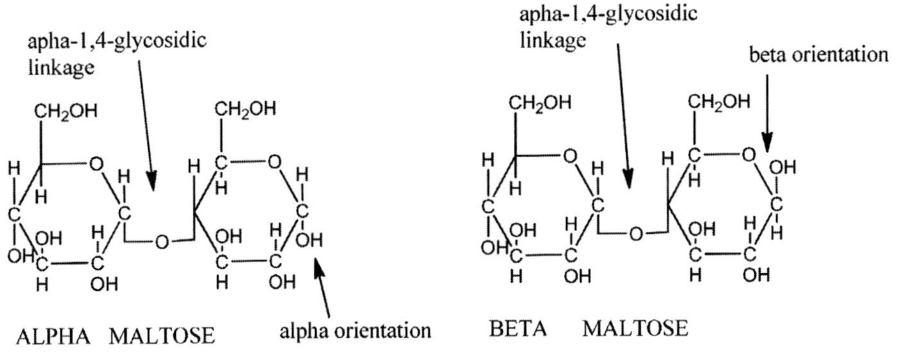

It is a white crystalline solid, soluble in water, and dextrorotatory in nature. Maltose sweetness is about 30% compared to sucrose. It does not occur in the body. Maltose is hydrolyzed by maltase (acts on alpha glycoside linkage and not on beta) and is found in the intestine Partial hydrolysis of starch and glycogen yields maltose.

C. Isomaltose

It is an isomer of maltose. Its chemical formula is $C_{12}H_{22}O_{11}$. It has a molar mass of 342.30 g·mol^{-1}. It is soluble in water. Its melting point is 98-160 °C. It is a disaccharide similar to maltose but with α-(1-6) linkage instead of α-(1-4). Both of the sugars are glucose and pyranose. It can turn into an open-chain form so that its functional group makes it a reducing sugar. It exists in glycogen, amylopectin, and certain bacterial dextrans. Isomaltose is used as a sweetening agent.

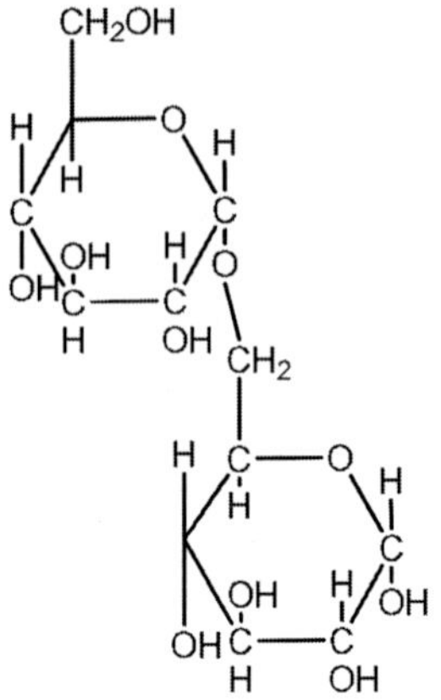

ISOMALTOSE

D. Cellobiose

It is a white crystalline solid, soluble in water, dextrorotatory, forms osazone and oximes, and shows mutarotation in an aqueous solution. Its formula is $(C_6H_7(OH)_4O)_2O$. It has two D-glucopyranose units linked by a 1-4-beta glycoside linkage. Like maltose, cellobiose also reduces sugar, but both differ in their linkages. It also exists in a reducing form, having a free aldehydic form. Maltose has an α-1, 4-glycosidic bond, whereas cellobiose has a beta 1,4-glycosidic bond. Cellobiose, having a beta-1,4 linkage, is resistant to hydrolysis by the human small intestine and, hence, reaches the colon undigested. Cellobiose has no taste.

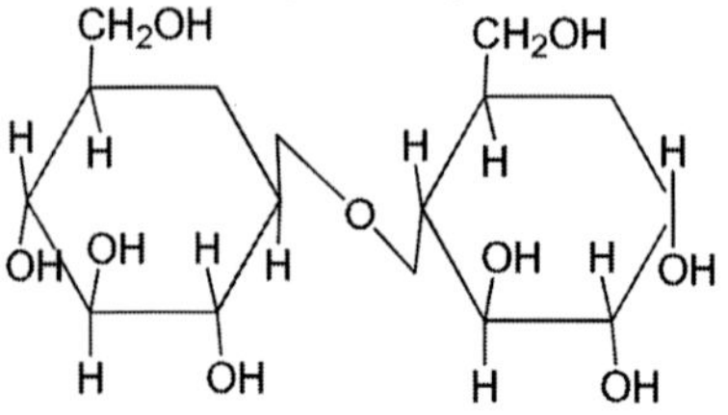

CELLOBIOSE

In animals, plant cellulose is digested by microbial cellulase to cellobiose, which, upon further hydrolysis, yields D-glucose units. Enzymatic or acidic hydrolysis of cellulose/cellulose-rich materials (cotton, jute, and paper) can yield cellobiose.

Gentiobiose

It occurs in rhizomes of several gentian species and has two D-Glucose pyranose units joined with a β (1-6) linkage. It is a reducing sugar. Gentiobiose may be related to maturation processes in seeds and fruits.

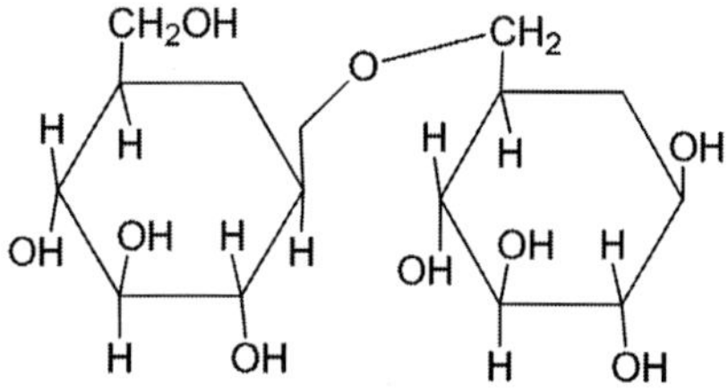

GENTIOBIOSE

The molecular formula is $C_{12}H_{22}O_{11}$. It is a white crystalline solid, soluble in water or hot methanol. Molecular weight is 342.30. It is obtained by a process of caramelization of glucose. **Caramelization** is a non-enzymatic oxidation of sugar used in cooking to produce a nutty flavor and brown color. Gentiobiose is bitter in taste.

F. Chitobiose

Chitobiose is a disaccharide amino sugar comprising glucosamine units linked by β-1,4 bonds. The N-acetylamino group is present on C-2 of chitobiose. Its molecular formula is $C_{16}H_{28}N_2O_{11}$.

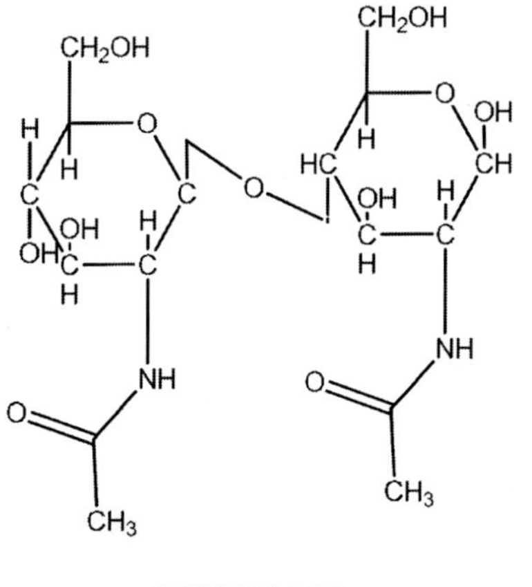

CHITOBIOSE

II. Non-reducing Disaccharides

A. Sucrose (Table Sugar, Cane sugar, Beet sugar)

It is formed by the union of one α-D (+)glucose and one β (-) D fructose by removing a water molecule. The glycosidic bond involves carbon 1 of the pyranose ring of glucose and carbon 2 of the fructose furanose ring, creating an α-1, β-2-glycosidic (head-to-head) linkage. As a result of this kind of linkage, the anomeric carbons of both component monosaccharide units are linked. Anomeric forms of sucrose are not possible. Sucrose contains no reactive free anomeric carbon atom and is non-reducing in nature. It does not form osazone or an oxime. Sucrose does not show mutarotation as the free aldehyde or ketone group is linked and unavailable. It is in one form, irrespective of solid or solution state, because it is non-reactive and is chemically stable.

Sucrose is dextrorotatory (+66.5), but upon hydrolysis, it breaks down into equal molecules of two monosaccharides (1:1 mixture) having opposite plane-polarised light rotations. The value for Fructose is more levorotatory (-92.40°) than glucose, which is dextrorotatory (+52.7°), and so the value of the hydrolyzed mixture of sucrose changes to levorotatory-19.90, e.g., {1/2(+52.7 +(-92.40)}=-19.90° from +66.5° dextrorotation. This change in the optical rotation of sucrose from dextrorotation to levorotation after its hydrolysis is known as inversion, and sucrose is known as inverted sugar. The enzyme sucrase (invertase) splits the molecule by adding a water molecule.

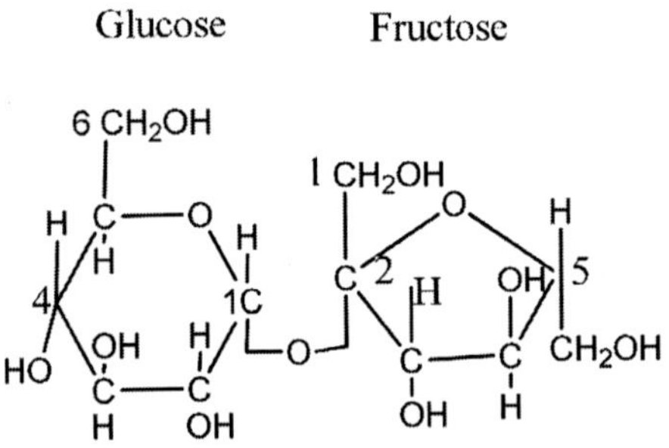

Sucrose
(alpha-glucopyranosside
beta-fructofuranoside)

The molecular formula of sucrose is $C_{12}H_{22}O_{11}$. It has a Molar Mass or Molecular Weight of 342.30 g/mol. It does not exist in the animal or human body. It is the product of photosynthesis. It is stored in sugar cane, sugar beet, pineapple, carrots, fruits, seeds, flowers, and roots. The nectar of the flowers is rich in sucrose, also found in honey. Sucrose is the sweetest among glucose, lactose, and maltose.

B. Trehalose or Mycose

Trehalose is a non-reducing disaccharide. Its molecular formula is $C_{12}H_{22}O_{11}$. It comprises two glucose units linked "head-to-head" in an α-(1→1)-linkage. It is distributed in bacteria, fungi, mushrooms, seaweed, lobster, shrimp, and plants, but not in mammalian cells.

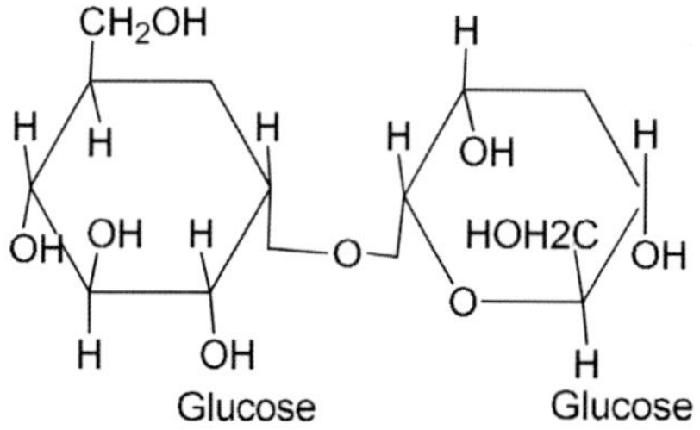

Trehalose

Blood sugar is trehalose in insects, including grasshoppers, locusts, butterflies, and bees. It is difficult to be hydrolyzed even by acids, but trehalase in fungi breaks food trehalose to release two glucose molecules. It is also used to prolong food shelf life. It has a molar mass of 342.30 g/mol. Trehalose is an antioxidant.

4.2 Trisaccharides

They are made up of three monosaccharides held together by two glycosidic bonds. As a result of regiochemistry (different bond combinations) and stereochemistry (alpha- or beta-), the trisaccharide formed is a diastereoisomer having different chemical and physical properties. They are usually present in plants and some fungi in small amounts. They are classified as reducing and non-reducing trisaccharides.

Trisaccharides	**Monosaccharides in Order of Their Linkage**
A. Non-reducing Trisaccharides	
Raffinose	Fructose-Glucose-Galactose
Mannotriose	Mannose-mannose-mannose
Melezitose	Glucose-Fructose-Glucose
Gentianose	Glucose-Fructose-Glucose
B. Reducing trisaccharides	
Maltotriose	Glucose-Glucose-Glucose

I. Non-Reducing Trisaccharides

A. Raffinose

It is a trisaccharide, a non-reducing sugar, found in plant seeds, leaves, stems, roots, sugar beet, cottonseed meal, and fungi. The glycosidic bonds are found at its anomeric carbon atoms. Rich sources of Raffinose are beans, cabbage, and broccoli. Raffinose is not digested in humans' stomachs and small intestines due to a lack of necessary enzymes for its breakdown. However, bacterial fermentation occurs in the large intestine.

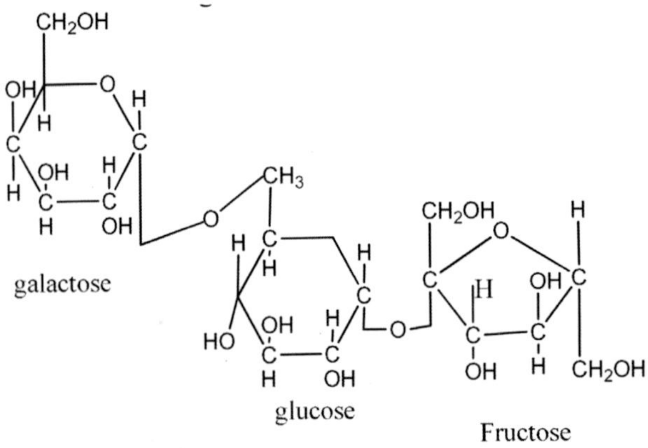

RAFFINOSE

It comprises galactose (pyranose), glucose (pyranose), and fructose. (furanose) units. Galactose is linked via α (1→6) to glucose, which is further connected via β (1→2) to fructose. Glucose acts as a link between galactose and fructose.

Galactose → α (1→6)→glucose→ β (1→2) --- fructose.

Hydrolysis of Raffinose.

- Raffinose has both α and β glycosidic linkages. As a result, it can be hydrolyzed by enzymes having α-glycosidic activity or by enzymes having β-glycosidic activity.
- Weak acid hydrolysis of raffinose and the action of sucrase yield one molecule each of fructose and melibiose.
- With α-glycosidic activity (α-galactosidase) enzymes, it is broken down into galactose and sucrose. Alpha maltase is used in the sugar industry to produce sucrose sugar from the raffinose of sugar beet.
- It is broken down to melibiose and fructose via enzymes having β-glycosidic activity (β-fructosidase).

- Herbivorous animals digest raffinose; hence, beetroots are used as food for cattle. But poorly utilized by human.
- Raffinose is entirely hydrolysed by bottom yeast because it contains both the enzymes sucrase and melibiase required for hydrolysis.

B. Gentianose

It is a reducing trisaccharide. It comprises β -D-glucopyranose, α-D-glucopyranose, and β -D-fructofuranose units. It forms polymers with other molecules, such as caproic acid. It is found in Gentiana lutea root, Gentiana orbicularis, and other organisms. It has anti-inflammatory, antioxidant, and antimicrobial activities.

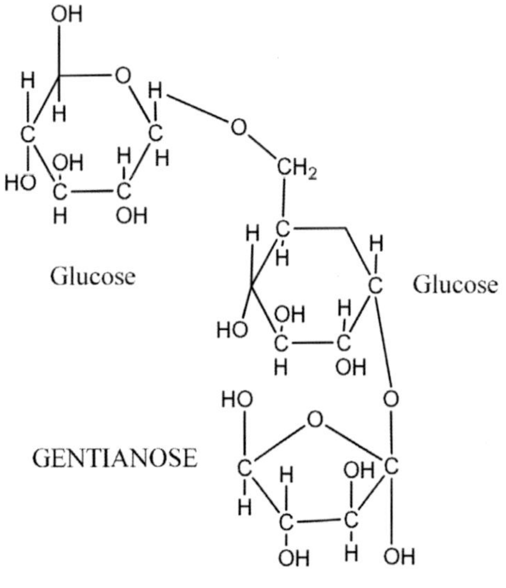

Fructose

On complete hydrolysis, it yields one molecule of fructose and two molecules of glucose. It contains beta-D-glucopyranosyl, beta-D-fructofuranosyl, and alpha-D-glucopyranoside units.

C. Melezitose (Melicitose)

It has wide industrial applications in the pharmaceutical, food, fisheries, broiler, and cosmetic industries. It is the primary trisaccharide in honeydew. It is made up of alpha-D-glucopyranosyl(1->3), -beta-D-fructofuranosyl(2->1)-alpha-D-glucopyranoside.

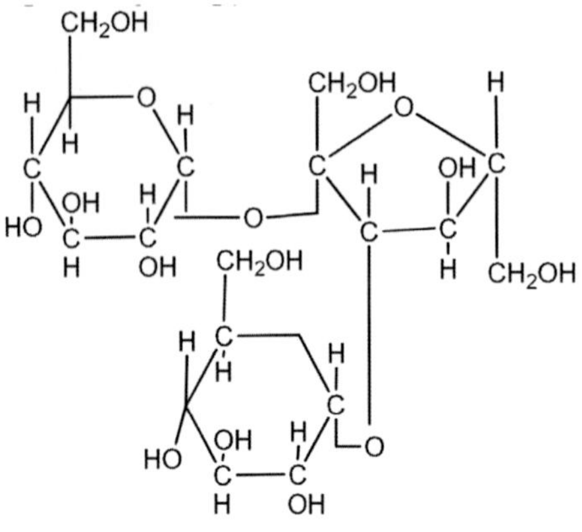

MELEZITOSE

D. Mannotriose

It is a non-reducing trisaccharide made up of three β-D mannose sugar units linked by β-1-4 glycosidic bond in a linear chain structure.It is no-reducing sugar because all three anomeric carbons of three units are involved in glycosidic linkages formed. It is formed from the hydrolysis of tetrasaccharide stachyose.

II Reducing trisaccharides

A. Maltotriose

It comprises three glucose molecules bound together with α-1,4 glycosidic bonds. It is a reducing trisaccharide. It is released by the digestive enzyme α-amylase (in human saliva and the small intestine) when it acts on amylose in starch. When the acid or enzyme amylases partially hydrolyze starch, the obtained products are maltose, maltotriose, and oligosaccharide dextrin.

Its chemical formula is $C_{18}H_{32}O_{16}$.

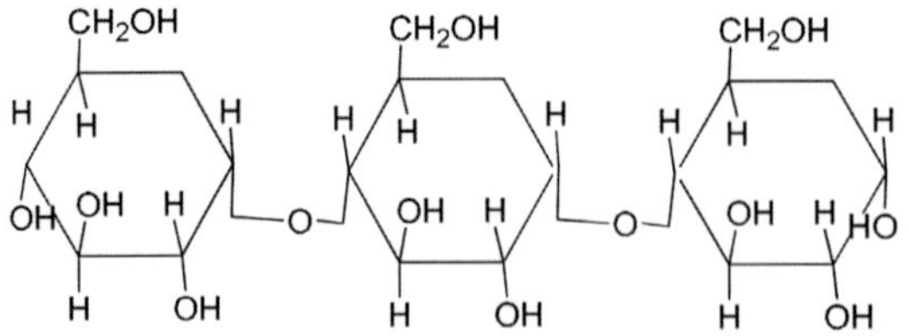

MALTOTRIOSE

It is used as a sweetener, 30% sweeter than sucrose

4.3 Tetrasaccharides

A tetrasaccharide comprises four monosaccharides, which are released upon hydrolysis of molecules. $C_{24}H_{42}O_{21}$ is the general formula of a tetrasaccharide.

I. Stachyose

It is a tetrasaccharide having four monosaccharides, of which two are galactose (pyranose) units, one glucose (pyranose) unit, and one fructose (pyranose) unit. It is a non-reducing sugar. Its formula is $C_{24}H_{42}O_{21,}$ and its molar mass is 666.578 g/mol. In this structure, glucose and fructose units are linked by C−1 of α− glucose and C−2 of β− fructose. The linkage of stachyose is arranged as follows:

galactose (α1→6) galactose(α1→6) glucose(α1→2β) fructose

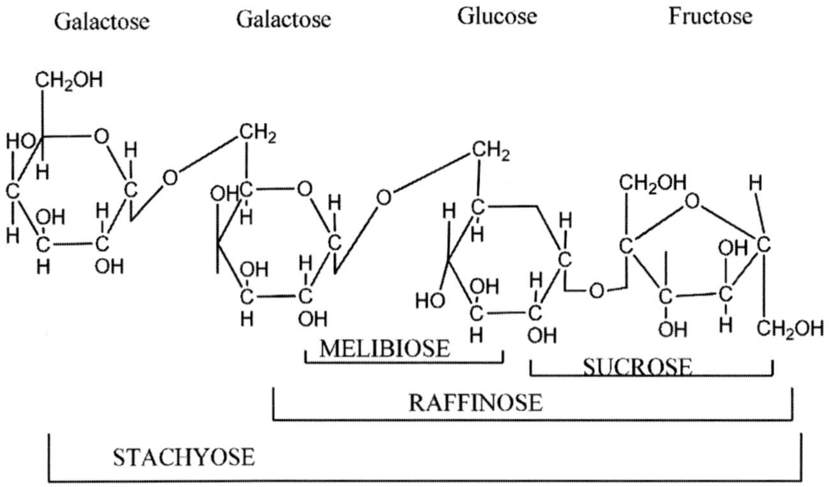

Stachyose is a normal metabolite in human milk, vegetables (various beans), cowpeas, and other legumes. It is a major storage sugar in woody plants, cucurbits, and legumes.

Pentasaccharides

I. Verbascose

This pentasaccharide exists in legumes and certain plants. Its molecular formula is $C_{30}H_{52}O_{26.}$ The structure of verbascose consists of five monosaccharide units, three galactose molecules, one molecule, and one fructose molecule. In this structure, glucose and fructose units are linked by C−1 of α-glucose and C−2 of β-fructose. It has a stachyose structure with one more α-D-galactose pyranose unit. Verbascose can be synthesized from stachyose. A transfer of galactosyl from raffinose to stachyose, catalyzed by the enzyme Galactosyltransferase (GGT), forms verbascose. It is soluble in water.

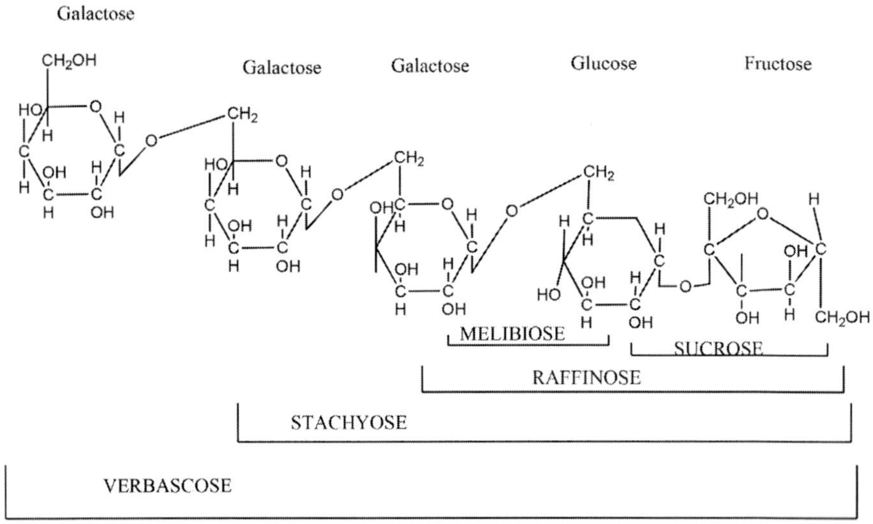

It is undigested by the human body. It is fermented by large intestinal bacteria, with the products formed being short-chain fatty acids and gases.

Key Takeaway

- Oligosaccharides comprise 2 to 10 units of monosaccharide molecules, which are released upon hydrolysis.
- Disaccharides are formed when glycosidic linkages covalently bond two monosaccharides together.
- During the union of two monosaccharide units, water molecules are eliminated, forming a glycosidic bond.
- Substances joined with glucuronic acid via glycosidic bonds increase water solubility (**glucuronidation**).
- Chitobiose is a disaccharide amino sugar comprising glucosamine units linked by β-1,4 bonds.
- What is the effect on the optical rotation of sucrose before and after hydrolysis.?
- Sucrose has an α-1, β-2-glycosidic linkage between glucose and fructose. As a result of this kind of linkage, the anomeric carbons of both component monosaccharide units are linked.

- Trehalose is a non-reducing disaccharide. It comprises two glucose units linked in an α-(1→1)-linkage.
- Trehalose is the major blood sugar in insects.
- Herbivorous animals digest raffinose; hence, beetroots are used as food for cattle. But poorly utilized by man.
- On complete hydrolysis, Gensianose yields one fructose molecule and two glucose molecules.
- Maltotriose comprises three glucose molecules bound together with α-1,4 glycosidic bonds. It is a reducing trisaccharide.
- Stachyose is a tetrasaccharide having four monosaccharides, of which two are galactose (pyranose) units, one glucose (pyranose) unit, and one fructose (pyranose) unit. It is a non-reducing sugar

Intext Questions

1. Write the names of three disaccharides with their monosaccharide sugars.
2. Why do some sugars have reducing properties?
3. What sugar is known as milk sugar? Write the monosaccharides present in it.
4. What is lactose intolerance?
5. How does maltose differ from isomaltose?
6. Draw the ring structures for cellobiose and lactose.
7. How is cellobiose obtained?
8. What is caramelization?
9. What is the major blood sugar in humans and insects?
10. Why does sucrose have no anomeric forms?
11. Name the monosaccharide units in maltose, maltotriose, and cellobiose.
12. Define the term oligosaccharide.

13 Choose one correct option for the following from the options given below (The answer is given in **bold** letters):

a A number of monosaccharides found in disaccharides

i Two ii Three

iii Four

b What is the effect of Glucuronidation on the solubility of a substance

i. **Increased** ii. Decreased

iii. Can't say anything

c A bond which held Glucose and Galactose together

i **β-1-4 glycosidic bond** ii α-1-4 glycosidic bond

iii β-1-3glycosidic bond

d Which of the following monosaccharides make maltose?

i Glucose and galactose ii **Glucose and glucose**

iii Glucose and fructose

e Which sugar is not digested in case of lactose intolerance?

i Galactose ii. Fructose

iii None of the above i. and ii.

f Which of the following is not a disaccharide sugar?

i Maltose ii **Galactose**

iii lactose

g Which type of reaction can split the disaccharides to release their component monosaccharides?

i Oxidation ii Reduction

iii Hydrolysis

h The linkage between sugar molecules and Gentiobiose?

i β-1,4 -linkage ii **β-1,6-linkage**

iii α-1,6-linkage

i Which sugar is known as inverted sugar?

i Maltose **ii Sucrose**

iii Lactose

j What number of hydrogen atoms are present in sucrose?

i 32 ii **22**

ii 42

k Which of the following is a non-reducing disaccharide sugar?

i. Gentianose **ii. Trehalose**

iii. Melezitose

l Maltose and Trehalose differ in their

i Number of monosaccharide units

ii Type of monosaccharide sugars

iii Type of linkages

m Major blood sugar in insects is

i Glucose ii Fructose

ii Trehalose

n Gentianose sugar is

i A reducing trisaccharide. ii A non-reducing trisaccharide.

iii A reducing tetrasaccharide.

o Which one of the following is an oligosaccharide?

i Stach ii Glycogen

ii **Lactose**

n The byproduct obtained during the formation of the glycosidic bond?

i Carbon dioxide ii Oxygen

iii **Water**

o Stachyose is a tetrasaccharide having four monosaccharides. The monosaccharide units present are

i Two galactose units, one glucose unit, and one fructose unit.

ii Two glucose units, one galactose unit, and one fructose unit.

iii One galactose unit, two glucose units, and one fructose unit.

Further Readings

Biomolecules: (Introduction, Structure, & Function). Carbohydrates. Suman Khowala, Deepak Verma, Samudra P. Banik. Drug Development and Biotechnology.Indian Institute of Chemical Biology.4, Raja S. C. Mullick Road.Kolkata-700032. INDIA.

https://www.verywellhealth.com/the-benefits-of-glucosamine-88809

Jain J.L., Sunjay Jain, Nitin Jain, Fundamentals of Biochemistry. S.Chand & Company Ltd. New Delhi.

Medical Biochemistry. Ethiopia public health initiative.

Textbook of medical biochemistry. Dr(Brig) MN Chatterjea and Rana Shinde

5

Polysaccharides (Glycans, Polyosides Poly-Carbohydrates, Polyholosodes)

Polysaccharides comprise several monosaccharide units (more than ten) linked by glycosidic bonds; therefore, they are high molecular weight carbohydrates. They are also called polyanhydrides of simple sugars. The most common monosaccharide unit of polysaccharides is D-glucose. The polysaccharides may differ in their monosaccharide units, molecular weight, chain nature, glycosidic bond types (alpha, beta), and linkages (1-2, 1-3, 1-4, or 1-6). They are tasteless, colorless, amorphous powders that are a little water-soluble. Some may form colloidal solutions. The polysaccharides provide structural stability (structural polysaccharides) to cells, prevent cells and tissues from drying out due to their water-binding properties, and reserve food for the future. Reserved polymeric carbohydrates are osmotically less active and can be stored in large quantities within the cell.

The right and left end of polysaccharides, respectively, are called the reducing and non-reducing ends. Based on the functional aspects, they are grouped in two ways.

- The first way depends on its digestibility—1. Nutrient polysaccharides are (digestible) starch, dextrin, and glycogen, and 2. Structural polysaccharides are (indigestible), chitin and cellulose.
- The second way, depending upon the constituent monosaccharide units, they are 1. Homoglycans (same units), and 2. Heteroglycans (different units).

5.1 Homoglycans

Homoglycans are those polysaccharides that, when subjected to hydrolysis, yield the same type of monosaccharide. Starch, glycogen, inulin (storage), cellulose, pectin, and chitin (structural) are homoglycans.

I. Starch

It is a glucose polymer, and alpha-linkages link its glucopyranose units. It is non-reducing in nature. It is found as insoluble starch granules within the plant cell. It is a reserved food material in cereals, legumes, potatoes, and other vegetables. Plants store starch in grains, roots, stems, leaves, rhizomes, etc. Different plant species have various shapes and sizes of starch grains.

The starch grain shapes differ and may be spherical, oval, lens-shaped, or irregular. Shapes of starch granules found in potatoes are oval or eccentric in shape. Wheat and potato starch granules are of A-type or B-type. Starch granules that are disc-like or lenticular are A-type. They are B-type starch granules if the shape is roughly spherical or polygonal. A-type starch granules are bigger than B-type starch granules. Rice contains polygonal and angular in shaped starch granules. Maize starch granules are angular in shape. In cereals, the starch grains are arranged in concentric layers. In potato starch, the arrangement is eccentric; comparatively, their granules are the largest, whereas rice starch granules are the smallest.

The raw starch has an outer cellulose layer, which makes it insoluble in cold water. This layer, when ruptured by heating in water, starch becomes soluble in water. Starch is a soft, amorphous white powder. It lacks sweetness. The specific rotation of starch is $+196^{o}$.

The saliva and pancreatic juice contain the alpha-amylase enzymes, which can digest the starch by breaking its alpha 1,4 linkages in a straight chain. The maltose and glucose units are released due to starch digestion by alpha-amylase. Salivary alpha-amylase is absent in the case of cows and buffaloes. Beta amylase is also an enzyme capable of digesting starch and is a plant enzyme.

Natural starches are made up of only glucose units, but these units in starch are arranged in two ways: as

A. amylose and

B. amylopectin.

All starches are made up of amylose and amylopectin. Their ratio can affect the rate and extent of starch digestion. The Amylose to Amylopectin ratio in starch is not the same, but it is different for different plants and even in plant varieties. Amylose content is 17-24% in bananas and potatoes, <1% in waxy maize, and 15-35% in rice. Amylopectin content is 76-83 % in banana and potato, is >99% in waxy maize, and is 65-85% in rice.

Amylose (10 to 20 %) comprises alpha-D-glucose units linearly arranged with α (1→4) linkages. Those having been placed in a branched manner with α (1→6) branch linkages make amylopectin (80 to 90 %) structure.

Starch yields glucose; hence, it is called glucosan.

A. Amylose or alpha amylose (1→4)-alpha-D-Glucopyranan

Amylose is a polysaccharide with a long, unbranched, straight chain structure containing 500 to 6000 alpha-D-glucose units with alpha 1-4 glycosidic linkages. Amylose in starch is found in the range of 15 to 20%. The molecular weight of Amylose ranges between 1,000,000 to 80,000. Hydrolysis of amylose by amylase at alpha 1,4 linkages yields mainly maltose units. Maltose is a disaccharide having an alpha-1,4 linkage between the two glucose units. The amylose part is responsible for the development of blue color with iodine.

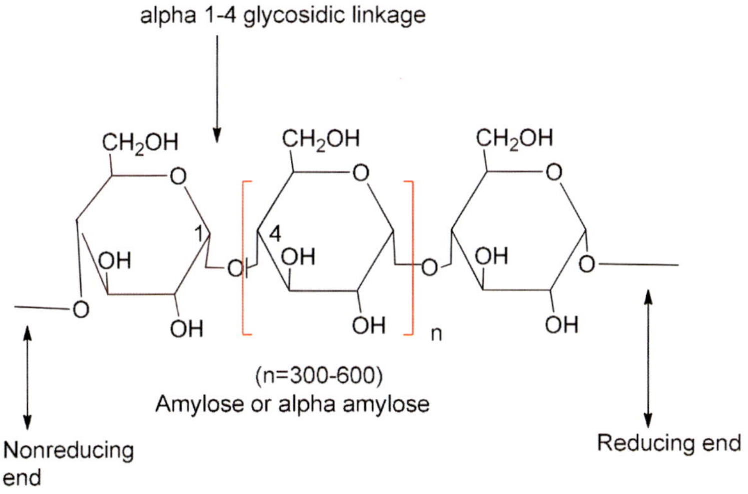

B. Amylopectin or isoamylose or beta amylose

Amylopectin is a polymer of several D-glucose units, about 80 to 85 % of the starch, arranged like branches in a tree. Amylopectin structure has alpha-D-glucose units arranged in two different ways. The alpha-D-glucose units of the main straight chain structure have 1-4 linkages like amylose and get attached to another chain to establish branch points through 1-6 linkages. Thus, amylopectin is a branched structure. Its molecular weight range is extensive, ranging from 50,000 to 10.00.000 Daltons. This means there is a very high number of glucose units of 300 -5,500 per molecule.

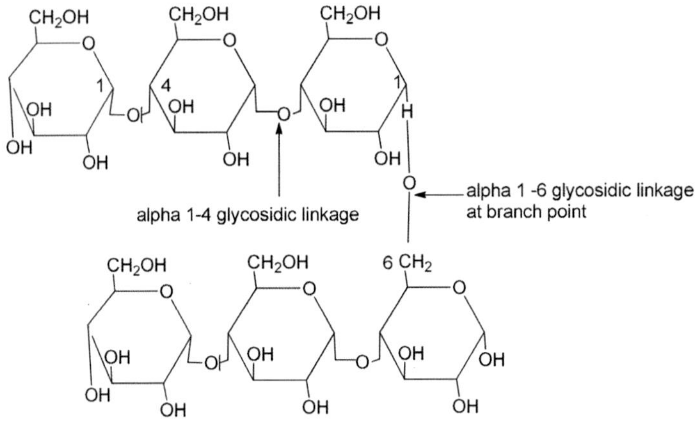

Amylopectin

Carbon 1, 4, and 6 of glucose are engaged in forming alpha 1-4 glycosidic linkage and forming alpha 1-6 glycosidic linkage at the branch point. Isomaltose is produced during the incomplete digestion of amylopectin. On average, the chain contains 24 glucose units.

Comparisons between amylose and amylopectin are under

- Amylose is an unbranched, simple structure. Its solubility is higher in water than that of amylopectin.
- Amylose, without swelling, is soluble in hot water. Amylopectin swells in water.
- Amylose does not form gel or starch paste, contrary to amylopectin.
- Amylose produces a typical blue color with iodine, whereas amylopectin forms a purplish color with iodine.
- Amylopectin can also be attacked, but not at 1-6 linkages, for which a debranching enzyme (alpha-1, 6 glucosidase) is required.

Starch hydrolysis by amylase enzymes

The types of amylases are alpha, beta, and gamma amylases.

Alpha amylase

Alpha-amylase is secreted in pancreatic juice and saliva. It is also found in humans, animals, plants, and microorganisms. They hydrolyze starch into maltose. They produce Dextrins (dextrinogenic) and can attack α-($1 \rightarrow 4$) bonds in the interior of starch molecules and related oligo- and polysaccharides. They produce Dextrins and then convert them into maltose. Manufacturing of beer and other alcoholic beverages prefers cereal alpha-amylases.

Beta amylase

It is found in microbes and plants. Unlike alpha-amylase, beta-amylase acts only at the non-reducing chain end to liberate beta-maltose. It is known to function by multiple attacks, producing several maltose molecules before the malto-oligosaccharide chain is released from the active site. It hydrolyzes starch into maltose.

Gamma amylase

It is found in microbes, animals, and plants. Gamma amylase cleaves both the alpha-1, 4-glycosidic and alpha-1-6-glycosidic linkages. It starts from the non-reducing end of the starch molecules, releasing glucose residues.

It also cleaves the alpha-1-6-glycosidic linkages.

II Dextrins

Dextrins are starch or glycogen hydrolysis products. They are of low molecular weight carbohydrates, smaller and less complex molecules than starch. The Molecular weight ranges between 4500 and 85000, depending on the number of $(C_6H_{10}O_5)n$ units in the polymer chain. The general formula of starch and dextrin is the same. However, both have different structures; dextrin is a less complex, smaller molecule. It comprises several glucose molecules bound together using alpha 1, 4, and alpha 1, 6 glycosidic links. The partial hydrolysis of starch releases Dextrins. Partial hydrolysis is carried out in different ways, including salivary amylase, dilute mineral acids, and heat. The Dextrins formed are amylodextrin, erythrodextrin, and achrodextrin, which give blue, red, and no color, respectively, with iodine. They have a faint, sweet taste. They form a strong adhesive paste with water. Dextrins are often used as a mucilage or adhesive for paper.

Starch and glycogen, both polysaccharides, have branch points at alpha -1,6 bonds, where alpha-1,4-glucanmaltohydrolase comes to its limit of hydrolysis of amylopectin or glycogen. So, the leftover fractions of the polysaccharide are limited dextrin.

III. Dextrans

They have glucose units linked together by α-(1-6) and α-(1-3) glycosidic linkages so that they have linear and branched chains.

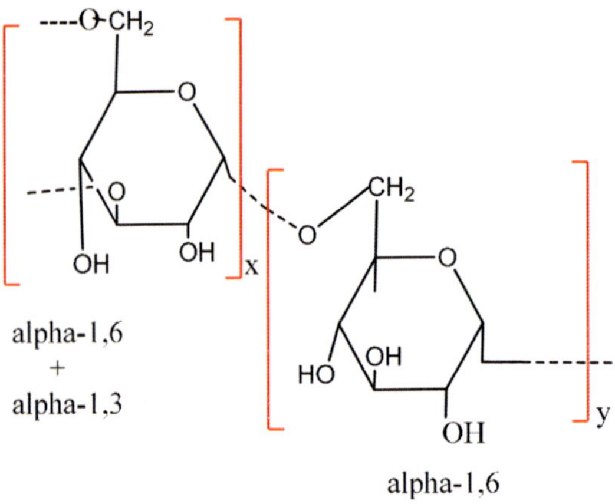

Dextrans

Applications of Dextrans

- It is used medicinally as an antithrombotic (reduces function) and decreases vascular thrombosis. It reduces erythrocyte aggregation and platelet adhesiveness due to an increase in electronegativity by the presence of dextrans.
- It decreases blood viscosity.
- It, as a volume expander, causes hemodilution and improves blood flow. Due to their larger size, Dextrans cannot pass out of the vessels and cause water retention due to their potent osmotic effect.
- Dextran is a plasminogen activator with thrombolytic features.
- It is used as an eye lubricant in eye drops.
- It is occasionally used as a volume expander to replace the lost blood when replacement blood is unavailable. Its use must be done carefully as it creates electrolyte disturbances, including hyponatremia.
- It can raise blood sugar levels.
- Dextrans are used for several different purposes, e.g., in cryopreservation, as well as for storing organs for transplantation and as carriers in vaccines.

IV Glycogen

It is stored in animals' liver (7% of wet weight) and muscles. It is reserved food for animals. It is formed from glucose, and the same process occurs upon acid hydrolysis, which yields glucose units.

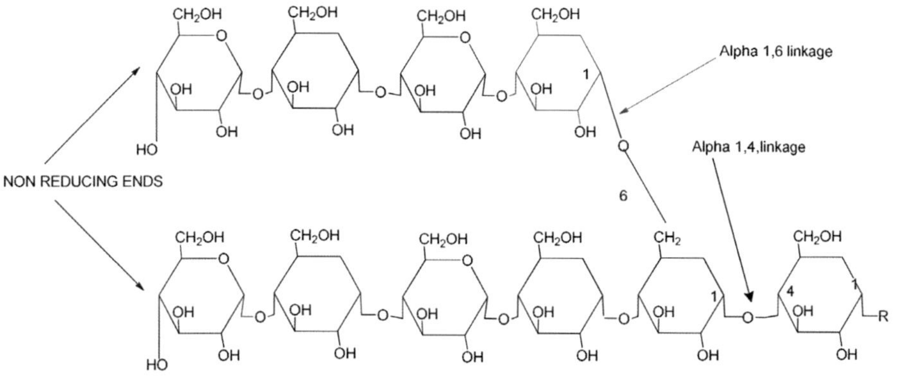

STRUCTURE OF GLYCOGEN

- It is also hydrolyzed by alpha-amylase. The structure of glycogen is similar to the structure of starch, but it contains more branch points, and branching is present after 8-10 glucose units. Phosphorylase enzyme also acts on glycogen and release glucose -1- phosphate and (glycogen) n-1 units. When hydrolysis is incomplete, it yields maltose. It is a non-reducing sugar. The red color given with iodine disappears upon boiling and reappears on cooling.

Physical Properties

- Glycogen particles are smaller than starch grains found in plants.
- Its color is white and is soluble in water (more than amylopectin).
- It is a highly branched and compact (like amylopectin) molecule.

V Cellulose

Cellulose is an odorless, solid and colorless compound. It is fibrous in nature, rigid in property, and white in color. It is not found in animals. It cannot dissolve in water and many organic solvents. It is the most abundant on earth. It lacks sweetness and gives no color with iodine. It absorbs water and increases the bulk. When concentrated acids (H_2SO_4, HCl) or concentrated alkali(NaOH) are used for the hydrolysis of cellulose, its monomer units released are glucose.

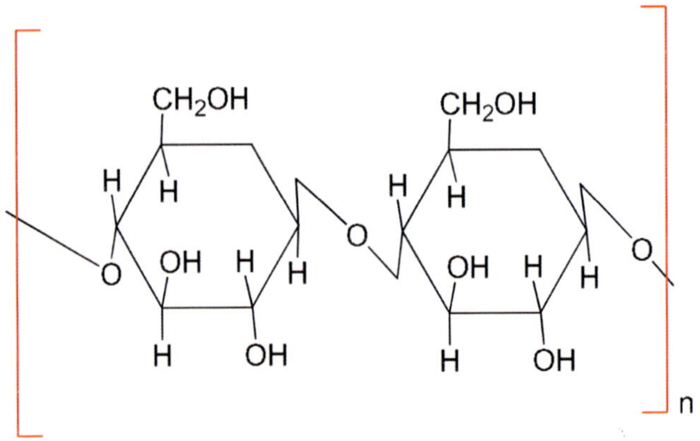

Cellulose

It is a polysaccharide structure of a plant cell wall. It is the most widely distributed plant polysaccharide for cell walls. It is present in the wood of the trees along with lignin. It gives tensile strength to the plants. Its molecular weight (2,00,000 to 20,00,000 Daltons) and glucose residue per molecule (1250 to 12,500 residues) have a wide range. A homopolysaccharide comprising glucose units (glucopyranose) bound by beta 1-4 glycosidic linkage and arranged in a linear structure. It contains several parallel chains that are united by hydrogen bonds. It is digested or hydrolyzed by cellulase (beta-glucosidase cleaves β-linkages) enzyme present in the rumen microbes, intestinal microbes of herbivores, wood-rotting fungi, termites and by a parasitic protozoan (Trichonympha). The animals cannot digest cellulose without an enzyme that cleaves β-linkages between glucose molecules. Cows, sheep, and goats can digest cellulose through microbial digestion. Some termites can digest cellulose. Wood, paper, cardboard, and cotton are all primarily cellulose.

Polymeric products are derived through the synthetic modification of cellulose by the esterification of free hydroxyl groups.

1. **Cellulose Nitrate** (nitrocellulose or guncotton). It is obtained when all hydroxyl groups of cellulose are replaced with nitrate groups (-ONO2) after its treatment with nitric acid (HNO3). It is an inflammable material.

   When partially nitrated, pyroxylin is obtained and used to manufacture various items like collodion, plastics, lacquers, and nail polish.

2. **Cellulose Acetate**. A reaction between cellulose and acetic anhydride, and an acid catalyst, synthesizes it. It can be converted to acetate rayon filaments for fabrics.

3. **Rayon.** Viscous rayon fiber can also be obtained. It has a super absorbent nature, doesn't trap heat, and wicks away sweat.

VI. Inulin (Fructans)

It is another storage form of homopolysaccharide and occurs as a reserve carbohydrate in many plant families, such as chicory, rye, onion, garlic, and dahlia bulbs. It is made up of fructose units and is also known as fructan. The fructose units of insulin are joined by a $\beta(2\rightarrow1)$ glycosidic/glucosidic bond; usually, the terminal group is glucose. The molecule is almost exclusively linear, with only a few percent branching.

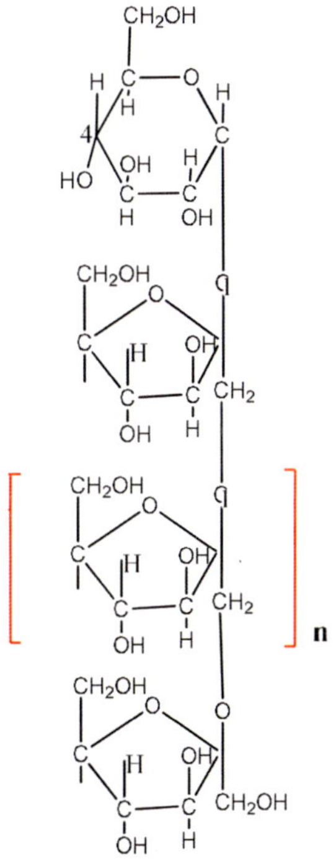

Inulin

It is used to measure the glomerular filtration rate and the renal clearance rate. Inulin is neither digested nor absorbed in the human stomach and can act as a prebiotic. It passes through the digestive system, except in the colon, where bacteria metabolize inulin, releasing significant amounts of carbon dioxide and methane. Inulin is a hypoglycemic agent.

Inulin comprises 30 to 35 fructose units per molecule. Its molecular weight is about 5000 Daltons. It is a non-reducing sugar and does not give any color with iodine.

Physical properties

- It is tasteless.
- It is a white powder in color.
- Insoluble in cold water.
- It dissolves easily in warm water, forming a colloidal solution that does not create a gel when cooling.

VII. Chitin

It is found in fungi, crustaceans, and insect exoskeletons (cuticle).

It is a homopolysaccharide comprising linear N-acetyl-D-glucosamine units held together by beta-1, 4 glucosidic linkages. It is closely related to cellulose.

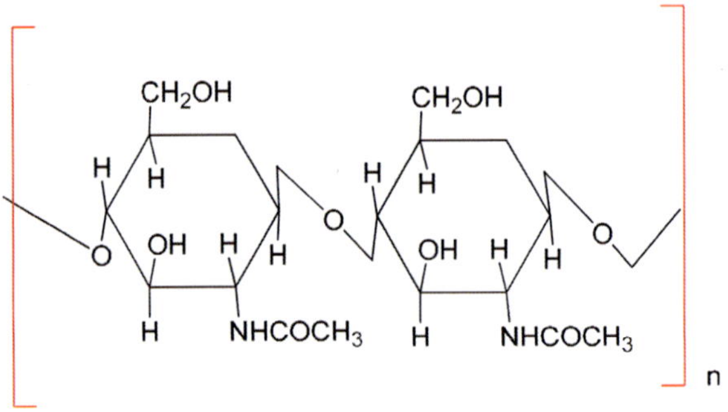

Chitin

Vertebrate animals do not digest chitin (like cellulose). In nature, Chitin is the second most abundant polysaccharide, next to cellulose. It is most favorable to make surgical threads due to their flexible properties and strong material. Its biodegradability properties accelerate wound healing.

5.2 Heteroglycans

These polysaccharides are made up of a mixture of two or more different monosaccharides, which are released upon their hydrolysis.

I. Mucopolysaccharides or Glycosaminoglycans (GAGs)

These polysaccharides are unbranched heteropolysaccharides having both the simple and derivatives of sugars (amino and uric sugars). Keratan sulfate contains galactose and not acid sugar.

They were first isolated from mucin and hence so named. They, therefore, are also known as Glycosaminoglycans (GAG). These are gelatinous substances. They support the structure of connective tissues or mucous membranes. They hold large quantities of water and are sticky, acting as lubricants. For example, hyaluronic acid, heparin, chondroitin -4- sulfate, and chondroitin -6- sulfate.

The mucopolysaccharides classification includes

A. Polycarboxylates

- Hyaluronic acid and chondroitin

B. Polysulfates

- keratan sulfates

C. Polycarboxy sulfates

- chondroitin 4 sulfates (chondroitin A)
- chondroitin 6 sulfates (chondroitin C),
- dermatan sulfate and
- Heparan sulfate

Common Mucopolysaccharides

Mucopolysaccharides	**components of disaccharide units**	**Linkages for disaccharides for repeated units**	
Hyaluronic acid or Hyaluronan	D-glucuronic acid+ N-acetyl-D-glucosamine	ß-1→3	ß-1→4
Chondroitin sulfate A	D-glucuronic acid+ N-acetyl-D-galactosamine-4-sulfate	ß-1→4	ß-1→3
Chondroitin sulfate B	L-iduronic acid+ N-acetyl-D-galactosamine-4-sulfate	Alpha-1→3	ß-1→4
Chondroitin sulfate C	D-glucuronic acid+ N-acetyl-D-galactosamine-6-sulfate	ß-1→3	ß-1→4
Dermatan sulfate	Iduronic acid, N-acetylgalactosamine,	β (1→3)	Only one with iduronic acid
Keratan sulfate	Galactose and others vary	β (1→4)	Very variable
Heparin	Glucuronic acid , Glucosamine	α (1→4)	Intracellularly
Heparan sulfate	Glucuronic acid , Glucosamine	α (1→4)	Extracellularly

A. Polycarboxylates— glycosaminoglycans

1. Hyaluronic acid (HA)

Hyaluronic acid (HA) is a polysaccharide with a high molecular weight. It is produced by specialized connective tissue cells and fibroblasts. It is found in the umbilical cord, nasal cartilage, vitreum, cutis, lymph of the thorax, and plasma. Synovial fluid and connective tissue(synovial membrane) contain HA in the highest concentration. It is quickly removed from the blood by the enzymatic action of hyaluronidase. It is a cell-cementing substance found in the eye's synovial fluid and vitreous humor. The structure comprises repeating units of glucuronic acid united to N-acetyl-d-glucosamine by beta-1,3. Beta-1,4 glycosidic linkage unites such repeating units. It acts as a shock-absorbing fluid for the joints. Hyaluronan helps in tissue repair and is a major component of the skin.

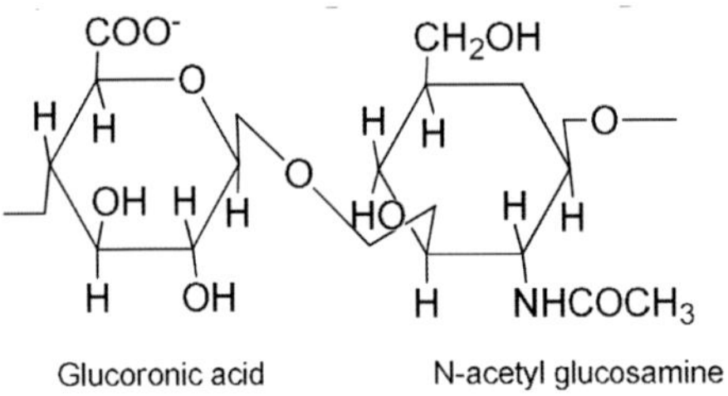

Hyaluronan Acid (HA) Unit

Clinical Utility

1. Rheumatic disease: Elevated serum HA levels indicate synovial inflammation (rheumatoid arthritis (RA).
2. Liver disease: Higher serum or plasma HA levels can be found when there is an increased production, with decreased **c**learance under various conditions:
    - Production of HA is more
    - Release of HA from damaged tissues
    - Removal of HA by the liver

2. Chondroitin

This mucopolysaccharide is found in the cornea and cartilage. Chondroitin occurs naturally in animal connective tissues. Its sulfation produces chondroitin sulfate. Since there are no major dietary sources of chondroitin, its supplements are preferred. Supplements are prepared from animal cartilage. Clinically, it reduces osteoarthritis pain, joint stiffness, and swelling.

B. Polysulfated Glycosaminoglycans

1. Keratan Sulfates (kerato sulfate)

It is a sulfated glycosaminoglycans and does not contain any uronic acid.

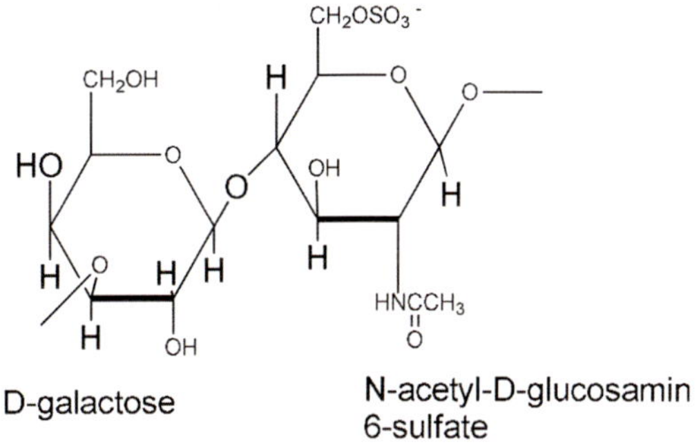

Keratan sulfate

There are two main types: one in the cornea (Type I), cartilages (Type II), and bone. Its polymer structure is linear and consists of repeating disaccharide units.

C. Polycarboxy Sulfates Glycosaminoglycans

1. Heparin

Heparin and heparan sulfate are sulfated glycosaminoglycans (GAGs), acidic mucopolysaccharides. It occurs as a proteoglycan (PG). Heparin consists of 15 to 100 sulfated two types of monosaccharides, alternating L-iduronic acid and D-glucosamine. These sulfated monosaccharide units are held by 1 to 4 links in linear or unbranched chains.

Heparin is formed from heparan sulfates. Heparan sulfates are released as heparin in blood after their modification in the mast cells (of the lungs, liver, and other cells). Heparin is an important anticoagulant. One disaccharide unit of heparin has more sulfate groups (3 number) than those in heparan sulfate(1 number).

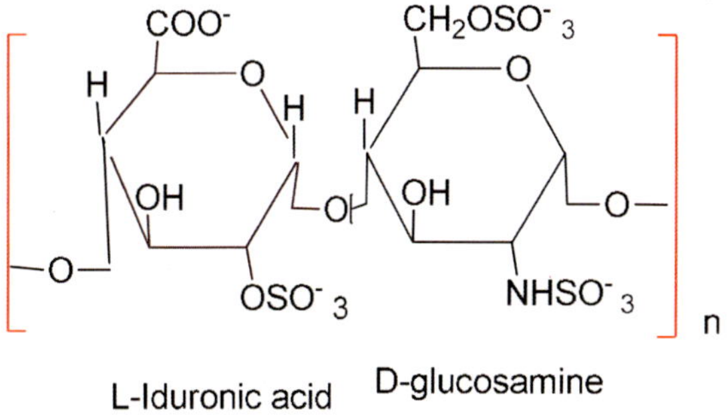

Heparin

2. Chondroitin Sulphate

It is a sulfated glycosaminoglycan (GAGs) made of a chain consisting of alternating sugars (N-acetylgalactosamine and glucuronic). It occurs as a proteoglycan due to the attachment of the chain. Two major chondroitin sulfates are A and C. The positions of their sulfate group are different in both of them. The sulfated group is at position 4 in the case of chondroitin sulfate A (chondroitin 4-sulfate) and is at position 6 in the case of chondroitin C (chondroitin 6-sulfate). The essential structural component of cartilage is chondroitin sulfate.

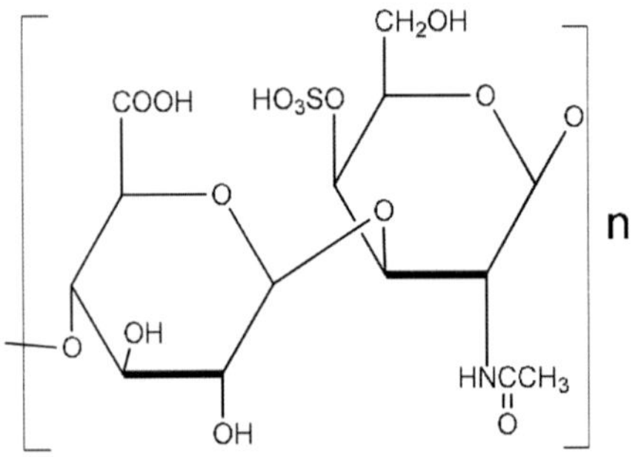

Chondroitin Sulfate

Chondroitin – 4 – Sulfate (chondroitin A)

It is present in tendons and synovial fluid. It comprises repeating units of disaccharide glucuronic acid and 4-sulfated galactosamine.

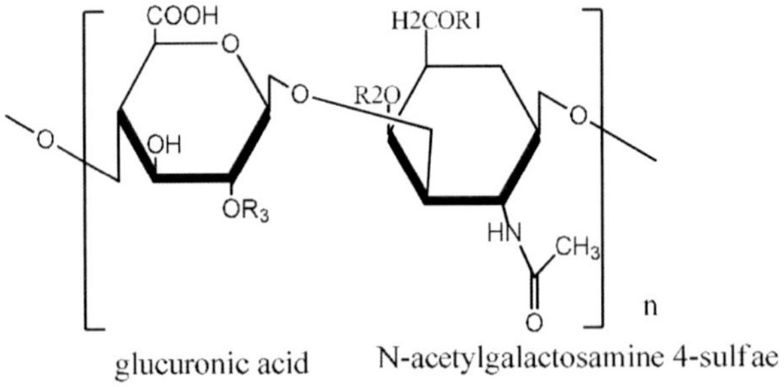

Chondroitin 4-sulfated
R1 is H, R2 is SO3H, and R_3 is H.

Chondroitin – 6 – Sulfate (Chondroitin C)

It is similar in structure to chondroitin -4- sulfate but possesses a sulfate group at six carbons of galacturonic acid.

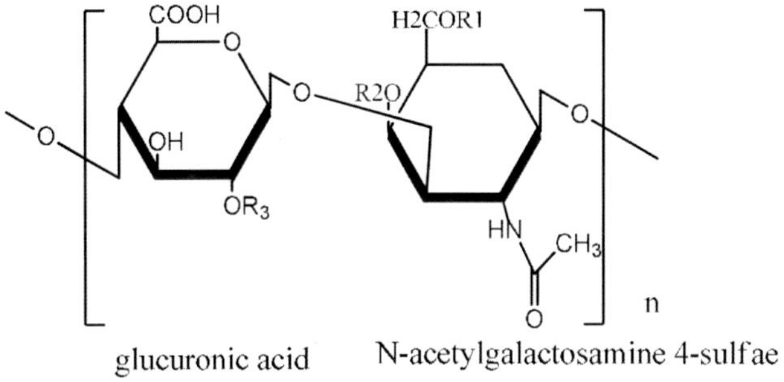

Chondroitin 6-sulfated

R1 is SO3H,R2 and R3 is H

3. Dermatan Sulfate (Chondroitin sulfate B)

It is a linear mucopolysaccharide structurally similar to chondroitin sulfate A (CS-A), but L-iduronic acid replaces the D-glucuronic acid. These two uronic acids differ only at C5. DS chains are made with the alternating arrangement of 50-200 units of disaccharide units consisting of L-iduronic acid (IdoUA) and *N*-acetyl-d-galactosamine (GalNAc) residues, held together with β 1→3 or 1→4 bonds. These two sugars are arranged in an alternate repeating disaccharide unit [L-IdoUA-β(1→3)-D-GalNAc4S-β(1→4)]. Galactosamine is sulfated in the C6 position, and iduronic acid is sulfated in the C4 position.

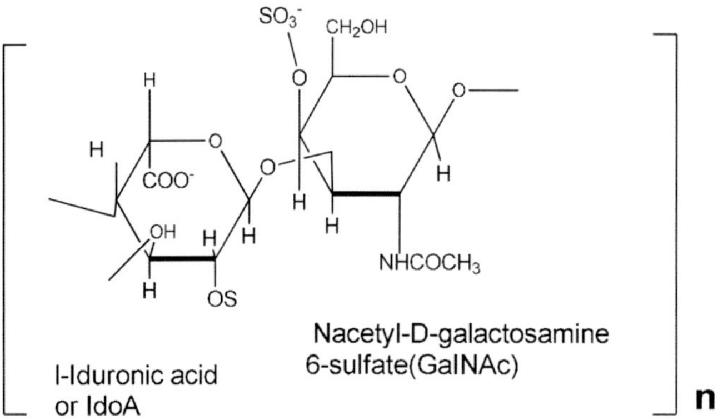

Dermatan sulfate

Chondroitin and dermatan sulfates are synthesized from chondroitin sulfate (CS) by epimerization of the glucuronic acid of CS to yield iduronic acid. Dermatan sulfate (DS) is covalently attached to the core proteins of proteoglycans (PGs). Dermatan sulfate (DS) is more in the skin and cartilage.

Therapeutic applications of Dermatan sulfate are

- Anticoagulant and antithrombotic activity.
- Dermatan sulfate, an essential component of the skin, can help repair wounds.
- Dermatan sulfate in infectious processes increases resistance to infectious diseases.
- Dermatan sulfate as a potential anticancer drug. Specifications.

D. Other Heteropolysaccharides

1. Agar or Agar Agar

Agar is a gelatinous polysaccharide employed in tissue culture. It is an unbranched **mucopolysaccharide**, a polymer of D-galactose 3-6 anhydrous L-galactose having sulfate esterification after every ten-galactose unit. It is obtained from some red algae or seaweed cell membrane species. Two classes of polysaccharides, agaropectin and agarose, are present in it, making it a heterogeneous mixture. It is used as a stabilizer, emulsifier, and thickener in the food industry. It cannot dissolve in cold water, but it dissolves in boiling water. It is used in icings, processed cheese, jelly sweets, and as a substitute for gelatin.

2. Pectins

It is a mucopolysaccharide and an essential constituent of plant cell walls. Chemically, they are known as polygalacturonides. Pectins form jellies in the presence of sugar and proper acid. They are found in the pulp of citrus fruits, apples, beets, and carrots. Pectins are methyl esters of pectic acid, which is a chain of 300 to 1000 alpha-D-galalctopyranosyluronic acid units in which galacturonic acid moieties are linked by **α**-1-4 glycosidic linkage. The most common side chain sugars are xylose, galactose, and arabinose.

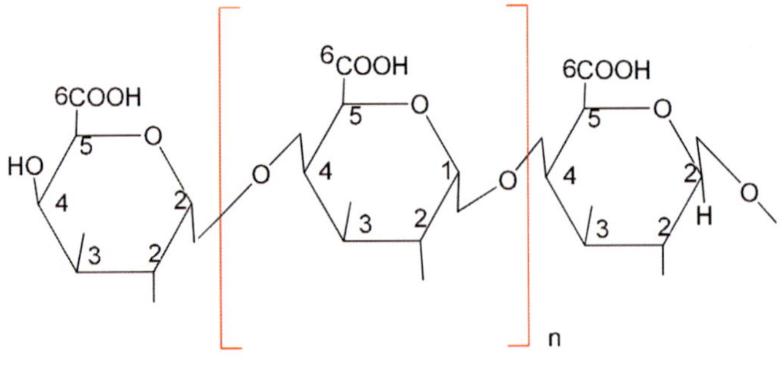

Polygalacturonic acid

Propectin is found in immature fruits, which is water-insoluble but gets converted to water-soluble pectin, which helps to ripen fruits, remain firm, and retain their shape. Pectin converts to pectic acids and soluble sugars (overripe fruit gets mushy). Protopectin

(Immature fruit, firm texture) →pectinic acids(mature)→

pectin(mature) →pectic acids (soluble sugars--overripe, mushy texture).

3. Gum acacia or Gum Arabic

It consists of a complex mixture of glycoproteins and polysaccharides, mainly polymers of arabinose and galactose. It is a higher polymer of Arabic acid. On hydrolysis, it yields galactose, arabinose, rhamnose, and glucuronic acid.

4. Agarose

The agar and agarose are not the same. Agar is derived from red algae. Agar, being gelatinous in nature, is used in the food industry as an ingredient. Agarose is obtained from agar (purification) when the agaropectin part is removed from the agar.

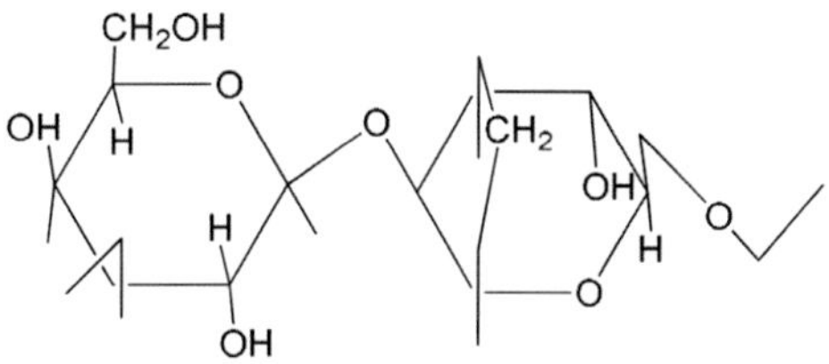

Agarobiose

Agarose is a linear polymer or polysaccharide of agarobiose(disaccharide) and is ordinarily used in gel electrophoresis. Monosaccharide units of agarobiose are D-galactose and 3,6-anhydro-L-galactopyranose linked by 1-4 linkages. Agarose is more expensive than agar.

5. Xylan

Xylan occurs in plant cell walls and some algae. It is a heteropolysaccharide with a variable structure that depends on the plant species. They are mainly backbone chains and side chains. Their backbone or main chain comprises monosaccharide D-xylose residues linked by a β-1,4 linkage. Glucose, arabinose, and acetylated sugar are various sugars found at the branch point.

E. Bacterial Cell Wall Polysaccharides

Pathogenic bacteria commonly produce a thick, mucous-like polysaccharide layer known as a capsule. Capsular polysaccharides consist of repeating subunits of one to six monosaccharides. There is enormous structural diversity. E. coli. Alone, it produces nearly two hundred different polysaccharides. Gram-positive and gram-negative bacterial cell wall chemical structures are different from each other. In the case of gram-positive bacteria, the cell wall's chemical structure is simpler than the chemical structures of gram-negative bacteria.

Mixtures of capsular polysaccharides (conjugated or native) are used as vaccines.

The bacterial cell wall maintains shape and structural integrity. Cell-surface polysaccharides provide structural components for the formation of biofilms. It also acts as a barrier between the cell wall and the environment and mediates host-pathogen interactions. The functions of bacterial cell wall polysaccharides are as follows:

1. Peptidoglycan (*murein or mucopeptide*)

The term peptidoglycan indicates that a molecule is derived from peptides and sugars (glycan). Short peptides cross-link a linear glycan strand. The rigid portion of the cell wall is due to peptidoglycan, known as murein. Peptidoglycan is found only in bacterial cell walls but not in human cells. Peptidoglycan (PG) is a heteroglycan polymer or polysaccharide that possesses a long chain of alternatively arranged two different types of derivatives of monosaccharide glucose, namely NAG (N-acetylglucosamine) and NAM (N-acetylmuramic acid). These two amino-sugar derivatives, NAM and NAG, are components of glycans. Besides long chains, they have cross-linking tetrapeptide side chains of four amino acids attached to NAM. The four amino acids are L-alanine, D-glutamine, L-lysine(or meso-diaminopimelic acid, DPA), and D-alanine. When cross-linked to other chains, the side chain provides a mesh-like structure that provides rigidity to peptidoglycans. This rigidity character helps to maintain cell shape and prevent osmotic lysis of cells. If the synthesis of peptidoglycans is inhibited, as in the case of antimicrobial agents, cell lysis occurs.

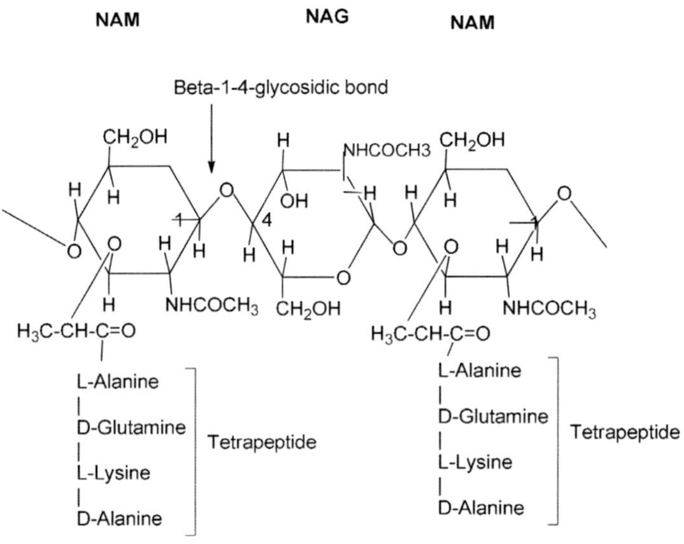

Peptidoglycan monomer arrangement

The cells make use of the L-isomeric form of amino acids. However, using the mirror image, D-amino acids in peptidoglycans protect the cell wall from the attack of proteases. The tetra peptides can be **directly cross-linked** to one another, with the D-alanine on one tetra peptide binding to the L-lysine/ DPA on another tetra peptide. In many gram-positive bacteria, a cross-bridge of five amino acids, such as glycine (peptide interbridge), connects one tetrapeptide to another. However, the cross-linking increases the strength of the overall structure.

The cell wall becomes insensitive to lysozyme in the presence of peptidoglycan. Peptidoglycans are involved in bacterial cell reproduction. About 90% of the dry weight of Gram-positive bacteria is contributed by peptidoglycans in addition to their higher cell wall thickness (20 to 80 nanometers). About 10% of the dry weight of Gram-negative bacteria is contributed by peptidoglycans, but they have a cell wall thickness of 7 to 8 nanometers. The peptidoglycans are found only in bacterial cell walls.

2. Lipopolysaccharides (LPS) or lipoglycans

It consists of a lipid and a polysaccharide. LPS is found in Gram-negative bacteria and provides strong immune responses in animals.

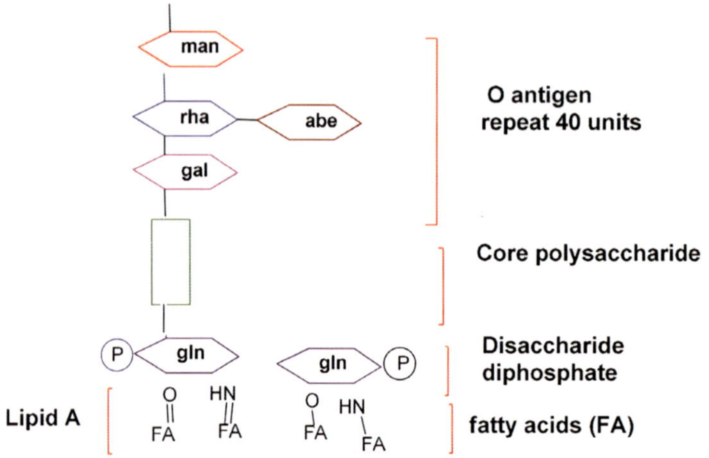

It also protects bacteria from environmental stress factors. It comprises three structural domains, e.g., Lipid A, core oligosaccharides, and the O antigen. Lipid A (hydrophobic portion) is an acylated β-1'-6-linked glucosamine disaccharide.

3. Alginate

Alginate is a natural exopolysaccharide that has two monomeric units, e.g., β-(1, 4)- linked d-mannuronic acid (ManAp or M) and α-(1, 4) linked l-guluronic acid (GulAp or G). Alginates are extracted from brown algae or produced by soil bacteria. The residues found are G residues (GGGGGG-- consecutive), M residues (MMMMMM-- consecutive), and M and G residues (GMGMGM- alternating residues).

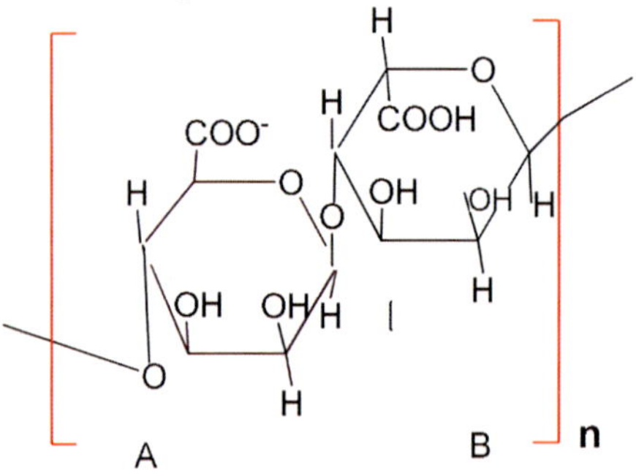

Ring A. beta-D-Mannuronic acid
Ring B. alpga-L-Guluronic acid.

Alginic acid

Food uses alginate as a stabilizer, viscosifier, and gelling agent. It is also used in various industries related to beverages, paper, printing, pharmaceuticals, and biotechnology. It is used to make dental impressions in dentistry and medicine to treat gastroesophageal reflux disease and stomach ulcers. In biomedicine, it is used as a hydrogel in wound healing, drug delivery, and tissue engineering applications. Treating indigestion, conditions like gastroesophageal reflux disease (GERD) and stomach ulcers.

Key Takeaways

- Polysaccharides comprise several monosaccharide units (more than ten) linked by glycosidic bonds; therefore, they are high molecular weight carbohydrates.
- The right end of a polysaccharide chain is called the reducing end, while the left end is called the non-reducing end.
- How are polysaccharides classified?
- Starch is a glucose polymer, and alpha-linkages link its glucopyranose units. It is non-reducing in nature.
- A-type starch granules are bigger than B-type starch granules.
- The raw starch is insoluble in cold water due to the outer cellulose layer on starch granules. This layer, when ruptured by heating in water, starch becomes soluble in water.
- All starches are made up of amylose and amylopectin.
- The amylose part is responsible for developing a blue color with iodine.
- The structure of glycogen is similar to that of starch, but it contains more branch points.
- Inulin is a homopolysaccharide of fructose units, also known as Fructans.
- Inulin is used to measure the glomerular filtration and renal clearance rates.
- Chitin is a homopolysaccharide comprising linear N-acetyl-D-glucosamine units united by beta-1, 4 glucosidic linkages.
- Fibroblasts and other specialized connective tissue cells produce hyaluronic acid. It is found in the highest concentrations in synovial fluid and connective tissue, such as the synovial membrane.
- In the mast cells of the lungs, liver, and other cells, heparan sulfate chains are modified further and released into the bloodstream as heparin. It has important anticoagulant properties.

- Agar is a gelatinous polysaccharide employed in tissue culture. It is an unbranched mucopolysaccharide. Two classes of polysaccharides, agaropectin, and agarose, are present in it, making it a heterogeneous mixture.
- Pectins are a mucopolysaccharide and an essential constituent of plant cell walls. Chemically, they are known as polygalacturonides.
- The agar and agarose are not the same. Agar is derived from red algae. Agarose is obtained when agaropectin is removed from agar.
- Pathogenic bacteria commonly produce a capsule, a thick, mucous-like polysaccharide layer.
- Gram-positive bacteria have simpler chemical structures in their cell walls than Gram-negative bacteria.
- The term peptidoglycan indicates that a molecule is derived from peptides and sugars (glycan). The two amino sugars, NAG(N-acetylglucosamine) and NAM(N-acetylmuramic acid), are the components of glycans.

Intext Questions

1. Write about the starch granules' properties.
2. Why is raw starch insoluble in cold water but becomes soluble upon heating the water?
3. What are homoglycans and heteroglycans?
4. What color with iodine is given by starch and glycogen?
5. How are glucose units arranged in starch?
6. How does dextran differ from dextrin?
7. How do glycogen and starch differ?
8. Humans cannot digest cellulose, but some animals do. Give the reason for this statement.
9. Which polysaccharide is used for kidney function tests? Why?
10. Which mucopolysaccharide is produced by fibroblasts, and is its highest concentration found in synovial fluid?
11. Which mucopolysaccharide has anticoagulant properties?
12. Which two classes of polysaccharides are present in agar?
13. How are agar and agarose obtained?

14. What does the term peptidoglycans indicate?
15. What is the expanded form of two amino sugars-NAM and NAG?
16. Choose one correct option for the following from the options given below (The answer is given in **bold** letters):

a The most common monosaccharide unit of polysaccharides is

i. **D-glucose.** ii. D-fructose

iii. L-glucose

b Polysaccharides having the same type of monosaccharides is

i. Heteroglycans ii. **Homoglycans**

iii. Both i. and ii.

c Stored polysaccharide in plants is

i. Glycogen ii. **Starch**

iii. Cellulose

d The place where glycogen is stored in animals.

i. Muscles ii. Liver

iii. **Both i. and ii.**

e The example of structural polysaccharide

i. Glucose ii. Starch

iii. **Chitin**

f A polymer has beta glucose units

i. Cellulose ii. Starch

iii. **Both i. and ii.**

g Polymer having alpha -D-glucose units

i. Cellulose ii. **Glycogen**

iii. Starch

h The polysaccharide that cannot be digested by human beings**, Cellulose**

i. Dextrins ii. Starch

i The monomer units present in inulin polysaccharides are

i. **Fructose** ii. Galactose

iii. Mannose

j The Starch gives a blue color with iodine; by nature, it is

i. Reducing ii. **Non-reducing**

iii. Can't say about the nature

k The starch contains

i. Only amylose ii. Only amylopectin

iii. **Both amylose and amylopectin.**

l The color given by starch with iodine is

i. **Blue** ii. Red

iii. Green

m Which is an unbranched structure?

i. **Amylose** ii. Amylopectin

iv. Both i. and ii.

n Amylopectin is a part of starch, a plant polysaccharide, whereas glycogen is an animal polysaccharide. The glucose units in both of them are linked by glycosidic linkage. Which linkage(s) is/are common in their structures?

i. Alpha 1-4 ii. Alpha 1-6

iii. **Both i. and ii.**

o How many classes of polysaccharides are present in agar? making it a heterogeneous mixture.

i. Four ii. Three

iii Two

p Which polysaccharide has anticoagulant properties?

i. Keratin sulfate ii. **Heparin**

iii. Chitin

q Which one of the following statements is true for the size of starch granules?

i. **A-type starch granules are bigger than B-type starch granules.**

ii. A-type starch granules are smaller than B-type starch granules.

iii The sizes of starch granules A-type and B-type are the same.

Further Readings

Belitz .H.-D., Grosch W., Schieberle P. "Food Chemistry" 4th ed. Springer, 2009

Dermatan sulfate: new functions from an old glycosaminoglycan. Janet M. Trowbridge, Richard L. Gallo. Glycobiology, Volume 12, Issue 9, 1 September 2002, Pages 117R–125R, https://doi.org/10.1093/glycob/cwf066.

Dumville J.C., Fry S.C. (2003). Gentiobiose: a novel oligosaccharide in ripening tomato fruit. Planta 216: 484–495. [PubMed] [Google Scholar]

Functional carbohydrates. Zhang, Guoqiang, and Qinhong Wang. (2017), pp.269-298. (https://www.researchgate.net/publication/345618278). https://www.naturalproductsinsider.com/formulation/understanding-starch-functionality

https://www.amjmed.com/article/0002-9343(69)90162-4/abstract

https://www.drugs.com/inactive/alginic-acid-5.html

https://www.naturalproductsinsider.com/formulation/understanding-starch-functionality.

https://www.slideshare.net/Kaviv2/mucopolysacchrides

Identification and characterization of a stachyose synthase gene controlling reduced stachyose content in soybean. Dan Qiu, Tri Vuong, Babu Valliyodan, Haiying Shi, Binhui Guo, J. Grover Shannon, and Henry T. Nguyen. Theor Appl Genet. 2015; 128(11): 2167–2176. Published online 2015 Jul 16. doi: 10.1007/s00122-015-2575-0. PMCID: PMC4624830. PMID: 26179337.

International University of Africa, Biochemistry Assignment on Polysaccharides Structure and Functions. Chirombo Idris, 2014

Isomaltulose Calories, Foods, Benefits, Dangers. (2017, January 6). Retrieved from ://www.nutrientsreview.com/carbs/disaccharides-isomaltose.html Link

Life Sciences, Fundamentals and Practice I. Pranav Kumar and Usha Mina. (2017). Sixth Edition.Pathfinder Publication. New Delhi.

Microbial alginate production, modification, and its applications. Iain D Hay 1, Zahid Ur Rehman, M Fata Moradali, Yajie Wang, and Bernd H A Rehm. Microb Biotechnol. 2013 Nov; 6(6): 637–650. Doi: 10.1111/1751-7915.12076. PMCID: PMC3815931. PMID 24034361.

Palmer RE, Anderson RL (1972) Metabolism of gentiobiose in Aerobacter aerogenes. Journal of Bacteriology 112, 1316-1320 [PubMed:4640505] [Abstract]

Pathophysiological Significance of Dermatan Sulfate Proteoglycans Revealed by Human Genetic Disorders. Shuji Mizumoto,Tomoki Kosho, Shuhei Yamada and Kazuyuki Sugahara. Pharmaceuticals. 2017, 10(2),34; https://doi.org/10.3390/ph10020034.

Stipanuk M.H., Caudill M.A. Biochemical, physiological, and molecular aspects of human nutrition. 3rd Edition. Elsevier Health Sciences, 2012

Trehalose and Trehalose-based Polymers for Environmentally Benign, Biocompatible and Bioactive Materials Naozumi Teramoto, Navzer D. Sachinvala and Mitsuhiro Shibata Molecules 2008, 13(8), 1773-1816;

6

Metabolism, Metabolic Pathways, and Electron Transfer Chain

6.1 Metabolism

Dietary components like carbohydrates, lipids, and proteins are continuously broken down and synthesized through various chemical reactions (exergonic or endergonic) that are concerned with energy. These processes not only break them down but also interconvert them for their synthesis. Dietary components, when utilized, are known as nutrients the body. The number of reactions occurring in living cells maintains the body cells and fluid's homeostasis and dynamic status. To keep the dynamic state of a cell, many reactions are going on to synthesize or break down the different types and sizes of molecules. The total of these reactions is known as metabolism.

The metabolism may be **anabolic or catabolic**. Further, it may be aerobic or anaerobic, depending on oxygen availability. These four terms are explained below:

I. Aerobic and Anaerobic Metabolism

- **Aerobic metabolism--** In aerobic metabolism, the carbohydrates (substrate) in the presence of oxygen are metabolized to release energy, carbon dioxide, and water. The released energy is converted to ATP by two processes in the mitochondria (the Krebs cycle and oxidative phosphorylation). In oxidation, the electrons released from a glucose molecule are transferred to reduced NAD+ and FAD. The oxygen (electron acceptor) is reduced to water. Water formation [from H2(g) and O2(g)] is an exothermic reaction. The energy released from the chemical bond of glucose is captured as ATP, and heat is released.
- **Anaerobic metabolism---**Anaerobic metabolism is the combustion of carbohydrates (substrate) without oxygen. Glucose is converted to lactate when limited amounts of oxygen (O_2) are available. No ATP is formed in ETC, even though NADH is oxidized to NAD+ during the

conversion of pyruvate to lactate. This metabolism occurs during short and intense exercises, providing the required energy by utilizing more glucose molecules than in aerobic conditions.

- **Anabolic metabolism:** It involves synthesizing biomolecules and various reactions to convert the smaller molecules to larger ones. Examples are the conversion of monosaccharides to polysaccharides, amino acids to protein, and fatty acids to lipids. The anabolic state of the body is indicative that the body is gaining or growing.

Catabolic Metabolism

In this process, the larger molecules are broken down into smaller or simpler units. Simpler or smaller units generated have two options: either to be oxidized for energy release or to be incorporated in anabolic reactions. If catabolism is more than anabolism, the body wastes away during fasting, disease, and old age.

6.2 Metabolic Pathways for Carbohydrates

A pathway through a series of chemical reactions can build or break down molecules in a cell for cellular processes. Glucose is the most preferred energy source of all organisms. The metabolic study helps to know about the enzyme(s) involved in a given reaction; one can learn about the type of a given reaction (exergonic or endergonic) and various intracellular controls that determine the speed of the pathway reactions.

I Pathways for Carbohydrate Metabolism

There are various pathways for carbohydrate metabolism for the synthesis and breakdown of various types of carbohydrates. Carbohydrates (oligosaccharides and polysaccharides) must first be broken down into smaller monosaccharides by respective enzymes, enabling them to enter into catabolic pathway(s). The different pathways for carbohydrate catabolism are listed below:

- Glycolysis: There are anaerobic and aerobic glycolysis. Aerobic glycolysis ends in pyruvic acid formation. In anaerobic glycolysis, lactic acid forms from glucose. This pathway is known as the Embden Meyerhof Pathway (EMP).
- Krebs cycle or Tricarboxylic acid cycle is the aerobic pathway.
- The PPP or Pentose phosphate pathway or Hexose monophosphate shunt pathway. It converts hexose sugar into pentose sugars and regenerates NADPH.
- The gluconeogenesis pathway converts noncarbohydrates, simple compounds like amino acids and fatty acids, into glucose molecules

through *de novo* synthesis.

- Glycogenesis is the process that converts excess glucose into glycogen for cellular storage in animals. This process regulates the blood sugar levels and controls the osmotic pressure of the cells.
- Glycogenolysis is a process that involves the breakdown of glycogen into glucose, which can be supplied to different body tissues as and when required.

II Catabolic Pathways for Glucose

Breakdown of Carbohydrates occurs through various pathways classified into two groups:

A. The pathways mainly for energy

i) Glycolysis or Embden-Meyerhof pathway,

ii) Conversion of pyruvic acid to acetyl CoA through decarboxylation

iii) Acetyl CoA oxidation in TCA cycle/citric cycle/Krebs cycle,

B. The minor pathways for oxidation, but not mainly for energy purposes.

i) Hexose monophosphate pathway (HMP).

ii) Uronic acid pathway

The body gets energy from its various activities, which are performed by the body at work or rest. The body's activities depend on its metabolic rate. Even at rest, vital organs perform their activities. Such organs are the heart, kidneys, lungs, and brain. The metabolism or energy expenditure at the body's resting state refers to the resting metabolic rate (RMR) required to maintain life. It is also referred to as the resting energy expenditure (REE). It is essential to maintain homeostasis in the body, which needs energy. The metabolism or energy expenditure rate under basal conditions is known as BMR. However, BMR and RMR differ slightly.

6.3 Mitochondrial Arrangement

During various metabolic pathways, the mechanisms of electron transfer and the formation of high-energy bonds take place. One can only understand these events better if the role of mitochondria is understood clearly. Mitochondria are unique in having distinct double membranes, unlike other cellular organelles, and have their DNA (genome) and ribosomes (70S). It produces high-energy bond ATP in a cell's well-organized series of chemical reactions. Furthermore, mitochondria store calcium for cell signaling activities and generate heat. It mediates cell growth and death.

A mitochondrion (plural form, mitochondria) is a round or oval shape; its size ranges from 0.5 to 10 μm; it is found in the cytoplasm of eukaryotic organisms. The number of mitochondria per cell varies from zero (red blood cells) to hundreds or even thousands (liver and muscle cells) per cell. Their number is directly related to cellular activities. (The eukaryote that lacks mitochondria is the oxymonad *Monocercomonoides* species).

The mitochondrion has a double membrane arrangement with the following components:

1. **Outer membrane.** Shuttling of pyruvate from the cytosol is enabled by the located transport proteins located in the outer membrane. The outer mitochondrial membrane freely allows small molecules to pass through. For the transport of the larger molecules, it contains special channels.
2. **The inner membrane.** It is the site for oxidative phosphorylation due to the presence of the electron transport chain and ATP synthase. Only very small molecules are permitted into the gel-like matrix (central mass) after crossing the inner membrane. The inner membrane is folded as cristae to increase the surface area.
3. **Cristae:** The inner membrane thrown into folds is known as cristae, which are functional dynamic components for capturing the released energy into high energy bonds. Cristae increase the surface area of the inner membrane, which accommodates a much greater amount of electron transport chain enzymes and ATP synthase so that efficient oxidation occurs in the mitochondrion. Cristae help mitochondria to produce more energy at a faster rate. The cristae house the protein components of the electron transport chain (ETC). The ETC helps to transfer electrons through a series of oxidation-reduction reactions for oxidative phosphorylation of ADP (adenosine diphosphate) to ATP to harness the valuable energy derived from glycolysis and lipolysis. Pyruvate and fatty acids produced in the cytosol get converted to acetyl-CoA in the mitochondrial matrix to get oxidized to CO_2 via the citric acid cycle (the central pathway of oxidative metabolism).
4. **Intermembrane space**. It is a small space between the inner and outer membranes of mitochondria. The outer membrane is permeable in nature, but the inner membrane is not very permeable. The space acts as a functional barrier between the cytosol and the matrix. It helps in maintaining the proton gradient required for oxidative phosphorylation. The hydrogen atom has 1 proton and 1 electron. When it loses its electron

and forms an H^+ ion, it is left with only one proton in its nucleus; it is referred to as the only proton.

5. Matrix. The mitochondrial matrix is a viscous fluid in nature, containing a mixture of enzymes and proteins. It contains ribosomes, inorganic ions, DNA of the mitochondrial genome, nucleotide cofactors, and organic molecules. TCA cycle enzymes are located here to oxidize many metabolites entering it, to generate various intermediates, energy, water, and carbon dioxide.

6. Protein Complexes and Carriers

There are water-soluble and lipid-soluble carriers. The lipid-soluble ones are four membrane-bound complexes embedded in the inner membrane. They are named Complex I, II, III, and IV. Except for complex II, the other three are proton pumps. The water-soluble electron carriers are Coenzyme Q and cytochromes.

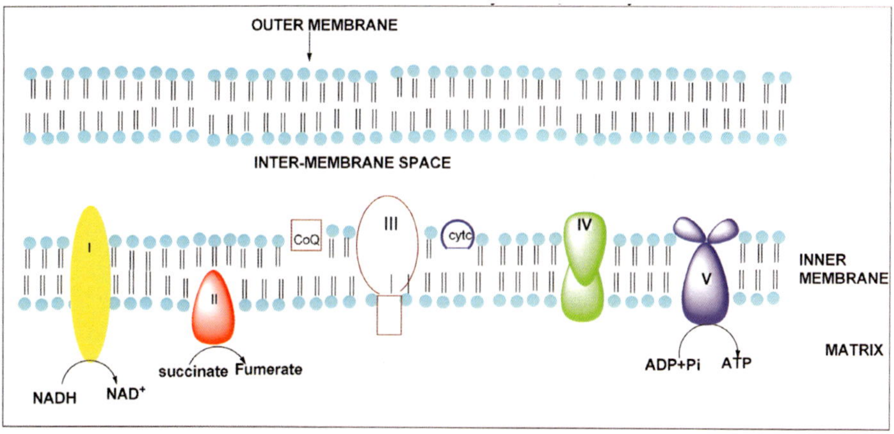

Fig. 6.1: Mitochondrial complexes

6.4 Electron Transport Chain (ETC) or Respiratory Chain and ATP Synthesis

A series of electron carriers collectively known by different names. They are known as the electron transport chain (ETC), the Cytochrome oxidase system, or the Respiratory chain. This chain provides a mechanism for transporting the hydrogen (proton) and electron stepwise from one acceptor to another and to the final hydrogen acceptor, which is molecular oxygen.

The hydrogen and electron acceptors are part of the oxidation-reduction machinery. Various metabolic pathways, like glycolysis, oxidation of pyruvic acid, and the citric acid cycle, release hydrogen (2H), which is accepted by appropriate coenzymes (in the electron transport chain) like NAD (nicotinamide adenine dinucleotide), FMN (flavin mononucleotide), CoQ (Coenzyme Q), UQ (ubiquinone), and cytochromes (cyt b,c, a, and a3).

The energy is released through the electron transport chain due to electron transfer. This energy pumps protons from the matrix to the cytosolic side. This will help generate an electrochemical gradient in the form of a proton gradient and a membrane potential. The gradient helps in ATP synthesis when the protons travel down their gradient into the matrix through the ATP synthase channel. Thus, electron transport and ATP production co-occur and are tightly coupled.

I. Electron Transport Chain Complexes

In several cellular processes (glycolysis, the TCA cycle, the beta-oxidation), the transfer of electrons and protons occurs to coenzymes (NAD and FAD). When electrons and a proton reduce NAD, it becomes NADH, and FAD becomes FADH2 and can take on two protons and four electrons to become $FADH_2$. The beginning of the first stage in the electron transport chain is the binding of these coenzymes to the protein complexes of the electron transport chain.

Mitochondria help in capturing the energy released during biological oxidation in such a way as to minimize the waste of energy, which is more in free oxidation. In free oxidation, energy is released at once at a very high temperature; all energy is lost as heat, light, and sound. Several moles of ATP can be generated if a large amount of released energy is captured during the redox reaction of one mole of the reduced cofactor. The released energy is so large that it is beyond the capacity of the biological system to convert all energy into ATP.

In biological oxidation, the energy is not transferred to sound and light, as happens in free oxidation. In order to minimize wastage as heat energy, mitochondrial ETC is such an arrangement in the biological system that releases a small amount of energy stepwise through an electron transport chain so that the energy is captured in the form of ATP. This is possible because of many enzymes and prosthetic groups at their active site of the chains. These prosthetic groups can be of two general types. One group carries electrons and protons(Flavoproteins, Quinone, NADH, and NADPH). In contrast, the second group only carries electrons (Cytochrome-heme as a prosthetic group, Iron-Sulfur proteins, or Fe-S (nonheme).

II. Cofactors

A. Nicotinamide Adenine Dinucleotide (NAD):

When the substrate releases hydrogen, it is accepted by the oxidized form of NAD+(Nicotinamide adenine dinucleotide) or NADP (nicotinamide adenine dinucleotide phosphate). After accepting hydrogen, both the NAD and NADP get converted to their reduced form, NADH and $NADPH_2$, respectively. They provide the electrons, or protons, to reduce oxidized compounds.

B. Flavin Adenine Dinucleotide (FAD)

The flavoproteins are flavin groups containing protein, namely FAD (flavin adenine dinucleotide) or FMN (flavin mononucleotide). Both flavoproteins serve as oxidoreductase enzymes and remove free radicals involved in oxidative stress.

The FAD converts to $FADH_2$ after accepting hydrogen atoms, even from succinic acid. Similarly, FMN converts to $FMNH_2$ (reduced form), accepting 2H from NADH2. Complex I or NADH dehydrogenase is a flavoprotein (FMN), while the flavoprotein of complex II is succinate dehydrogenase (FAD).

III. Mobile Electron Carriers

A. Coenzyme Q (CoQ) or Ubiquinone

It is , located between the flavoproteins and cytochrome. Under anaerobic conditions, it is in an oxidized form known as quinone in mitochondria. In reduced form, it exists as hydroquinone. It is small and lipid soluble, so it is easily diffusible within the lipid bilayer of the inner membrane of mitochondria. It carries both protons and electrons. The complete reduction of ubiquinone occurs in two steps with the help of two electrons and two protons. Polyisoprenoid side chains occur in coenzyme Q, ranging from 6 to 10. The number is written after the letter Q. Coenzyme Q10 means it has ten isoprenoid side chains.Q refers to the quinone group. The coenzyme CoQ, when fully oxidized, is ubiquinone,Q. When CoQ is fully reduced it is called ubiquinol,QH_2, but when partially reduced, it is converted to semiquinone or ubisemiquinone. Coenzyme Q is not a protein-bound prosthetic group.

B. Cytochromes

Cytochromes are heme proteins, large complexes, and are involved in electron transport in mitochondria. About 30 cytochromes have been identified. They are differentiated by spectroscopic spectra. The mitochondria contain cytochromes classified as cytochrome b ($b_{566 \text{ and }} b_{562}$), c (c and c_1), and a (a and a_3). porphyrins. Cytochromes contain heme (a tetrapyrrole compound with Fe iron) prosthetic groups. Heme prosthetic groups of cytochromes a and b are not covalently bound to their associated proteins, but in the case of cytochrome c, they are covalently bound to their associated protein. Iron exists in interconvertible form as ferrous, (Fe^{2+}) or ferric (Fe^{3+}) forms, as Fe can accept or donate an electron, and hence the atom is reduced or oxidized.

IV. Protein complexes of mitochondrial ETC and steps in the movement of electrons

There are different types of complexes (I, II, III, IV, V) in mitochondria. Except for complex II, all others are proton pumps. Complex II is not a proton pump because it consists of succinate dehydrogenase directly connected to FADH. The passage of electrons maintains a proton gradient, moving protons from the matrix to the intermembrane space of mitochondria. When protons move back from the intermembrane space to the matrix, they generate ATP from the ATP synthase. The NADH at complex I produces 4 H^+ ions through ETC, 3 from complex III, and 2 from complex IV.

There are five protein complexes in the mitochondrial ETC (electron transport chain) or mitochondrial respiratory chain, as follows:

- Complex I : NADH-ubiquinone oxidoreductase
- Complex II: succinate-ubiquinone oxidoreductase
- Complex III: Ubiquinone cytochrome- c oxidoreductase
- Complex IV: cytochrome-c oxidase
- Complex V: ATP synthase

A. Complex I or NADH ubiquinone oxidoreductase or NADH dehydrogenase or Step 1. Electron movement to Ubiquinone (CoQ) through NADH.

Complex I is the largest multi-subunit complex of the respiratory chain. It is the first enzyme of the mitochondrial electron transport chain. It performs several functions related to oxidation and reduction, in addition to proton transport. It does NADH oxidation, ubiquinone reduction, and proton transport to create proton-motive force across the inner membrane of the mitochondrion. Glycolysis and the citric acid cycle generate NADH is oxidized to NAD. It has NADH dehydrogenase and flavoprotein (FPN), with FMN as the prosthetic group, and eight non-heme iron-sulfur (Fe-S) centers or clusters that help in the transport of electrons.

Step 1. Electron Movement to Ubiquinone (CoQ) through NADH.

When NAD accepts hydrogen, it converts to NADH in the mitochondrial matrix. It binds complex I to unload electrons and protons. The electrons via complex I are transferred from NADH (NADH dehydrogenase) to FMN(intermembrane space). The Fe-S center receives electrons from FMN (one electron to one Fe-S center). The CoQ (ubiquinone) receives electrons from the Fe-S center (one at a time), then CoQ is converted first to semiquinone and then ubiquinol ($CoQH_2$). Electron from coenzyme Q, is transferred to cytochromes b, c_1, c, a, and a_3. In this process, the proton is released. Electron transfer moves 4 electrons from the matrix.

Complex I pump protons with the energy generated due to the transfer of electrons. With this energy, Complex I can pump four hydrogen ions (protons or H^+) across the membrane from the matrix into the intermembrane space. This will generate and maintain the potential gradient such that the protons remain low in the matrix and high in the intermembrane space. Hydroxides are left behind in the matrix. The sequence of the hydrogen transport system:

- NAD → FMN → CoQ → Cyt.b → Cyt. c1 → Cyt.c → Cyt. a → Cyt. a_3

As the hydrogen atom or electron passes down the chain, there is simultaneous oxidation of one coenzyme and reduction of another at each step. $NADH_2$ and $FMNH_2$ lose hydrogen and get oxidized to form NAD and FMN, respectively. Upon accepting the hydrogen, NAD and FMN again get reduced to get converted to $NADH_2$ and $FMNH_2$ respectively. After transfer from the cytochrome a_3, an electron and a proton combine to produce hydrogen, which is accepted by the oxygen molecules, so that water is formed.

(NADH + H+) + CoQ + 4 H+(in matrix)

$$\downarrow$$

NAD+ + CoQH2 + 4 H+(out-intermembrane)

Thus built, the electrochemical potential difference is required to produce ATP. The electron transfer mechanism is as follows:

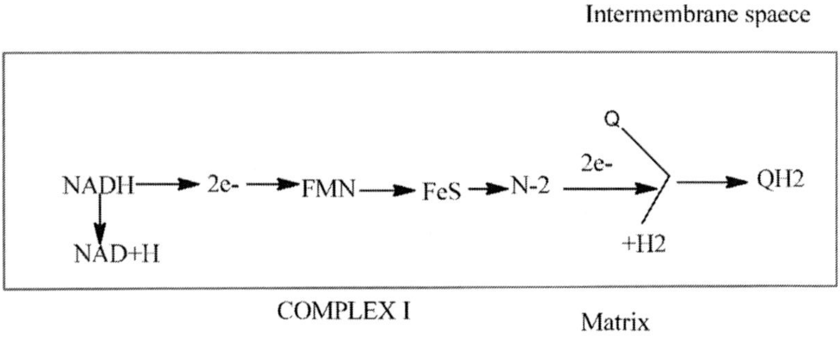

Fig. 6.2: The electron transfer mechanism from complex I to the formation of QH_2.

B.Complex II (Succinate Dehydrogenase or Succinate-CoQ Reductase) or Step 2. Transfer of Electrons from $FADH_2$ to CoQ

ComplexII is simpler and smaller than complex I. It converts succinate to fumerate, releasing the two electrons, which are transferred to FAD and then to the Fe-S center. Finally, it is transferred to coenzyme Q (ubiquinone) and reduced to ubiquinol.

This complex doesn't translocate H+ across the membrane, as does Complex I. Complex II is not a proton pump because it consists of succinate dehydrogenase and flavoprotein with the FAD prosthetic group and Fe-S centers. Here, FAD oxidizes to form $FADH_2$. The $FADH_2$ electrons are directly received at Complex II, bypassing the first complex.

The first two complexes are connected to the third complex via ubiquinone (Q). A hydrophobic membrane core permits the small molecule (lipid-soluble) to move freely. After receiving electrons, the Q molecule converts to QH_2 (ubiquinol). These electrons are delivered from QH_2 to complex III. Ubiquinone(Q) receives the electrons from two sources e.g., NADH and $FADH_2$.

The electrons received from NADH are via Complex I and those received from $FADH_2$ are via Complex II. The electrons generated from FADH2 are directly received by complex II. Therefore, FADH2 electrons bypass the first complex and fail to energize it, so complex I cannot pump the protons. Since the ATP generated depends on the protons pumped, it is evident that fewer ATP molecules are generated from FADH2.

The carriers Q molecule (ubiquinone) exist between complexes I and III, and between II and III. $FMNH_2$ (complex I) and $FADH_2$ (complex II) donate electrons to Coenzyme Q . Coenzyme Q subsequently transfers electrons to complex III.

Step 2: Transfer of Electrons from $FADH_2$ to CoQ

- The conversion of FAD to $FADH_2$ is a reduction reaction that occurs when there is oxidation of succinate to fumarate
- $FADH_2$ electrons enter to ETC with the help of complex II.
- Through a series of Fe-S centers, the electrons reach to CoQ
- Proton pump is absent in complex II and, therefore, doesn't pump any protons across the membrane, like complex I.

The overall reaction is as follows:

Succinate + $FADH_2$ + CoQ → Fumarate + FAD^+ + $CoQH_2$

C. Complex III or Ubiquinol-cytochrome c Oxidoreductase or Cytochrome bc1 Complex.

Step 3: Transfer of Electrons from CoQH2 to the cytochrome bc1 Complex and to Cytochrome c

Complex III comprises cytochrome b (two different non-heme groups), c (single heme group) proteins, a specific Fe-S center, and a Riske center (2Fe-2S center).

Cytochrome c transfers the electrons to complex IV. The electrons are passed to cytochrome c when complex III pumps protons through the membrane. Cytochrome c can accept only one electron from Q even though Q carries pairs of electrons.

- Complex III receives electrons arriving from both complex I and complex II. Complex III sends its electrons to complex IV.
- CoQ(ubiquinone) is converted to Ubiquinol (CoQH2) after receiving electrons. The formed CoQH2 delivers electrons through cytochrome b and c1 to cytochrome c to convert it back to CoQ (the iron center Fe^{3+} -oxidized form is converted to reduced form Fe^{2+}). This completes the Q (quinol) cycle. Simultaneously, protons are pumped from the matrix into the intermembrane space, creating proton gradients.
- This cycle is called the Q (quinol) cycle when electrons are transferred from QH2 to cytochrome c.

The Q Cycle Occurs in Two Stages

A. First half Q cycle

- First, QH2 (ubiquinol), which carries two electrons, binds to complex III.
- The released two electrons from QH2 get distributed so that:

  A. Rieske iron-sulfur protein receives one electron. From here, the electron is transferred to cytochrome c1 and then to reduce cytochrome c. At this time, two protons are released into the intermembrane space.

  B. The second electron is captured from ubiquinol by cytochrome b and then transferred to ubiquinone (Q). It is converted to semiquinone radical (QH+ -) after partial reduction.
- After oxidation or after the transfer of both electrons, the ubiquinone molecule gets oxidized before it is released back to the inner mitochondrial membrane. Now, complex III is ready to take up a second new ubiquinol to capture electrons.

B. Second Half Q cycle

- As stated above, a second new molecule of QH2 binds to complex III, and one electron is used to reduce cytochrome c. Two protons are released into the intermitochodrial space.
- The second electron goes to cytochrome b and then to semiquinone, which changes to ubiquinol (QH2). This QH2 is released into the inner mitochondrial membrane.

Such a Q cycle uses two QH2 and two cytochrome c. Each QH2 carries two electrons, but each cytochrome accepts only one electron. Protons ($2H^+$) are pumped to the intermembrane space at every Q cycle. A total of 4 protons are pumped out using the energy of electron transfer. The Q cycle increases (doubles) the efficiency of complex III for proton pumping.

D. Complex IV or cytochrome c oxidase.Step 4. Electrons pass from cytochrome c to molecular oxygen.

Cytochrome proteins c, a, a3 (cytochrome c oxidase or COX), two heme groups, and three copper ions arranged as two copper centers constitute this complex.

Its heme group, Fe^{3+}(oxidized form), accepts electrons from cytochrome c to become Fe^{2+}. The cytochromes tightly hold an oxygen molecule between the iron and copper ions until two electrons are ultimately reduced. The combination of reduced oxygen with hydrogen ultimately forms H_2O. The proton gradient is established during this process, as four protons are translocated across the membrane. Proton translocation is possible by the energy of the electrons going through the ETC.

2 cytochrome c(reduced) + $\frac{1}{2}O_2$ + 4 H^+(matrix) → 2 cytochrome c(oxidised) + 1 H_2O + 2 H+(intermembrane)

Step 4. Electrons pass on from cytochrome c to molecular oxygen pass on

- Complex IV receives electrons transferred from cytochrome c.
- The electron gained by complex IV (cytochrome oxidase) is transferred to molecular oxygen.
- During this process, the Fe^{2+} (reduced form) of the cytochrome c is converted to Fe^{3+} (oxidized form). Protons get pumped across the membrane using the released energy.
- For every four electrons passing through this complex, four substrate H^+ from the matrix convert O_2 to H_2O.
- The return of protons from the intermembrane space to the matrix results in the formation of ATP by ATP synthase.

V. ATP synthase is also known as Complex V.

Protons trapped in the intermembrane space pass to the matrix through the proton channels present in the complex V; this will form ATP using proton gradient energy. Protein synthase helps the protons return to the mitochondrial matrix. The ATP synthase creates a hydrophilic pathway across the inner mitochondrial membrane. Protons, when flowing through the ATPase, allow the synthesis of ATP. Oxidation of NADH generates 3 moles of ATP. $FADH_{2,}$ upon oxidation, generates 2 moles of ATP. The redox potential of hydrogen is lowest (-0.42), while it is highest in the case of oxygen (+0.82). All other substances have redox potential between these values of hydrogen and oxygen. When electrons flow from lower redox potential substances to substances having higher redox potential, the transfer of electrons yields energy. A reducing agent compound in one reaction becomes an oxidizing agent in another reaction. Thus, depending on the reaction type, a given compound agent can behave like a reducing or oxidizing agent.

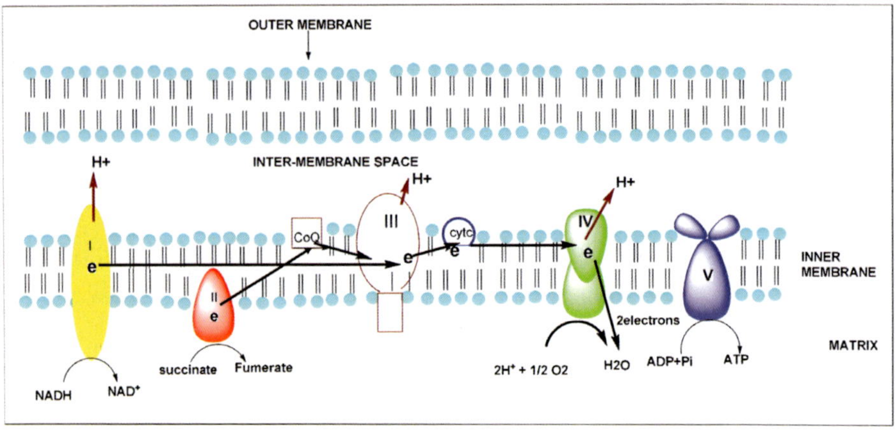

Fig. 6.3: Path of electrons and protons with formation of water and ATP in ETC.

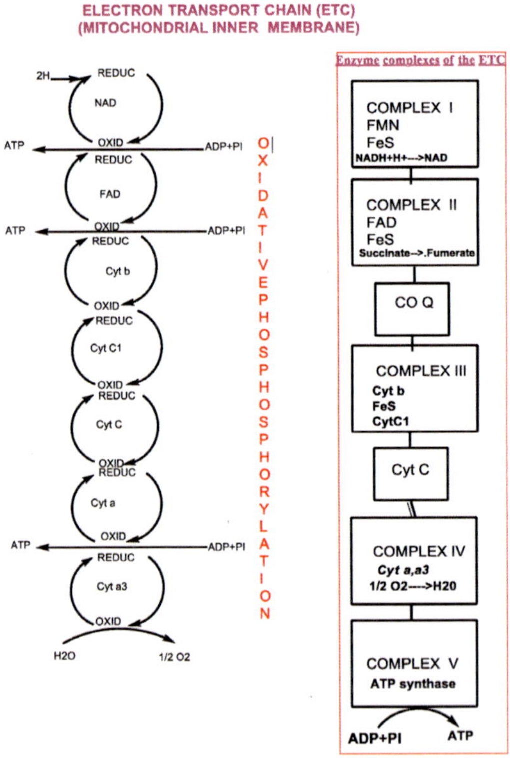

Fig. 6.4: Mitochondrial ETC and enzyme complexes

Biological oxidation-reduction or redox reaction

In a chemical reaction, one reactant can lose electrons, whereas the other accepts electrons. The reactant that loses electrons is called the oxidized, and the other reactant that accepts electrons is called the reduced. This type of reaction is oxidation-reduction or redox reaction, which always co-occurs. Such a reaction in a biological system is a biological oxidation-reduction reaction or redox reaction. In plants (autotrophs), sugar photosynthesis is an example of a redox mechanism wherein CO_2 is reduced, a water molecule is oxidized, and oxygen is released. The sugar formed is consumed by heterotrophs (animals and humans) break it down and capture stored energy, involving redox reactions.

When reactants undergo oxidation, they combine with oxygen or a hydroxyl group, remove hydrogen, or lose electrons; the reaction is called an oxidation reaction.

When reactants undergo reduction, they lose oxygen or a hydroxyl group and gain hydrogen or electrons; the reaction is called a reduction reaction. In general, the electrons are transferred from the reducing agent to the acceptor of electrons, which is the oxidizing agent. The tendency to release or accept electrons depends on the redox potential, an electrochemical concept measured in volts. The redox potential can be positive (has a higher affinity for electrons-O_2) or negative (lower affinity for electrons-NADH) than $H_{2.}$ A substance that can be in an oxidized form and a reduced form is called a redox couple.

Oxidation (electron loss) requires energy, and reduction yields energy (electron gain). A redox process involves dehydrogenation (oxidation) and hydrogenation (reduction).

6.5. Inhibitors, Uncouplers, and Other Factors Influence the Electron Transport Chain

The electron transport chain functions can be altered in many ways, such as using inhibitors, uncouplers, and other factors.

The inhibitors' and uncouplers' mechanisms of action influence ETC function in different ways. The inhibitors block the oxidation process, reducing oxygen demand and ATP generation. Uncouplers break the mitochondrial membrane, so ATP production is decreased, but there is an increase in the consumption of oxygen.

I. Inhibitors

The inhibitors bind to different components and block the specific electron carriers of ETC. This binding of inhibitors to the electron carriers of ETC is irreversible, so the oxidized state cannot be converted to a reduced state. Such a situation stops the synthesis of ATP due to the blocks of redox reactions. Examples of inhibitors with related information are tabulated below :

Table 1. Inhibitors and related information

Inhibitors	**Related information**
Rotenone	It has different properties, like insecticidal(broad-spectrum), acaricidal, and pesticidal. It inhibits complex I of the mitochondria, leading to the depletion of the formation of ATP due to the impaired process of electron transfer. It forms ROS, leading to oxidative stress and apoptosis in cells. It is now used only for fish management. It is a natural alkaloid (rotenoid) obtained from certain tropical plants' roots, leaves, seeds, and bark. Its molecular formula is $C_{23}H_{22}O_{6.}$ Its MW is 394.42.
Derris elliptica	It inhibits complex I of ETC through the HADH-CoQ reductase. It is an insecticide. It is used as a poison to paralyze the fish for accessible collection.
Amobarbital (Amytal)	Amytal is a barbiturate. It inhibits the NADH-CoQ reductase complex. It is used for the treatment of anxiety and anxiety disorders, seizures, and insomnia, and for the induction sedation during procedures.
Demerol	Demerol also inhibits complex I. It is used to cure severe pain.
Piericidin A	It inhibits complex I of mitochondria and NADH dehydrogenase. It is an antibiotic of microbial origin. It is an Antibody Drug conjugate (ADC) cytotoxic. It has antibacterial, cytotoxic, anti-cancer, and insecticidal activity.
Antimycin	An antibiotic that blocks complex III of ETC and inhibits the flow of electrons from cyt. b to cyt. c1.
Cyanide	It blocks the complex IV of the ETC, blocks cytochrome c oxidase, and binds with the iron of this protein complex. It prevents cells from using oxygen, causing rapid death.
Carbon monoxide	Like cyanide, it inhibits Complex IV (cytochrome c oxidase). It binds to the heme groups within Complex IV, obstructing oxygen binding. Carbon monoxide competes with oxygen for the binding to hemoglobin. It has a 200-fold more significant affinity to form carboxyhemoglobin and interferes with oxygen release from oxyhemoglobin.
Azide	It blocks the complex IV cytochrome oxidase in ETC. It reacts with the ferric form (Fe3+) of complex IV of the electron transport chain.
Hydrogen Sulphide	It is a respiratory inhibitor that blocks complex IV of the electron transport chain. It is toxic.
Oligomycin and Dicyclohexyl-carbodiimide (DCCD).	Inhibitors of complex V, ATP synthase complex (Phosphorylation inhibitor) in ETC, so ATP synthesis is prevented. It stops the proton channel, which prevents the formation of ATP.

Inhibitors	Related information
Atractyloside (a plant glycoside) or Bongkrekate (an antibiotic from a mold)	It is a transport inhibitor inhibiting the ATP-ADP translocase, which forms a tight coupling between oxidation and phosphorylation. Therefore, oxidation cannot proceed via the respiratory chain without simultaneous phosphorylation of ADP.

II. Uncouplers

The high-energy bond ATP is synthesized in mitochondria, chloroplasts, and various microorganisms with energy-transducing membranes. The ATP synthesis requires two coupling reactions: electron transport and phosphorylation. When this coupling reaction is uncoupled by an uncoupler or uncoupling agent, then there is no formation of ATP. Because the uncoupling of reactions produces energy by redox reactions, the energy cannot be used by phosphorylation. Thus, uncouplers do not inhibit electron flow and ATPase activities but fail to synthesize ATP. So, the released energy from electron transport remains untapped and is released as heat during uncoupling. This kind of heat release is known as nonshivering thermogenesis.

Natural uncoupler thermogenin is a protein found in brown adipose tissue of newborn and hibernating animals.

ATP is synthesized from ADP and Pi when H^+ enters the mitochondria via the H^+-ATPase. Uncouplers disrupt the mitochondrial membrane and reduce ATP production but increase oxygen consumption.

The uncoupling agents are lipid-soluble (hydrophobic) and bind with hydrogen ions, and they carry them back to mitochondria during the electron transport. This will stop generating a proton gradient.

Examples of uncouplers

Chemical Uncouplers

- 2,4-Dinitrophenol, dicumarol
- CCCP (carbonyl cyanide m-chlorophenylhydrazone) is a potent mitochondrial uncoupling agent.
- FCCP (p-trifluoromethoxy carbonyl cyanide phenyl hydrazone).

Physiological uncouplers

It includes long-chain fatty acids, Thyroxin, Thermogenin (protein), and Calcium ions.

III. The factors influencing the electron transport chain (ETC)

Various factors influence the ETC functions are

- **pH**: The optimum pH for some processes can vary. The pH of intact mitochondria remains alkaline through the electron transport chain (ETC) activity. In resting intact mitochondria, pH values range from 7.2 to 8.2 in different cell types. A low pH causes mitochondria to pump out H+ to the intermembrane space, leading to ATP production by ATP synthetase. However, a pH that is too low can negatively affect the process. At an acidic pH of 3 or 9, enzyme protein denaturation due to disarrangement of the bonds occurs so that the enzyme's function is impaired. Extreme pH levels disarrange the bonds that hold the protein in its operational shape. The mitochondrial matrix pH is about 7.8, higher than the pH of the intermembrane space, 7.0-7.4.
- **Temperature**: The rate of perchlorate removal can vary with temperature. Mitochondria are up to 15 °C hotter than their external environment in living cells. The kinetic energy of a molecule increases with an increase in temperature. Here, ATP production is increased due to the frequent collisions between the enzyme and substrate. Cell and cell membrane damage occur at too high or too low temperature. If prolonged exposure occurs to extreme temperature ranges, it can kill the cell.
- **Electron acceptors**: They can be inorganic or organic in nature. They are oxidizing agents. The oxygen is the final electron acceptor in mitochondria. If it is absent, the ETC will not be complete and will stop, so ATP will not be produced. The electron acceptors in anaerobic conditions are nitrate, nitrite, ferric iron, magnesium, sulfate, and carbon dioxide.

  Organic molecules as fumarate. NAD+ and FAD each can accept 2 electrons.
- **Inhibitors**: They are molecules that slow down or stop chemical reactions. Rotenone reacts with NADH dehydrogenase (complex I) so that oxidation of NADH to NAD. When NAD is not formed, substrate reactions involving NAD have stopped, e.g., glutamate and pyruvate.
- **Uncoupling agents**: the molecule disrupts oxidative phosphorylation, dissociating the ATP formation reaction by ATP synthase in ETC.
- **Disruption of the phospholipid bilayer**: It can be possible with higher temperatures, mild alkaline hydrolysis, and some detergents. This can cause protons to flow more freely, weakening the electrochemical gradient and preventing ATP production. Disruption of the phospholipid bilayer affects the ion balance inside the cells, and proteins can leak.

- **Lack of oxygen**: When the final electron acceptor, oxygen, in ETC is if unavailable, the ETC will not work, and ATP synthesis will fail.
- **Cyanide poisoning**: Cyanide blocks cytochrome a3, which, therefore, cannot reduce oxygen. This causes the breakdown of the proton gradient and to stops ATP synthesis.
- **The number of hydrogen ions pumped through the membrane**: It is essential to pump four hydrogen ions from the matrix into the intermembrane space to establish and maintain the hydrogen ion gradient between these two compartments.
- **The shuttle of electrons across the membranes of the mitochondria**: Electron transporters of ETC shuttle electrons from NADH and FADH2 to molecular oxygen. The NADH generated from glycolysis cannot quickly enter mitochondria because glycolysis occurs in the cytoplasm. NADH cannot directly cross the mitochondrial membrane to deliver electrons, but can do so through a "shuttle system."

6.6 Carbohydrate Digestion in Non-Ruminants, Ruminants, Poultry and Variation in Blood Glucose Levels.

I. Digestion of carbohydrate

A. Non-ruminants

Carbohydrate digestion occurs differently in non-ruminants, with simple stomachs than in ruminants with compound stomachs.

The digestion of carbohydrates in simple stomach animals, including humans, occurs due to various enzymes secreted by the digestive system. Besides monosaccharides (glucose and fructose) and disaccharides (sucrose, lactose, maltose), starch is the only polysaccharide that simple stomach animals digest.

Starch digestion starts from the mouth's salivary gland using the salivary α-amylase (ptyalin) enzyme to produce smaller molecules, but time is the limitation. This enzyme acts only in an alkaline pH as it enters the stomach and faces acidic pH, and its action is inhibited here. There is no further digestion of carbohydrates in the stomach, but it occurs in the intestine. When cooked, polysaccharides break glycosidic linkages, so digestion becomes more accessible. In the small intestine, pancreatic amylase breaks α-1,4 glycosidic linkages of carbohydrates (starch and glycogen) to produce disaccharides; the action of pancreatic amylase is similar to salivary amylase. Both plant and animal polysaccharides, respectively starch and glycogen, are broken down into maltose, isomaltose, oligosaccharides, and alpha-limit Dextrins. Maltose is digested by maltase to yield two glucose molecules: lactose by lactase to produce glucose and galactose and sucrose by sucrase to yield glucose and fructose molecules. Their monosaccharide molecules, the glucose from the intestine, enter intestinal cells by cotransport, then enter the portal venous blood by facilitated transport. The entry of glucose into cells from the blood is done by facilitated transport and is also helped by insulin. Glucose is co-transported with sodium. If excess, sodium is pumped from cells to out actively using a sodium pump.

B. Ruminants

The ruminant animals differ from simple stomach nonruminants (including man) in that their stomach is also known as polygastric due to many compartments like rumen, reticulum, omasum, and abomasum. Grass is the chief diet of herbivores, which includes nonruminants and ruminants.

The structural carbohydrates, including cellulose, are not digested by non-ruminants. In horses (non-ruminant herbivores), the digestion of cellulose occurs in the large intestine and not in the stomach. Enzymes secreted by microbes and protozoa help in carbohydrate digestion in the ruminant stomach. The end products of the digestion of carbohydrates like cellulose and hemicellulose are not glucose but different acids in ruminants like cows, buffaloes, sheep, and goats. (The grains or non-fibrous food can provide glucose as the end product). These acids produced are acetic acid (CH3-COOH)to the extent of 60% to 70%, propionic acid (CH3-CH2-COOH) to the extent of 15 to 20%, and butyric (CH3-CH2-CH2-COOH) to the extent of 10% to 15%. Other products produced in small quantities are valeric acid and caproic acid. The formed lactic acid is converted to propionic acid immediately. The production of VFA depends on the food taken by animals.

In ruminants, the rate of digestion for soluble sugars(hexoses, sucrose, and fructose) is high and gets fermented rapidly. Starches are fermented less rapidly. Lignin's presence decreases the digestibility of cellulose and hemicellulose. A major part of cellulose digestion occurs in the rumen (70%), and the remaining is digested in the large intestine(30%).

fructasan→fructose →volatile fatty acids in rumen

hexoses and pentose→volatile fatty acids in rumen

cellulose →glucose →volatile fatty acids in rumen

hemicellulose →hexoses and pentosan(xylose)→ volatile fatty acids in rumen

lignin→mostly undigestible

Soluble carbohydrates like starch and cellulose produce more propionic acid and butyric acid, but a lower proportion of acetic acid. Here, the concentration of propionic and butyric acids formed are equal. If animals take more hay, acetic acid production is higher than propionic acid.

As a result of different end products(VFA) of carbohydrate digestion in ruminants, the blood glucose level in cows is nearly half that found in human. The volatile fatty acids are produced in different proportions depending on the nature of the diet. As a result, their tissues have to use acetate and convert it to acetyl-CoA to enter the TCA cycle. Thus, ruminants do not get glucose directly from carbohydrate digestion, but it is supplied by gluconeogenesis from acetate. VFAs are also used for milk fat and milk sugar lactose synthesis. Butyrate is metabolized to β-hydroxybutyrate in the rumen wall, whereas acetate and propionate are transported to the liver. Propionate is the primary source of blood glucose. Glucose is also formed by gluconeogenesis from amino acids undergoing deamination. Starch, which escapes ruminal fermentation or digestion, enters the small intestine and can be digested into glucose, contributing to blood glucose. Lactose, a disaccharide sugar (glucose+galactose), is found in milk. Ketone bodies are formed if sufficient sugar is unavailable, and ketosis occurs.

C. Poultry

Avian has a peculiar digestive system divided into the oral, esophageal, gastric, pancreatic, and intestinal regions, and digestion is similar to that in mammals.

Birds lack a soft palate and hence raise their head to swallow food. Corn, wheat, and other cereal grains are the main carbohydrate diet of poultry. Starch is the main carbohydrate in the food of poultry. The saliva secreted lubricates the food. The ingested starch is hydrolyzed by alpha-amylase at a pH optimum of 6.7.

The esophagus extends from the pharynx to the crop, a storage organ or region. The crop contains salivary amylase, and limited fermentation occurs here.

The gastric region consists of the proventriculus and gizzard. Fluid is secreted by the proventriculus. The gizzard is the organ for mastication and contains grit pieces. Starch digestion by alpha-amylase continues. Lactose and molasses ferment here. Lactose is converted to lactic acid here. Molasses increases fermentation and VFA production.

The pancreatic region contains pancreatic secretion of insulin and glucagon by islets of Langerhans as an endocrine portion, and pancreatic amylase is secreted by the exocrine part of the gland.

The intestine has a small intestine part and a large intestine part. The small intestine contains the duodenum, starting at the distal end of the gizzard, followed by the jejunum and ileum. The small intestine is the major site for carbohydrate digestion and absorption of monosaccharides. Starch granules are digested in the intestine by pancreatic amylase. The large intestine contains the cecum and colon.

II. Variations in blood glucose levels

Variations in blood glucose levels vary depending upon the factors that help enter and exit blood glucose, including the actions of different enzymes and hormones. Both the entry and exit of blood sugar take place simultaneously. The rate of entry and exit of blood glucose maintains the blood glucose level. If this process is disturbed, it will also disrupt the blood glucose level, leading to normal levels of hyperglycemia or hypoglycemia.

- Route for entry of glucose into the body: The entry of glucose into the body occurs via various routes, such as the digestion of food and the absorption of glucose released from the intestine, and the release of glucose from glycogenolysis in the liver. Glucose is also produced by gluconeogenesis.
- Routes for major uses of glucose from the blood: Tissues use glucose for oxidation and to get energy for work. It is converted to glycogen in the liver and other tissues in excess. In adipose tissues, it is stored as fat if in excess. It is used to synthesize milk sugar and lactose in lactating mammary glands. It is used for the biosynthesis of glycolipids and mucopolysaccharides and is converted to ribose for nucleic acid synthesis. In the case of glucose disorders like diabetes mellitus, it is excreted into the urine.

The liver and kidneys are essential for the regulation of blood glucose concentration.

Liver cells are permeable to glucose, and more glucose enters the liver and is converted to glycogen when blood glucose rises beyond the normal level. Glucose is converted to glucose-6-phosphate by glucokinase. The activity of the enzyme glucokinase is not inhibited by glucose-6-phosphate within the hepatic cells. However, in extrahepatic tissues, glucose-6-phosphate inhibits glucokinase.

Kidney glomeruli filtrate contains glucose even in normal conditions. The filtered glucose is essential to the body and is actively reabsorbed by the renal tubules. There is a limit to the maximum absorbance of glucose by the renal tubules; when it is exceeded, glucose is excreted in urine (glucosuria). If blood glucose is within the physiological limit, it is completely reabsorbed and does not enter the urine. The term renal threshold of glucose is used for the lowest glucose level in the blood at which glucose appears in the urine. Its value changes with the species of animals. It is 180mg/dl for men, 180-200mg/dl for dogs, 160 to 180 mg/dl for horses, and 280-290 mg/dl for cats. The transport maximum for glucose (TmG) is the maximum rate of tubular glucose reabsorption per minute (350mg/dl).

6.7 Convergence of Metabolic Pathways

The great Nobel Prize-winning scientist Hans Krebs described the citric acid and urea cycles concerned with metabolism. Metabolism (anabolism or catabolism) is studied separately for metabolites of food components (carbohydrates, protein, and fats). However, in vivo, such separate metabolism does not occur for different food components. The different metabolites are processed simultaneously so that the end product of one pathway can become the substrate for the other interconnected paths. This type of arrangement will avoid excessive accumulation or acute shortage of any one or more specific food metabolites in the body, which is very likely to favor the condition of homeostasis. Therefore, metabolism comprises the total number of chemical reactions in the cell.

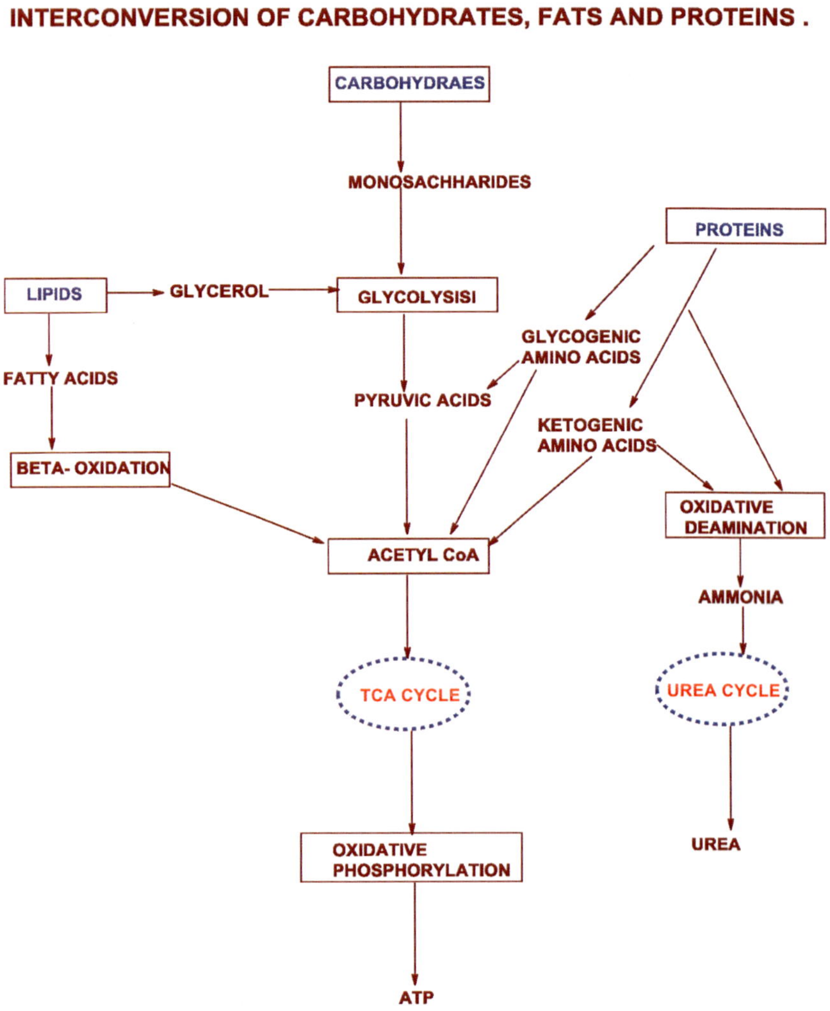

Fig. 6.5: Schematic diagram for the interconversion of carbohydrates, proteins, and lipids through the TCA cycle.

Key Takeaway

- The metabolism can be viewed from different aspects. It may be **anabolic or catabolic** type. It may be aerobic or anaerobic, depending on oxygen availability. In aerobic metabolism, the substrate in the presence of oxygen is metabolized to release energy, carbon dioxide, and water.

- In oxidation, the electrons released from a glucose molecule are transferred to reduced NAD+ and FAD. The electron acceptor oxygen is reduced to water. The energy released from the chemical bond of glucose is captured as ATP, and heat is released.
- Glucose is converted to lactate when limited amounts of oxygen (O_2) are available.
- Anabolism involves the synthesis of biomolecules and various reactions to convert the smaller molecules to larger ones. Catabolism consists of the breakdown of larger molecules into smaller or simpler units.
- The gluconeogenesis pathway converts noncarbohydrates, simple compounds like amino acids and fatty acids, into glucose molecules through *de novo* synthesis.
- Glycogenesis is the process that converts excess glucose into glycogen for cellular storage in animals. Glycogenolysis is a process that involves the breakdown of glycogen into glucose,
- The metabolism or energy expenditure at the body's resting state refers to the resting metabolic rate (RMR) required to maintain life. It is also referred to as the resting energy expenditure (REE). The metabolism or energy expenditure rate under basal conditions is known as BMR.
- Mitochondria are unique organelles and have distinct double membranes.
- The number of mitochondria per cell varies from zero (red blood cells) to hundreds or even thousands (liver and muscle cells) per cell. The inner membrane is thrown into folds known as cristae to increase its surface area. The cristae house the protein components of the electron transport chain (ETC).
- There are four lipid-soluble membrane-bound complexes embedded in the inner membrane. They are named Complex I, II, III, and IV. Except for complex II, the other three are proton pumps. The water-soluble electron carriers are Coenzyme Q and cytochromes.
- A series of electron carriers is collectively known as the electron transport chain (ETC), the Cytochrome oxidase system, or the Respiratory chain. The ETC provides a mechanism for transporting the hydrogen (proton) and electron stepwise from one acceptor to another, and the final hydrogen acceptor is molecular oxygen.
- The first stage in the electron transport chain is binding NAD and FAD coenzymes to the electron transport chain's protein complexes (I, II, III, and IV). Complex I, or NADH ubiquinone oxidoreductase, or NADH dehydrogenase is involved in the movement of electrons to Ubiquinone (CoQ) through NADH.

- Flavoproteins, Quinone, NADH, and NADPH carry electrons and protons. In contrast, Cytochrome-heme is a prosthetic group, and iron-sulfur proteins or Fe-S (nonheme) only carry electrons.
- Polyisoprenoid side chains occur in coenzyme Q, ranging from 6 to 10.
- Complex II (succinate dehydrogenase or succinate-CoQ reductase) transfers electrons from $FADH_2$ to CoQ. Complex III or ubiquinol-cytochrome c oxidoreductase, or cytochrome bc1 complex, transfer electrons from $CoQH_2$ to the cytochrome bc1 complex and to the cytochrome c
- Complete reduction of ubiquinone occurs in two steps with the help of two electrons and two protons. The first two complexes are connected to the third complex via ubiquinone (Q). The Q molecule (lipid-soluble) can move freely.
- The redox potential of hydrogen is lowest (-0.42), while it is highest in the case of oxygen (+0.82). All other substances have redox potential between these values of hydrogen and oxygen.
- The end products of the digestion of carbohydrates like cellulose and hemicellulose are not glucose but different acids in ruminants. The blood glucose level in cows is nearly half that found in men, so acetate gets converted to acetyl-CoA to enter the TCA cycle.
- The inhibitors' and uncouplers' mechanisms of action influence ETC function in different ways.
- The inhibitors bind to different components and block the specific electron carriers of ETC. This binding of inhibitors to the electron carriers of ETC is irreversible,
- Various inhibitors bind with different complexes.
- Complex I is bound by Rotenone, Derris elliptical, Demerol, and piecing A.
- Antimycin occupies complex III.
- Complex IV is occupied by Cyanide, carbon monoxide, azide, and hydrogen sulfide
- Complex V gets bound by oligomycin and DCCD.
- Carbohydrate digestion occurs differently in non-ruminants with simple stomachs than in ruminants with compound stomachs.
- When cooked, polysaccharides break glycosidic linkages, so digestion becomes more accessible.

- The entry of glucose into cells from the blood is done by facilitated transport and is also helped by insulin.
- The ruminant animal's stomach is known as polygastric due to many compartments like rumen, reticulum, omasum, and abomasum.
- A major part of cellulose digestion occurs in the rumen (70%), and the remaining is digested in the large intestine(30%).
- Soluble carbohydrates like starch and cellulose produce more propionic acid and butyric acid, but a lower proportion of acetic acid.
- As a result of different end products(VFA) of carbohydrate digestion in ruminants, the blood glucose level in cows is nearly half that found in human.
- Ruminants do not get glucose directly from carbohydrate digestion, but it is supplied by gluconeogenesis.
- Ketone bodies are formed if sufficient sugar is unavailable, and ketosis occurs.

Intext Questions

1. How does glucogenesis differ from glycogenesis?
2. What substrates are required for the process of glucogenesis and gluconeogenesis?
3. Which types of metabolism commonly occur at a young age and old age?
4. Explain the term BMR.
5. What is the complete form of ETC?
6. What is the relation between the activity of a cell and the number of mitochondria in a cell?
7. How many lipid-soluble membrane-bound complexes are embedded in the inner membrane of mitochondria? Give their name.
8. What is the stage in the electron transport chain during metabolism?
9. Why does a wastage of released energy occur during the reoxidation of the reduced cofactor?
10. What are resting metabolic rate and resting energy expenditure?
11. Name the carriers for protons and electrons present in mitochondria.
12. Write the complete form of NAD, FAD, and NADP.

13. What is the importance of Complex I and Complex II?
14. Why is the blood glucose level in ruminants lower than that of humans?
15. Choose one correct option for the following from the options given below (The answer is given in **bold** letters)

a. The metabolism may be

i. Anabolic ii. or catabolic

iii. **Both i. and ii.**

b. Generally, aerobic metabolism cannot occur without

i. **Oxygen** ii. Nitrogen

iii. Carbon dioxide

c. The breakdown of chemical bonds releases

i. Energy ii. Heat

iii. **Both i. and ii.**

d. Which of the following processes is related to the biosynthesis of biomolecules

i. **Anabolism** ii. Catabolism

iii. None of the i. and ii.

e. Which of the molecules are converted to glucose in gluconeogenesis

i. Glucose ii. **Amino acids**

iii. Fructose

f. The inner membrane of the mitochondrion is thrown into folds to increase its surface area, known as

i. Riste ii. **Cristae**

iii. Istae

g. The process that converts excess glucose into glycogen

i. Gluconeogenesis ii. **Glycogenesis**

iii. Glucogenesis.

h. Which of the following is a double membrane-bounded organelle?

i. Ribosome
ii. **Mitochondrion**
iii. Endoplasmic reticulum

i. The number of lipid-soluble membrane-bound complexes embedded in the inner membrane is

i. Two
ii. **Four**
iii. Six

j. Mitochondria have four lipid-soluble membrane-bound complexes; one does not act as a proton pump like the other three, and that is

i. **Complex II**
ii. Complex III
iii. Complex I

k. The solubility nature of the electron carriers Coenzyme Q and cytochromes is

i. **Water-soluble**
ii. Lipid-soluble
iii. Both i. and ii.

l. The final hydrogen acceptor molecule in ETC is

i. Carbon dioxide
ii. **Oxygen**
iii. Nitrogen

m. What is the function of the proton pump from the following?

i. To pump oxygen
ii. **To pump hydrogen**
iii. To pump carbon dioxide

n. A side chain units in coenzyme Q which ranges from 6 to 10 in number

i. **Polyisoprenoids**
ii. Fatty acids
iii. Amino acids

o. Which directional movement of protons generates ATP?

i. **Movement of protons from the mitochondrial intermembrane space to the matrix.**
ii. Movement of protons from the mitochondrial matrix to the intermembrane space
iii. Whenever protons move in and out of the mitochondrial matrix to the intermembrane space.

p. The process of conversion of NADH to NAD^+ is

i. **Oxidation** ii. Reduction

iii. Shuttle mechanism

q. Number of proton pumps in the single respiratory chain on the inner mitochondrial membrane.

i. 4 ii. **3**

iii. 1

r. A number of protons are transferred from the matrix to the intermembrane space when 2 electrons move to oxygen from NADH.

i. 8 ii. **10**

iii. 12

s. The terminal complex of oxidative phosphorylation in mitochondria.

i. **Cytochrome c oxidase** ii. Cytochrome a

iii. Cytochrome b

t. In complex II, the reductant and oxidant in ETC are

i. The Reductant is Succinate, and the oxidant is CoQ

ii. **The Reductant is ubiquinone(Q) and the oxidant is succinate**

iii. The Reactant is fumarate, and the oxidant is cytochrome c

u. Which statement is valid for the Fe-s centers in mitochondrial complexes

i. **Complex II contains both heme and iron-sulfur proteins**.

ii. Complex I contains only heme proteins.

iii. Complex IV contains zn-s centers

v. Molecular oxygen oxidizes one of the complexes found in ETC

i. **Complex IV** ii. Complex III

iii. Complex I

w. Which complex transfers electrons from succinate to ubiquinone

i. Complex I ii. **Complex II**

iii. Complex III

x. Identify the inhibitor from the following:

i. FCCP ii. CCCP

iii. **Antimycin**

y. Identify the uncoupler from the following:

i. **Dicumarol** ii. Rotenone

iii. Cyanide

z. The end products of carbohydrate digestion in the rumen of ruminants are —

i. **VFAs** ii. Glucose

iii. Both of the above.

Further Readings

A Textbook of Biochemistry. Berry, A. K. (2009), sixth edition, Emkay Publishers, Delhi.

A textbook on animal physiology. A, K, Berry. (1997). Sixth Edition. Emkay Publication, Delhi.

Biochemistry. Naik, Pankaj (2009). Second Edition, Jaypee Brothers, New Delhi.

Carbohydrate fractions of legumes: uses in human nutrition and potential for health. Guillon F. and Champ M.M.J. (2002). British Journal of Nutrition. PMID: 12498630. DOI: 10.1079/BJN2002720. https://pubmed.ncbi.nlm.nih.gov/12498630/

Cell Biology. E.D.P. De Robertis, Wiktor W. Nowinski, and Franciso A. Saez (1970). Fifth Edition, WB Saunders Company. Philadelphia, London, Toronto.

Cell Physiology. Arther C. Giese (1968), Third Edition, WB Saunders Company. Philadelphia, London, Toronto.

Fundamentals of biochemistry. J.L.Jain, Sunjay Jain and Nitin Jain(2005). First edition., S.Chand & Company Ltd. New Delhi.

H.Terada (1990). Uncouplers of oxidative phosphorylation. Environmental Health Perspectives. Jul.87:213-218. Doi:10.1289/ehp.9087213.

http://localhost:8080/jspui/handle/123456789/2796

https://easybiologyclass.com/what-is-q-cycle-in-the-electron-transport-chain/

https://microbenotes.com/electron-transport-chain/

https://microbenotes.com/electron-transport-chain/#a-complex-i-mitochondrial-complex-i

https://microbenotes.com/electron-transport-chain/#electron-transport-chain-complexes.

Life Sciences. Fundamentals and Practice I. Pranav Kumar and Usha Mina. (2017).Pathfinder publication. New Delhi.

NELSON, L., COX, M.M. 2017.Lehninger Principles of Biochemistry, 7th Edition. Pdf. W.H. Freeman and Sapling Learning.

Pretest Physiology. James P. Ryan and Ronald F. Tuma (1996), Eighth Edition, McGraw-Hill, New York, Milan, Montreal, New Delhi.

Principles of physical, organic, and biological chemistry.John R. Holum. (1969), John Wiley & Sons, Inc., New York, London, Sydney, Toronto.

Textbook of Biochemistry for Medical Students. Vasudevan, D. M.; Sreekumar, S.; Vaidyanathan, Kanan (2019). Ninth Edition, Jaypee Bros. Medical Publication, New Delhi.

Textbook of Veterinary Biochemistry. Dhanotiya R.S.(2006).Reprint of the second edition. Jaypee Brothers, New Delhi.

The Equine Microbiome: Diet, Disease, and Behaviour. Holly Heartz, MSESS, RD. IAAABC FOUNDATION JOURNAL. https://journal.iaabcfoundation.org/the-equine-microbiome-diet-disease-and-behaviour/

7

Glycolysis or Embden-Meyerhof-Pathway (EMP)

Glycolysis involves a series of reactions to extract energy from glucose to release two molecules with three carbons called pyruvate. The process is not dependent on oxygen but can occur under aerobic and anaerobic conditions, releasing different end products. Under aerobic conditions, the pyruvate formed is then oxidized in the mitochondria to CO_2 and H_2O. The energy released in glycolysis is in the form of ATP (High Energy Bond). All the reaction steps occur in the cytoplasm, as it is the extramitochondrial pathway. In anaerobic conditions and the absence of mitochondria (RBC), the end product of glycolysis is lactic acid.

The net reaction of Glycolysis (anaerobic) may be summarized as under

Glucose + 2ADP + 2Pi $\rightarrow$ 2 Lactate + 2ATP

7.1 Steps in EMP Cycle or Glycolytic Pathway

Table 7.1: Steps and enzymes of glycolysis or EMP cycle.

No	Reactions and energy account	Enzyme
I	Formation of Glucose-6-phosphate through phosphorylation of glucose (Committed step) ---use of 1ATP (endergonic reaction)	Glucokinase or Hexokinase. It is one of the critical enzymes.
II	Isomerization of Glucose-6-Phosphate to Fructose-6-Phosphate	Phosphohexokinase
III	Phosphorylation of F-6-P to Fructose 1,6-Biphosphate ----use of 1ATP (endergonic reaction)	Phosphofructokinase—It is one of the critical enzymes.
IV	Cleavage of Fructose 1,6-biphosphate	Aldolase
V	Interconversion of the Triose Phosphates	Triose phosphate isomerase
VI	Oxidative phosphorylation of Glyceraldehyde-3-phosphate to 1,3-Bisphosphoglycerate ----2 molecules generate 2 NADH equivalents to 6 ATP	Glyceraldehyde 3-phosphate dehydrogenase
VII	Conversion of 1,3-Biphosphoglycerate to 3-Phosphoglycerate ---2 molecules generate 2ATP	Phosphoglycerate kinase

No	Reactions and energy account	Enzyme
VIII	Conversion of 3-phosphoglycerate to 2-phosphoglycerate	Phosphoglycero mutase
IX	Dehydration of 2-Phosphoglycerate to Phosphoenolpyruvate	Enolase
X	Conversion of Phosphoenol Pyruvate to Pyruvate—2 molecules involved, so the generation of 2ATP	Pyruvate kinase

ADD THIS COLOUR FIG 7.1

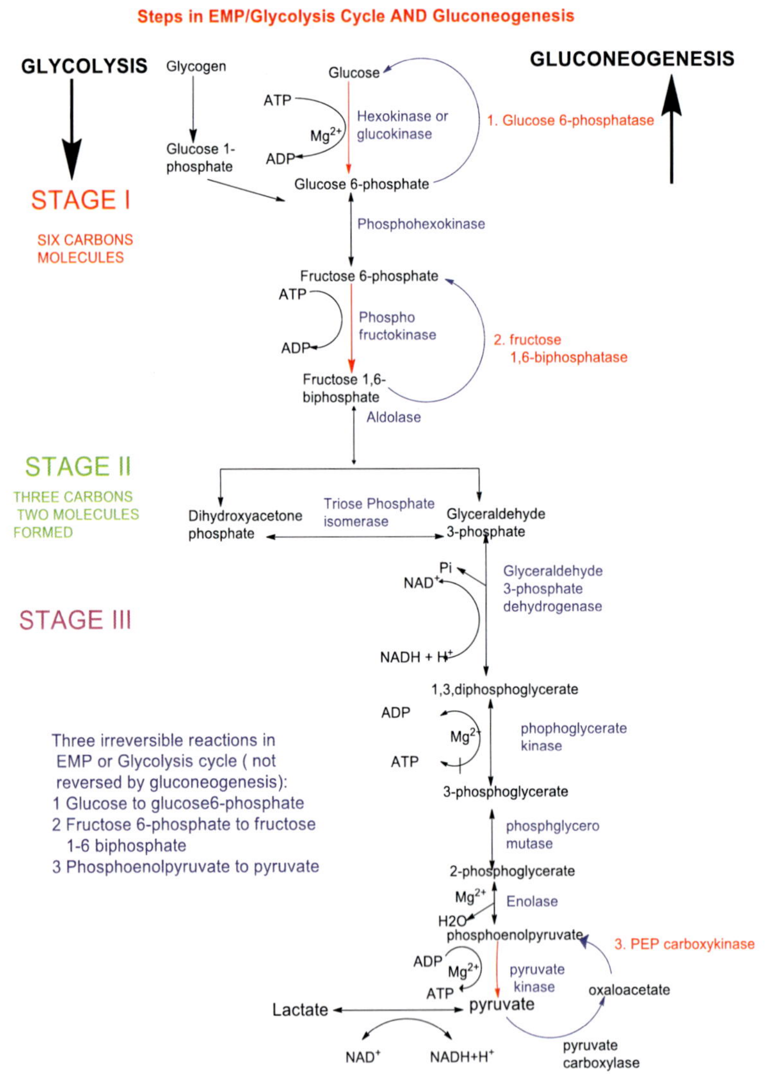

Fig. 7.1: EMP/Glycolysis cycle and Gluconeogenesis steps.

I. Uptake and Phosphorylation of Glucose (one ATP utilized)

Glucose uptake in different cells, from the simplest single-cell bacterium to the highly specialized mammalian cell, is facilitated by GLUTs (glucose transporters) present in the plasma membrane. Upon entering the cells, glucose is rapidly phosphorylated by glucokinase (GK) or hexokinase and further metabolized. **Glucokinase** facilitates the addition of phosphorus at carbon no 6 of glucose, converting it to glucose-6-phosphate. It is found in the cells of the liver and pancreas of most vertebrates. This is the first priming reaction.

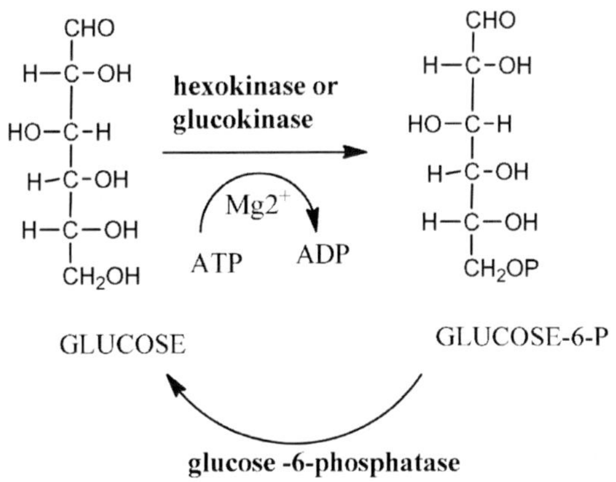

When glucose-6-p is formed, ATP is converted to ADP in the presence of Mg2+.

The above reaction is **irreversible** as different enzymes are required for forward and backward reactions.

II. Isomerization of Glucose-6-Phosphate to Fructose-6-Phosphate

This reversible reaction converts aldehyde phosphorylated sugar to ketone phosphorylated sugar with a P group on carbon no.6. Both these sugars differ at the first two carbons. Glucose 6-phosphate has an aldehyde group (CHO) at the first carbon and a primary alcohol group at the second carbon. Fructose 6-phosphate has a primary alcohol group at the first carbon and a keto group or carbonyl group (C=O) at the second carbon. The enzyme required is Phosphohexokinase

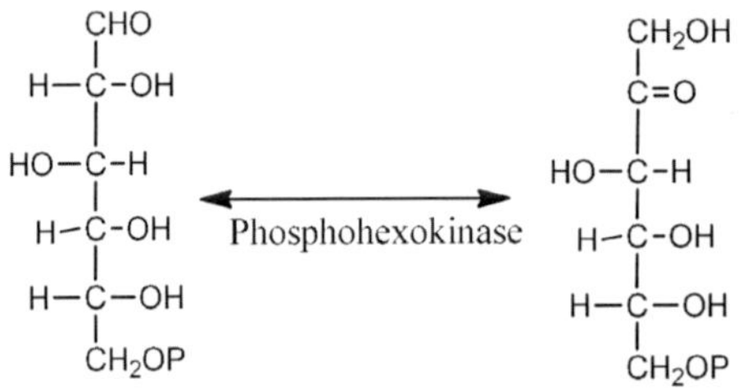

GLUCOSE-6-P FRUCTOSE -6-P

In the above reaction, two changes are noted. **Oxidation** occurs at a second carbon atom having a secondary alcohol group (CHOH) to form a keto-group (C=O). When **reduced**, the aldehyde group (CHO) at the first carbon atom is converted to a primary alcohol group ($-CH_2OH$).

III Phosphorylation of F-6-P to Fructose 1,6-Biphosphate

This *is the second priming reaction.* The enzyme transfers the phosphorus group from ATP to fructose-6-P, forming ADP and fructose 1-6-biphosphate. This reaction is irreversible and can run in a backward direction with the help of fructose-1-6-biphosphatase enzyme. PFK is stimulated by the presence of fructose-6-phosphate, AMP, and ADP but is inhibited by ATP and citrate

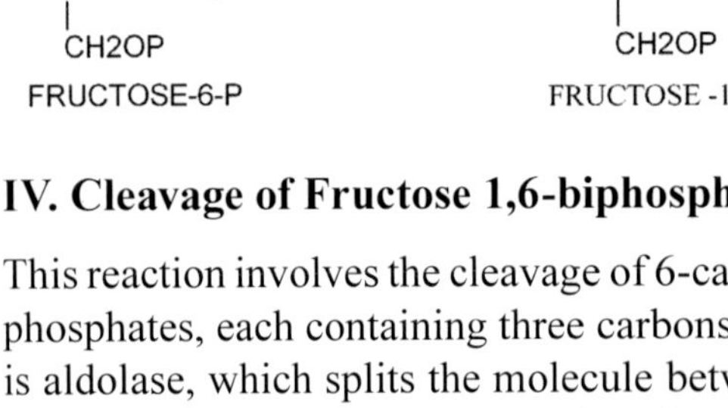

FRUCTOSE-6-P FRUCTOSE -1-6-BIPHOSPHATE

IV. Cleavage of Fructose 1,6-biphosphate

This reaction involves the cleavage of 6-carbon glucose phosphate to two sugar phosphates, each containing three carbons (two trioses). The enzyme required is aldolase, which splits the molecule between carbons 3 and 4. The products formed are dihydroxyacetone phosphate (ketone) and glyceraldehyde-3-P (aldehyde).

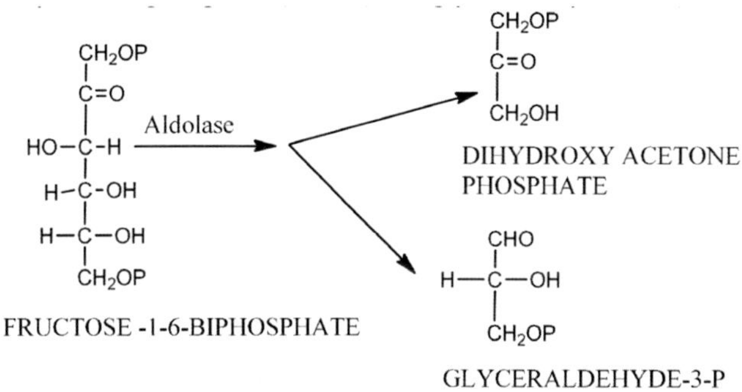

V. Interconversion of the Triose Phosphates

The presence of the enzyme Triose phosphate isomerase does the inter-conversion of triose phosphate, namely dihydroxy acetone phosphate and glyceraldehyde phosphate If the reaction has to go in the forward direction, then two glyceraldehyde-3-P are formed.

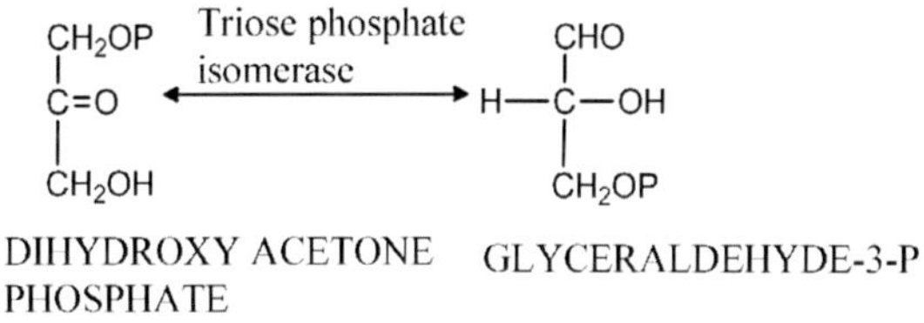

VI. Oxidative Phosphorylation of Glyceraldehyde-3-P to 1, 3-Bisphosphoglycerate

This reaction is characterized by oxidation and phosphorylation. It is a reversible reaction. One phosphorus is added to carbon 1 of the glyceraldehyde-3-P, forming 1, 3-biphosphoglycerate, and NADH.

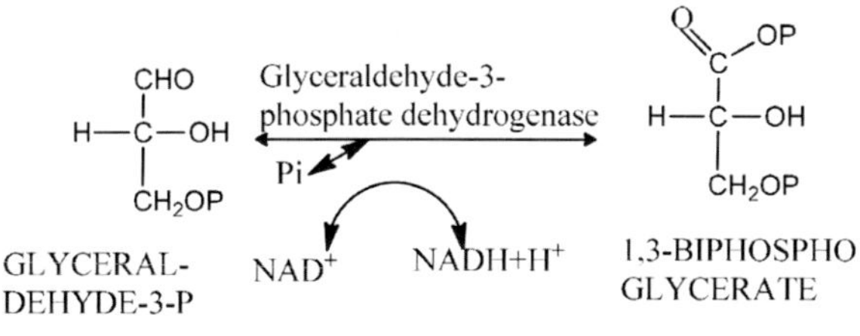

Aldehyde group, during its oxidation, releases hydrogen atoms or electrons that are accepted by a coenzyme NAD+ and reduced to NADH + H^+.

Note: In R.B.C., 1,3 Bi (di) phosphoglycerate is converted to 2,3-diphosphoglycerate by the enzyme diphosphoglyceromutase. This compound lowers the oxygen affinity of hemoglobin, releasing oxygen to peripheral tissues in case of high altitude. Fetal hemoglobin has high oxygen affinity due to less 2, 3-diphosphoglycerate than an adult.

VII. Conversion of 1,3-Biphosphoglycerate to 3-Phosphoglycerate

This is the first ATP-forming substrate-level phosphorylation. The enzyme required is phosphoglycerate kinase.

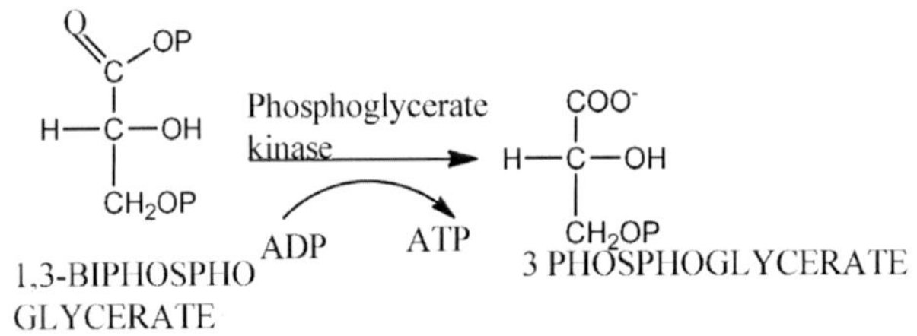

VIII. Conversion of 3-Phosphoglycerate to 2-Phosphoglycerate

It takes place in the presence of phosphoglycerate mutase.

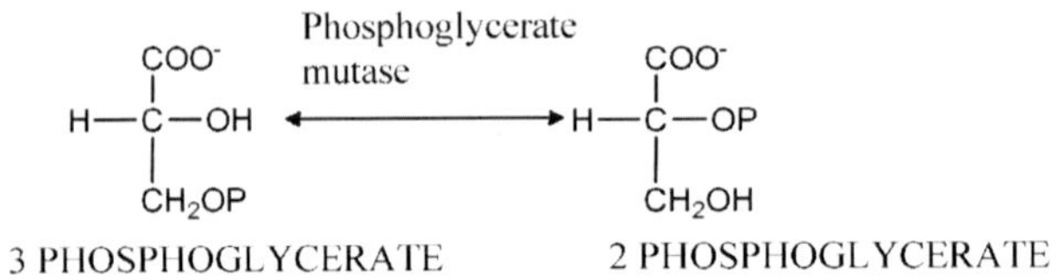

IX. Dehydration of 2-Phosphoglycerate to Phosphoenolpyruvate

This dehydration reaction, with the help of enolase, removes a water molecule from 2-phosphoglycerate to form Phosphoenolpyruvate.

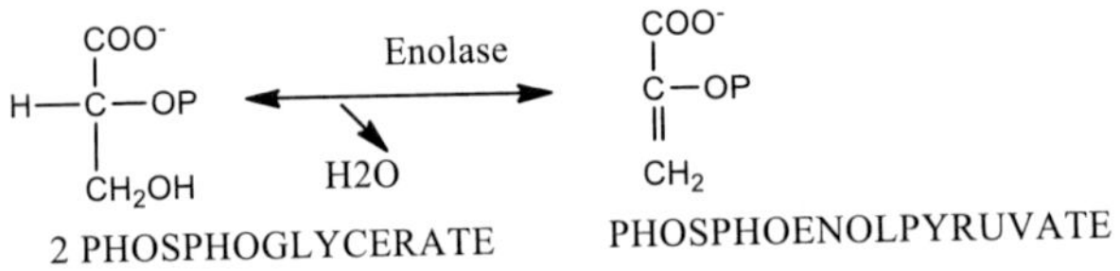

X. Conversion of Phosphoenol Pyruvate to Pyruvate

This reaction generates ATP, and the enzyme required is pyruvate kinase.

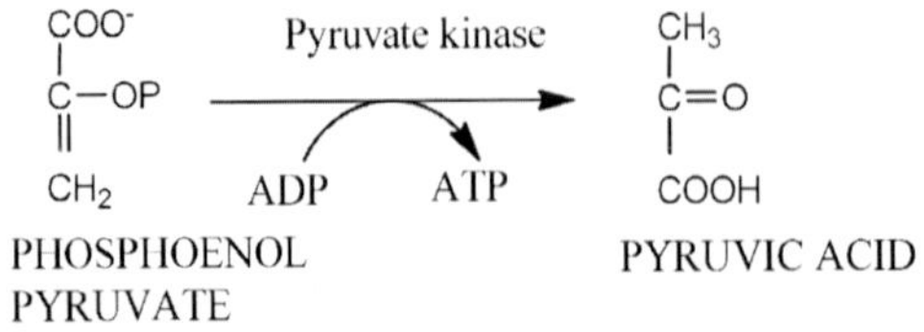

Pyruvic acid is metabolized in two ways. When there is a sufficient oxygen supply to living cells (aerobic condition), pyruvic acid is metabolized through the TCA/citric acid / Krebs cycle to yield more energy. When the oxygen supply to the tissues is insufficient or is lacking (anaerobic condition), it ferments to produce **lactic acid**. The body uses this lactic acid for gluconeogenesis.

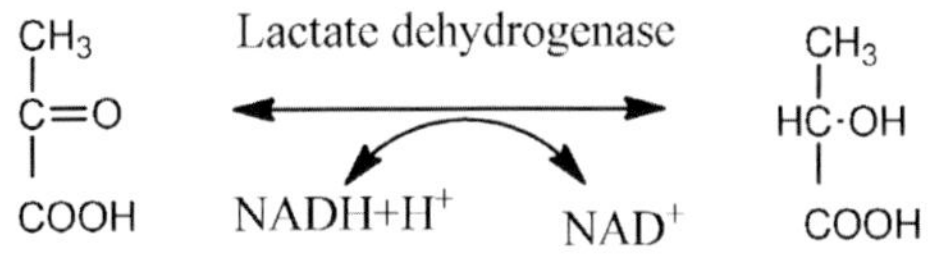

PYRUVIC ACID LACTIC ACID

NADH2 loses its hydrogen atoms or electrons it is accepted by pyruvate, forming lactate and NAD^+. This generated NAD+ is used by the reaction catalyzed by glyceraldehyde-3-phosphate dehydrogenase, which converts glyceraldehyde -3-p to 1, 3-biphosphoglycerate in the glycolysis reaction.

7.2 Utilization of Galactose and Fructose

- **Galactose utilization:** Galactose gets converted to galactose 1-phosphate in the presence of glucokinase and ATP. Galactose 1-phosphate is further converted to glucose 1-phosphate and, then to glucose 6-phosphate by the enzymes galactose-1-P-uridyl transferase and phosphoglucomutase, respectively. Glucose 1-phosphate can either be stored as glycogen or enter glycolysis through conversion to glucose 6-phosphate. The rate of conversion of galactose to glucose 6-phosphate is slower compared to the conversion of glucose to glucose 6-phosphate.
- **Fructose utilization:** Fructose, when acted upon by Fructokinase, is converted into fructose 1-phosphate. Fructose 1-phosphate subsequently gets converted into glyceraldehyde and dihydroxyacetone phosphate by aldolase B. These products or metabolites enter glycolysis. Fructose metabolism bypasses the PFK1-catalyzed reaction, a major regulatory step in glycolysis.

7.3 Factors Influencing Glycolysis

- Rate of enzyme synthesis—Induction of enzyme synthesis increases enzyme synthesis, whereas repression of enzyme synthesis –reduces the rate of enzyme synthesis.
- Rate of phosphorylation and dephosphorylation.
- Allosteric regulation (inhibition by the product) influences the enzyme activity. Glucose -6-p provides allosteric feedback inhibition.

  - Phosphofructokinase is allosterically activated and inhibited. It is activated by F-6-P and AMP and inhibited by ATP, citrate, and H+.
  - Pyruvate kinase is allosterically inhibited by ATP, fatty acids, alanine, and acetyl-CoA. and activated by F-1-6 diphosphate.
- Irreversible reactions in glycolysis.

  a. hexokinase, glucokinase c. Phosphofructokinase and d. pyruvate kinase.
- Hormonal regulation—The main hormones regulating blood glucose levels are Insulin and Glucagon secreted by the pancreas gland. The glucose levels are decreased by Insulin and increased by Glucagon.
- Inhibitors of glycolysis—Arsenate (ATP is not formed from 1-3diphosphoglycerate), iodoacetate (inhibits G-3-P-D), fluoride (inhibits enolase).
- AMP-activated protein kinase (AMPK): It is an important enzyme in all mammalian cells as a cellular energy sensor and for its pivotal role in energy metabolism. It is activated by an increase in a cell's AMP/ATP or ADP/ATP ratio, starvation, hypoxia, and toxins' inhibition of the mitochondrial respiratory chain. The AMPK switches off anabolic pathways (e.g., protein and lipid synthesis) and switches on catabolic pathways (e.g., glycolysis and fatty acid oxidation).

7.4 Glycolysis in Erythrocytes or RBC—the Rapaport-Lumbering Cycle or Shunt (for Synthesis of 2, 3-Bisphosphoglycerate)

Erythrocytes metabolize excessive amounts of glucose through the glycolytic pathway. This will generate more ATP than required and cannot be used by erythrocytes. To reduce ATP production, a diversion pathway in glycolysis occurs in erythrocytes. Therefore, a bypass reaction occurs in erythrocytes so that 1, 3-biphosphoglycerate (1.3 BPG) formed is not converted to 3-phosphoglycerate. Instead, it is converted to 2, 3-biphosphoglycerate (2, 3 BPG) in the presence of the diphosphoglyceromutase enzyme. In this step, ATP is not synthesized as the ATP-generating phosphoglycerate kinase reaction is bypassed. On the other hand, the formation of 2, 3-diphosphoglycerate is also required for hemoglobin function, which helps quickly release oxygen in the tissues. The oxygen affinity of hemoglobin is decreased by 2, 3 BPG. It increases oxygen dissociation from oxyhemoglobin when tissue demand is higher (as in the case of high altitude/hypoxic conditions/postoperative anemia).

The reaction is as follows.

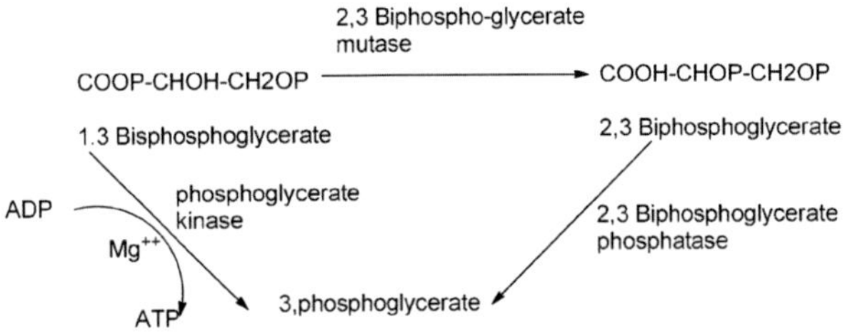

In the above glycolytic reaction, 1,3- Bisphosphoglycerate(1,3-BPG) gets converted to 2,3-BPC by the enzyme 2,3-bisphosphoglycerate mutase. The 2,3-BPC gets hydrolyze and lose one phosphate group by the enzyme bisphosphoglycerate phosphatase to form 3-phosphoglycerate.

In Adult hemoglobin is Hb-A1 type, and in fetal, its type is Hb-F. Both types of hemoglobin have different affinities to oxygen due to different concentration levels of 2,3-BPG. When the concentration of 2,3-BPG is high, affinity to O_2 is less as in the case of Hb-A1 type, and unloading/dissociation is more. The concentration of 2,3-BPG is low in fetal Hb-F, the affinity to O_2 is higher, and the unloading/dissociation is lower.

During blood storage, Inosine is preferred to be added as it can be converted to restore the depleted 2, 3-diphosphoglycerate in the blood bank. Conversion of Inosine to 2,3DPG occurs through the HMP shunt.

7.5 Points to Remember

- It is the primary cycle for glucose metabolism, which can operate under aerobic and anaerobic (RBCs and skeletal muscles are in heavy oxygen demand) conditions. It also facilitates the oxidation of galactose and fructose.
- Tissues like the cornea, lens, and erythrocytes lack mitochondria; hence, glycolysis is a major pathway for ATP synthesis.
- It is also an essential pathway for the brain, dependent on glucose for energy.
- It provides pyruvic acid to run the TCA/KREB cycle (through conversion to Acetyl CoA).
- It provides glycerophosphate, which is required to form triacylglycerol in lipogenesis.
- In RBCs, the cycle generates 2, 3 DPG necessary for oxygen transport.

- Under aerobic glycolysis, there is no need for LDH as the product formed is pyruvic acid, with the generation of more ATP in the cell cytoplasm.
- During anaerobic conditions, when glucose splits into two lactate molecules, overall, two molecules of ATP are gained.
- In anaerobic conditions, when the pyruvic acid is converted to lactic acid by lactate dehydrogenase, it converts NADH2 to NAD. The anaerobic reaction occurs in RBC or erythrocytes (without mitochondria) and during heavy muscular exercise.
- Skeletal muscles work under aerobic (resting) and anaerobic (exercise) conditions. Muscle glycogen cannot yield glucose as receptors for glucagon are absent, and it does not contain G-6-phosphatase. Epinephrine stimulates glycogenolysis in skeletal muscles.
- Cancer cells or tumor cells proliferate and produce more lactic acid. More lactate is used for gluconeogenesis at the expense of more energy and contributes to loss of body weight (cancer cachexia)
- Malignant tumor cells have a higher glycolytic rate than normal cells of their origin. High aerobic glycolysis is helpful in the clinical diagnosis of the condition. Targeting a reduction in mitochondrial metabolism by reducing glycolysis can starve the malignant cells.
- Sodium fluoride (NaF) is the preferred anticoagulant for blood when blood is collected for blood glucose estimation. Fluoride anticoagulants inhibit the enzyme enolase present in erythrocytes, as glycolysis is the only major pathway or source to meet the energy requirement. If NaF is not added to blood, then there is a possibility of getting a laboratory report of low blood sugar due to the utilization of sugar through uninhibited enolase enzyme activity in a glycolytic process.
- The four regulatory enzymes of glycolysis are 1. Hexokinase 2, glucokinase 3, phosphofructokinase, and 4 pyruvate kinase.
- Hexose kinase deficiency leads to a low level of the precursor of 2,3-DPG in red blood cells. Red blood cells will have a high oxygen affinity to release less oxygen for tissues.
- Pyruvate kinase deficiency leads to low production of ATP and reduces the activity of the Na+ and K+ stimulated ATPase ion pump.

Key Takeaway

- Glycolysis involves a series of reactions to extract energy from glucose to release two molecules with three carbons called pyruvate.
- Glucose uptake in cells is facilitated by GLUTs (glucose transporters) present in the plasma membrane.

- Allosteric regulation is found in the EMP cycle.
- Tissues like the cornea, lens, and erythrocytes lack mitochondria; hence, glycolysis is a major pathway for ATP synthesis.
- It is also an essential pathway for the brain, dependent
- The endergonic reactions in the EMP cycle are a. formation of Glucose-6-phosphate through phosphorylation of glucose phosphorylation of F-6-P to Fructose 1,6-bisphosphate.
- The exergonic reaction is a. Oxidative phosphorylation of Glyceraldehyde-3-phosphate to 1,3-Bisphosphoglycerate generate NADH

  b. Conversion of 1,3-Bisphosphoglycerate to 3-Phosphoglycerate c. Conversion of Phosphoenol Pyruvate to Pyruvate.
- The four regulatory enzymes of glycolysis are 1. Hexokinase 2, glucokinase 3, phosphofructokinase, and 4 pyruvate kinase.
- The enzymes concerned with the endergonic reactions are

  a. Glucokinase or Hexokinase and b. Phosphofructokinase.
- The enzymes concerned with the exergonic reactions are a. Glyceraldehyde 3-phosphate dehydrogenase b. Phosphoglycerate kinase and c. Pyruvate kinase.
- In R.B.C., there is a formation of 2, 3-diphosphoglycerate required for hemoglobin function. It helps quickly release oxygen in the tissues.
- During blood storage, Inosine is preferred to be added as it can be converted to restore the depleted 2, 3-diphosphoglycerate in the blood bank. Conversion of Inosine to 2,3DPG occurs through the HMP shunt.

Intext Questions

1. What is the full form of EMP?
2. What do you mean by the term glycolysis? Where does it occur?
3. What is the difference between hexokinase and glucokinase?
4. What is the difference between enolase and aldolase?
5. Why is the glucose converted to glucose 6-phosphate once free glucose enters the cell?
6. Write the endergonic reactions of the EMP cycle.
7. Write the exergonic reactions of the EMP cycle.
8. Which high-energy bonds participate in the EMP cycle?

9. In which reaction of the EMP cycle does isomerization occur?

10. Which steps in the EMP cycle are irreversible in nature?

11. Erythrocytes lack mitochondria. Which reaction product in erythrocytes favors oxygen dissociation?

12. Choose one correct option for the following from the options given below (The answer is given in **bold** letters)

a Which enzyme will act on fructose?

i. Glucokinase ii. **Hexokinase**

iii. Enolase

b The EMP cycle operates in

i. Nucleus ii. Mitochondria

iii. **Cytosol**

c R.B.Cs. are oxygen carriers and supply oxygen to all cells. The nature of glycolysis occurring in them is

i. Aerobic. ii. **Anaerobic.**

iii. Sometimes aerobic and sometimes anaerobic.

d Which enzyme is required for the conversion of glucose 6-phosphate to glucose?

i. Glucokinase ii. **Glucose phosphatase**

iii. Hexokinase

e Reaction type which converts glucose to fructose

i. Reduction ii. **Isomerization**

iii. Oxidation.

f Phosphofructokinase reaction is involved in the conversion of

i. **Fructose 6-phosphate to fructose -1-6-biphosphate**

ii. Fructose to fructose 6-phosphate

iii. Glucose to fructose

g Triose phosphate isomerase converts

i. Dihydroxy acetone phosphate to glyceraldehyde 3-p

ii. Glyceraldehyde 3-p to dihydroxy acetone phosphate

iii. **Both i and ii.**

h Conversion of glyceraldehyde-3-P to 1, 3-biphosphoglycerate converts

i. **NAD^+ to $NADH_2$.** ii. FAD to $FADH_2$

iii. $FADH_2$ to FAD

i How many phosphoglycerates are formed from one glucose molecule?

i. Three ii. **Two**

iii. Four

j The enzyme involved in converting 3 phosphoglycerates to 2 phosphoglycerates

i. **Phosphoglycerate mutase** ii. Phosphoglycerate kinase

iii. Enolase

k Enolase enzyme forms

i. Phosphoenol pyruvate ii. 2 phosphate glycerate

iii. **Both i. and ii.**

l The formation of pyruvic acid by pyruvate kinase is

i. **Exergonic reaction** ii. Endergonic reaction

iii. Substrate level reaction

m Lactate dehydrogenase converts

i. Pyruvic acid to lactic acid ii. Lactic acid to pyruvic acid

iii. **Both i. and ii.**

n Glycolysis is a major pathway for ATP synthesis in

i. Cornea ii. Lens

iii. **Both i. and ii.**

- o Which statement is not true for EMP cycle
  - i. **It is located in the inner membrane of mitochondria**
  - ii. It converts glucose to glucose-6- P
  - iii. Two triose phosphates are formed in the cycle.

Further Readings

AMPK's mechanisms in maintaining ATP balance during energy metabolism. Rong Ke, Qiao, Cong Li, Lingyu Luo.Deqiang Huang, 2017..

Biochemistry. Naik, Pankaj (2009). Second Edition, Jaypee Brothers, New Delhi.

Biochemistry By U.Satyanarayana. Edition no.3.

Glycolysis. Chandel NS. 2021. Cold Spring Harb Perspect Biol .2021 may, 3:13(5) : a040535 doi 10.1101/cshperspect.PMID.

Harper's illustrated biochemistry (30th ed.). Rodwell, V. W., Botham, K. M., Kennelly, P. J., Weil, P. A., & Bender, D. A. (2015). New York, N.Y.: McGraw-Hill Education LLC.

Life Sciences. Fundamentals and Practice I. Pranav Kumar and Usha Mina. (2017).Pathfinder publication. New Delhi.

Medical biochemistry By M.N.Chatterjea, Edition 8.

Qiao, Cong Li. Lingyu Luo, Deqiang Huang. 2017. Mechanisms of AMPK in the maintenance of ATP balance during energy metabolism.

Textbook of Biochemistry for Medical Students. Vasudevan, D. M.; Sreekumar, S.; Vaidyanathan, Kanan (2019). Ninth Edition, Jaypee Bros. Medical Publication, New Delhi.

8

Tricarboxylic Acid Cycle (TCA)/ Citric Acid Cycle/ Krebs Cycle and Shuttle Mechanisms

The Hans Krebs cycle, known by different names like the citric acid or **tricarboxylic acid cycle**, was named after the scientist **Hans Krebs.** The **citric acid** cycle is named after the citrate product formed in the cycle. The name **tricarboxylic acid** is given as citrate itself, which is a tricarboxylic acid due to the presence of three carboxyl functional groups (-COOH).

The tricarboxylic acid (TCA) cycle occurs in aerobic conditions and within mitochondrial ETC, involving various sequential oxidoreductase reactions. The cycle requires Acetyl-CoA (C2 molecule) and oxaloacetate molecule (C4 molecule) to begin with the formation of a citrate molecule (C6 molecule). Acetyl-CoA is generated from the pyruvate (glycolysis), the fatty acids (lipolysis), or by oxidative deamination of specific amino acids. The complete oxidation of a glucose molecule results in the end products like energy (ATP), CO_2 and H_2O. A molecule of Acetyl-CoA has two carbons. It is formed from a three-carbon pyruvic acid molecule generated from the glycolysis of a six-carbon glucose molecule. Acetyl-CoA is the substrate for this cycle and is an energy-rich molecule having a thioester bond between coenzyme A and acetic acid.

8.1 Formation of acetyl-CoA from Pyruvic Acid

Pyruvate is obtained by glycolysis or the oxidation of glucogenic amino acids. The pyruvic acid produced in the cytoplasm crosses the inner membrane of the mitochondrion to enter the mitochondrial matrix, where it undergoes oxidative decarboxylation, forming Acetyl-CoA. Acetyl-CoA carries (carrier molecule) the acetyl group to the citric acid cycle. It also links various metabolic products with the TCA cycle. Such products can be from the EMP cycle, beta-oxidation, and oxidative deamination of amino acids. Oxidative decarboxylation of pyruvic acid needs a complex enzyme system known as pyruvate dehydrogenase complex. This complex system comprises three enzymes and five co-enzymes working in cooperation. Oxidative decarboxylation of pyruvic acid is called a link reaction because this reaction links the glycolysis pathway and the citric acid cycle.

- Enzymes—pyruvate dehydrogenase(E1), dihydrolipoyl acetyl transferase (E2) and Dihydrolipoyl dehydrogenase (E3).
- Co-enzymes—TPP, FAD, NAD, CoASH, and lipoic acid. TPP, Lipoic acid, and FAD are prosthetic groups. NAD and CoASH are co-substrates.

I. Steps-wise action of the PDH complex

(A) Decarboxylation (E1, formation of hydroxyethyl-TPP)

A carbonyl group gets converted to CO2 and is then removed from the pyruvate molecule by its decarboxylation in the presence of TPP (the coenzyme of pyruvate decarboxylase), with the formation of a hydroxyethyl derivative of TPP (the coenzyme of pyruvate decarboxylase).

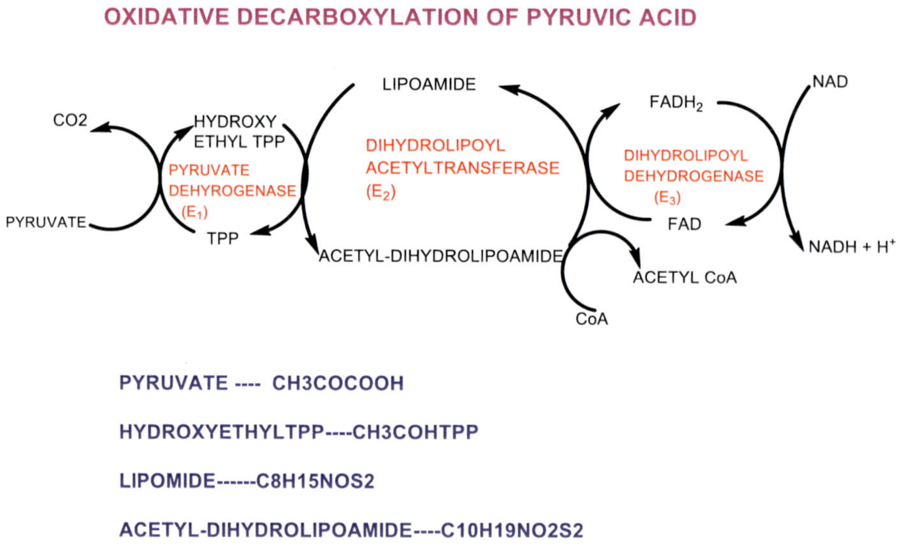

Fig. 8.1: Decarboxylation of pyruvic acid with E1, E2, and E3 enzymes.

ADD THIS COLOUR DIAGRAM

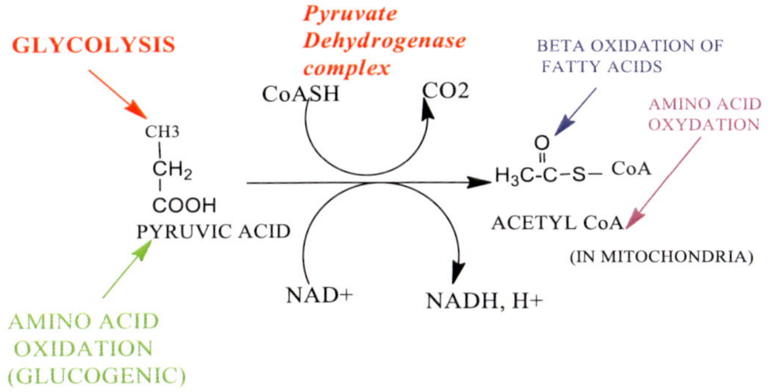

Fig. 8.2: Decarboxylation of pyruvic acid (Simplified)

(B) Oxidation (transfer of acetyl group to lipoamide).

The hydroxyethyl group attached to TPP, upon oxidation, forms an acetyl group. This formed group is transferred to the lipoamide group of the protein complex enzyme E2 to form acetyl-dihydrolipoamide in the presence of dihydrolipoyl acetyltransferase.

(C) Transfer of acetyl group from acetyl lipoamide to CoA

The acetyl group from acetyl lipoamide is then transferred to Coenzyme A to form the Acetyl-CoA. It is catalyzed by E2 and E3 enzymes.

(D) Oxidation of dihydrolipoamide to lipoamide (E3, FAD, NAD^+).

Oxidation of dihydrolipoamide leads to the formation of lipoic acid (regenerated). The FAD present in this reaction gets converted to $FADH_2$(reduced form) and again is oxidized to FAD by dihydrolipoyl dehydrogenase and NAD+ (oxidized form). The NAD+ accepts electrons as part of a specific reaction in glycolysis. In the presence of oxygen, NADH (reduced form) donates electrons and regenerates NAD+ again for glycolysis. The donated electrons pass into the electron transport chain (ETC), regenerating NAD+ again for glycolysis. The formed Acetyl-CoA is utilized when it enters the TCA cycle.

The overview of reactions of the pyruvate dehydrogenase complex

Pyruvate + TPP + lipoic acid + CoA + FAD^+ + $NAD^+ \longrightarrow CO_2$ + Acetyl-CoA + $NADH+H^+$

Allosteric inhibition of the pyruvate dehydrogenase complex reaction occurs due to factors like the high ratio of ATP to ADP, NADH to NAD+, and Acetyl-CoA to CoA.

8.2 Steps of Tricarboxylic Acid Cycle (TCA)/ Citric Acid Cycle/ Krebs Cycle

This cycle involves a series of sequential chemical reactions to release energy through the oxidation of acetyl-CoA, which is derived from food components (carbohydrates, fats, and amino acids). In the case of carbohydrates, a monomer glucose molecule gets converted to acetyl-CoA via pyruvic acid oxidation. This process yields two acetyl-CoA molecules from one glucose molecule. Because two acetyl-CoA are formed, the cycle proceeds twice per glucose molecule, yielding twice the products.

This cycle represents a critical aerobic reaction in mitochondria sequentially to oxidize acetyl-CoA to generate CO_2, H_2O, and energy from NADH.H+ and $FADH_2$. The $NADH_2$ and $FADH_2$ are oxidized to NAD and FAD through the respiratory chain for reuse. This cycle also provides a pathway for other molecules to enter and produce molecules required for various biosynthetic processes, depending upon the body's demand. The TCA Cycle is located in different places, including prokaryotes and eukaryotes. It is in the matrix of eukaryotic cells, but in the cytosol of prokaryotic cells. The pyruvate is formed in the cytoplasm, but enzymes for its metabolism are located in mitochondria. Therefore, pyruvate has to be brought into mitochondria for its metabolism. The mitochondrial matrix is essential for the TCA cycle as it is where the required enzymes are available, except for succinate dehydrogenase, which is a part of protein complex II (on the inner membrane) in the electron transport chain.

Table 8.1: Steps and enzymes catalyze the reactions of the TCA cycle.

No	Reactions and energy account	Enzyme
1	**Condensation reaction.** Acetyl-CoA (C2) joins with oxaloacetate (C4) to form citrate (C6). It is an irreversible reaction.	Citrate synthase.
2	**Dehydration (H2O removed) and Hydration reaction (H2O is added).** Citrate is (C6) converted first to cis-Aconitate and then to isocitrate (C6--an isomer of citrate).	Aconitase
3	**Oxidation and decarboxylation reaction.** Isocitrate is converted to alpha-ketoglutarate (a C5 molecule). The release of CO2 occurs. One NADH molecule is formed. Release of one H^+ ion. **This reaction is a rate-limiting step.**	Isocitrate dehydrogenase
4	**Oxidation reaction.** Alpha-ketoglutarate is converted to succinyl-CoA. A second molecule of NADH and carbon dioxide is produced, release of H^+ ion.	Alpha-ketoglutarate dehydrogenase

No	Reactions and energy account	Enzyme
5	• **A substrate-level phosphorylation.** Succinyl CoA is then converted to succinate (C4 molecule). One GTP molecule is produced. Release of CoA-SH occurs.	Succinyl-CoA synthase
6	**A redox reaction/oxidation reaction.** Succinate is converted into fumarate (C4 molecule). A molecule of $FADH_2$ is produced.	Succinic Dehydrogenase
7	• **Hydration reaction.** Fumarate is converted to malate (C molecule). H2O is added.	Fumarase
8	• **Oxidation reaction.** Malate, upon oxidation, is converted to oxaloacetate. This reaction regenerates the oxaloacetate molecule. • The third NADH molecule is produced. Release of H+ ion.	Malate dehydrogenase

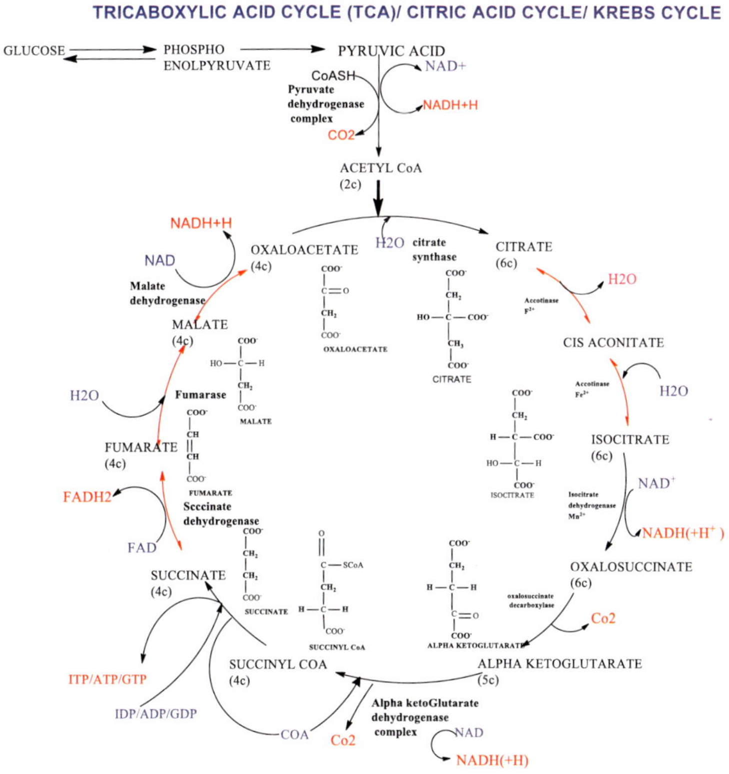

SUMMARY OF TCA CYCLE

MOLECULE	UTILISED	RELEASED
H2O	3	1
NAD	2	O
NADH(+H)	0	2
FAD	2	O
FADH2	0	2
CO2	0	2
COA	1	0

Fig. 8.3: Tricarboxylic acid cycle

I Different steps in one complete round of the cycle

A. Condensation reaction. This reaction is marked by the condensation of Acetyl-CoA (C2 molecule) with oxaloacetate (C4 molecule) to form citrate (C6 molecule) and free co-enzyme A molecules. The enzyme required is **citrate synthase.**

This step is irreversible and highly exergonic in nature. Negative feedback of ATP controls this reaction. If oxaloacetate is available in more, then more acetyl-CoA gets metabolized.

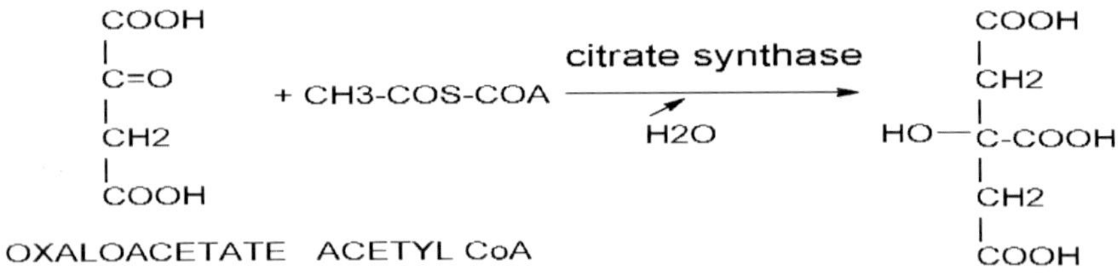

It is also possible to form oxaloacetate from pyruvate through carboxylation by pyruvate carboxylase. This reaction contributes to anaplerosis, as it helps replenish the oxaloacetate pool without using citrate. This is how citrate can be spared for lipid synthesis.

B. Dehydration (H2O removed) and Hydration reaction (H2O is added). Citrate is converted to isocitrate (an isomer of citrate). Here, citrate loses one water molecule (dehydrated) to *cis*-aconitate. When the *cis*-aconitate

If it is hydrated, it will form isocitrate (instead of citrate). The enzyme required is **Aconitase** in both of these reactions.

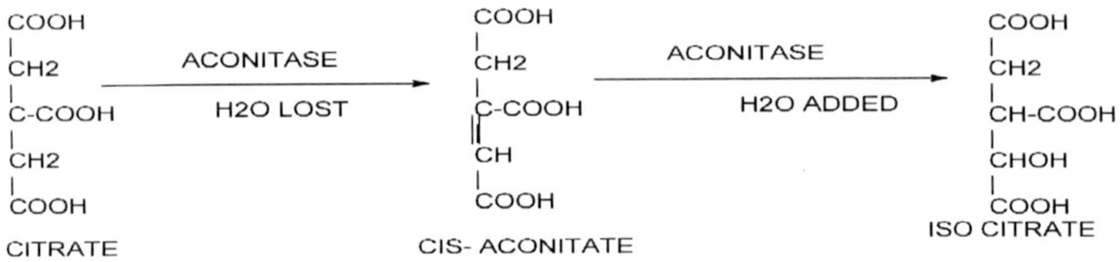

C. Oxidation and decarboxylation reaction. Isocitrate in the presence of NAD+ and the enzyme isocitrate dehydrogenase is converted to oxalo-succinate with the formation of NADH. In the subsequent step, the oxalo-succinate in the presence of the same enzyme loses carbon dioxide, forming alpha-ketoglutarate (C5 molecule).

In this reaction, NADH is formed in the first step, and CO2 is formed in the second. This step is under the control of negative feedback from NADH. The isocitrate dehydrogenase is an allosterically regulated enzyme; hence, this reaction is a rate-limiting step.

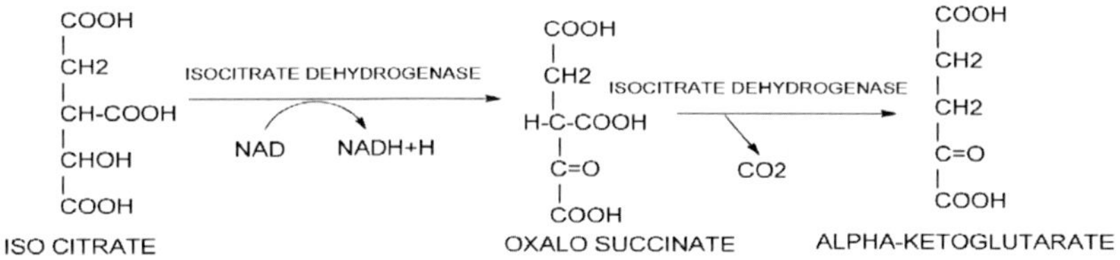

D. Oxidation reaction. In this reaction, succinyl-CoA is obtained due to the oxidation of alpha-ketoglutarate. A second molecule of NADH and CO2 is produced. The enzyme alpha-ketoglutarate dehydrogenase complex catalyzes this reaction. This enzyme is regulated by ATP, succinyl-CoA, and NADH feedback inhibition.

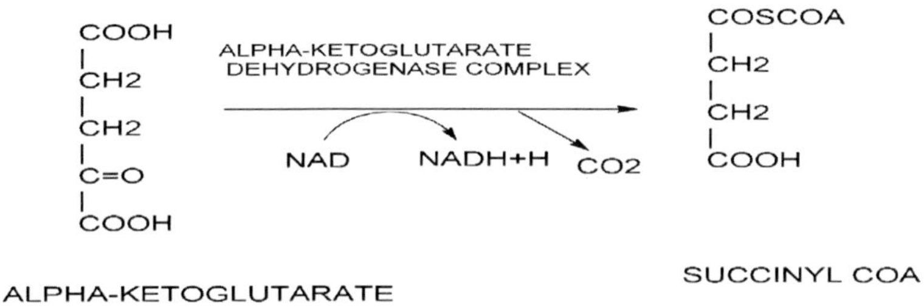

E. A substrate-level phosphorylation. Succinyl CoA is converted to succinate (a 4-carbon molecule), producing one GTP molecule. This reaction requires succinyl-CoA synthase. This reaction is a substrate-level phosphorylation event where coenzyme A is replaced, and a high-energy bond from GDP to GTP is formed.

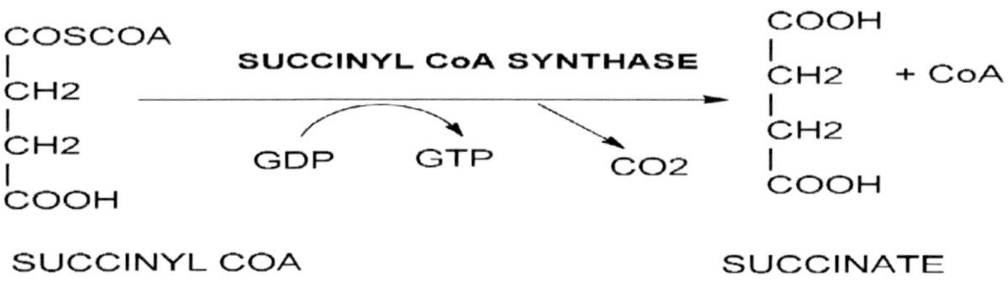

F. A redox reaction/oxidation reaction. This is a redox reaction where the oxidation of succinate in the presence of FAD+ forms a fumarate molecule. As a result of the transfer of two hydrogen atoms to FAD+, it is converted to $FADH_2$ which transfers electrons to the electron transport chain directly.

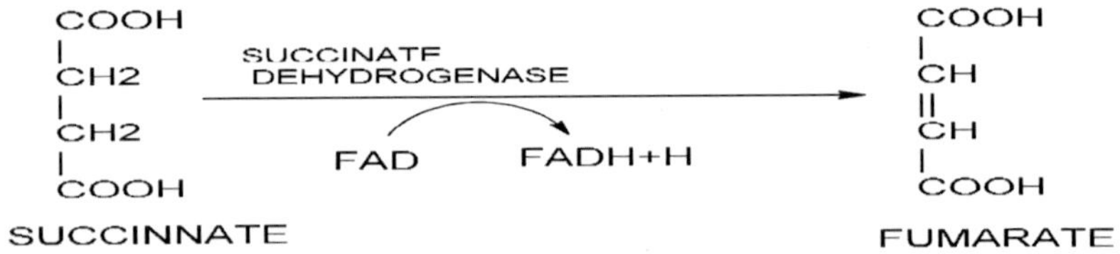

G. Hydration reaction. When a water molecule is added to Fumarate, the latter is converted to malate (C4 molecule). This reaction needs the fumarase enzyme.

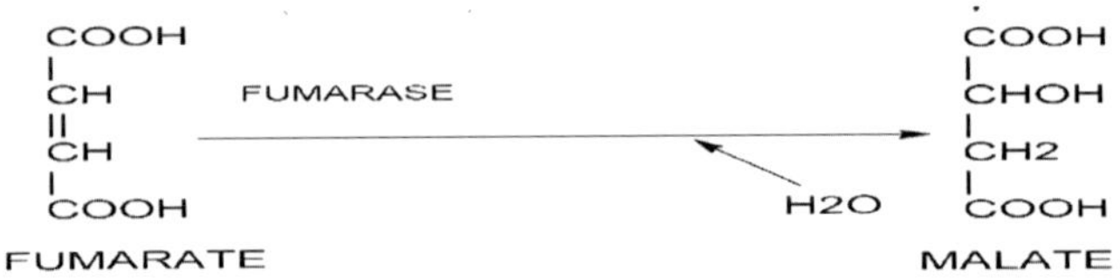

H. Oxidation reaction. Malate, upon oxidation, is converted to oxaloacetate using NAD+. The third molecule of NADH is also produced.

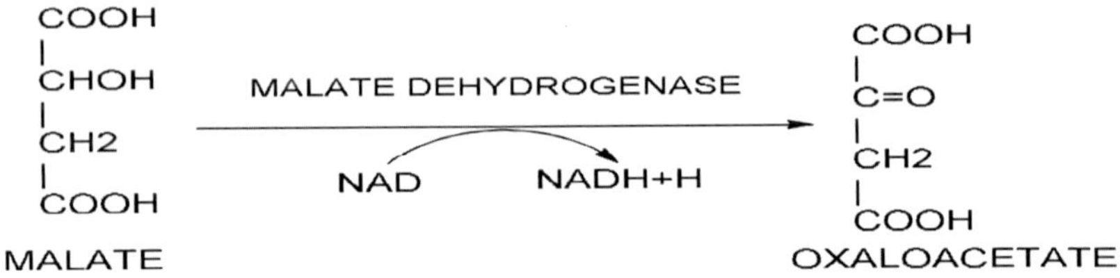

Oxaloacetate is regenerated with the help of this last reaction.

In the TCA cycle, both the anabolic and catabolic processes occur, so it is an amphibolic pathway.

II Details of energy production from one glucose molecule (aerobic condition)

Table 8.2: Energy accounts for one glucose molecule.

	Reaction	**ATP Account**
A	**Glycolysis Emp Cycle**	
	Glucose to Glucose-6-P by hexose kinase ATP to ADP	-1ATP
	Fructose 6-P to fructose -1, 6-bis-P through phosphofructokinase	-1ATP
	Glyceraldehyde-3-P to 1,3-BIS-Phosphoglycerate through Glyceraldehyde -3P-dehydrogenase 2NADH x 3ATP	+6ATP

	1,3-BIS-Phosphoglycerate to 3-P-glycerate through the enzyme phosphoglycerate kinase 2 ATP (due to two molecules of 1,3-BIS-Phosphoglycerate)	+2ATP
	P-Enol pyruvate to pyruvate through. pyruvate kinase (due to two molecules of 1,3-BIS-Phosphoglycerate)	+2ATP
	NET GAIN ATP (10-2)	+8ATP
B	**Oxidative Decarboxylation of Pyruvic Acid**	
	Pyruvate (2) to Acetyl-CoA (2) through the Pyruvate dehydrogenase complex Two pyruvate molecules from one glucose molecules 2NADH x 3ATP	+6ATP
C	**TCA Cycle**	
	1 Isocitrate (2) to oxalosuccinate (2) through Isocitrate dehydrogenase 2NADH x 3ATP	+6ATP
	Two alpha-ketoglutarate (2) to succinyl CoA (2) through the Alpha-ketoglutarate dehydrogenase complex 2NADH x ATP	+6 ATP
	3. succinyl-CoA (2) to succinate (2) through succinate thiokinase 2GTP x 1ATP	+2ATP
	4. succinate (2) to fumarate (2) through Succinate dehydrogenase 2 FADH2x 1 ATP	+4ATP
	5. Malate (2) to oxaloacetate (2) through Malate dehydrogenase 2 NADH x ATP	+6ATP
	Total ATP per Molecule of Glucose (Aerobic Condition)	38 (30+8)
	Energy Production from One Glucose Molecule (Anaerobic Condition): Pyruvate gets Converted to Lactate	+2ATP ONLY

III. Brief account of ATP formation during complete oxidation of a glucose molecule.

Table 8.3: Summary of ATP generation upon complete glucose oxidation.

Source	Stage of reactions	ATP from reduced coenzymes through ETS	Substrate-level phosphorylation or Direct ATP formation	Total ATP
Glycolysis (EMP)	Glucose(6C) to 2 pyruvates (3C)	6 ATP (2NADPH)	2 ATP	8
Link Reaction	2 Pyruvate(6C) to 2 Acetyl-CoA (4C) + 2CO2	6ATP (2NADPH)	0	6
TCA Cycle	2 Acetyl-CoA (4C) to 4CO2	18 ATP (6NADH) 4ATP (2FADH2)	2 ATP from GTP	24
Total	Glucose (6C) to 6CO2	34	4	38

Note

- ATP is Continuously Formed and Hydrolysed in the Body.

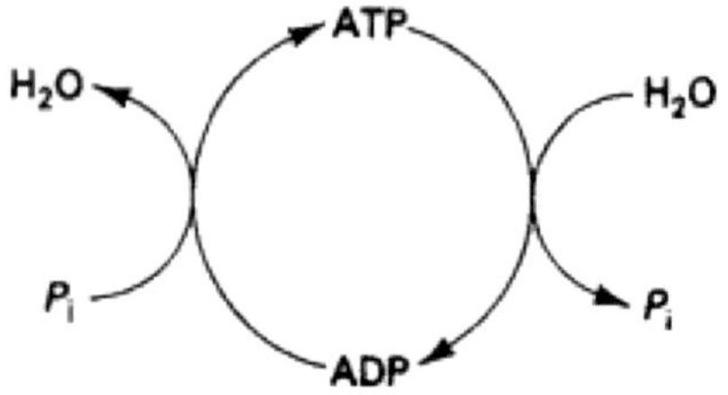

- One glucose molecule, upon complete oxidation, generates 38 molecules of ATP. In the above diagram, it can be seen that when one ATP molecule is formed, H2O is also formed. Therefore, 38 water molecules are also produced because phosphodiester bonds exist between ADP and Pi (inorganic phosphate), which involves splitting off a water molecule.
- One molecule of glucose 6c generates two molecules of pyruvate in glycolysis. Two pyruvates get converted to two acetyl-CoA. The TCA cycle has to run twice to oxidize two pyruvates /two Acetyl-CoA.

8.3 Regulation of the TCA Cycle

- When citrate inhibits phosphofructokinase of glycolysis, the availability of pyruvate and acetyl-CoA molecules gets reduced. They are required to enter the TCA cycle.
- When elevated, Calcium levels in the mitochondrial matrix will accelerate the activities of many enzymes in the TCA cycle.
- Allosteric regulation by metabolites is observed in the regulation of the TCA cycle, which is primarily determined by product inhibition and substrate availability.
- Various cycle metabolites provide negative feedback via the respective enzyme.
- ATP/ADP ratio and NADH/NAD ratio also influence the rate of the TCA cycle.
- ATP, Acetyl-CoA, and NADH inhibit citrate synthase.
- Isocitrate dehydrogenase is inhibited by both ATP and NADH.
- Succinyl CoA, NADH, and ATP inhibit alpha-ketoglutarate dehydrogenase.
- Oxaloacetate inhibits SDH enzyme.
- Four water-soluble B complex vitamins participating are

1. Riboflavin in the form of FAD, a cofactor in dehydrogenases (alpha-ketoglutarate dehydrogenase and succinate dehydrogenase).
2. Niacin, as the coenzyme in the form of NAD.
3. Thiamin (vitamin B1) as coenzyme TPP for the decarboxylation reaction and
4. Pantothenic acid is a part of coenzyme A in the form of acetyl-CoA, or Succinyl-CoA.

8.4 Importance of the TCA Cycle

- It serves as a common pathway for the complete oxidation of acetylCoA derived from the components of the food.
- It is the cycle that provides a major energy source for the cells.
- It is an amphibolic cycle as it provides the number of intermediates for the anabolic pathway involved in the biosynthesis of various molecules. Succinyl CoA is used for porphyrin and Hb synthesis. Oxaloacetate is used for gluconeogenesis, amino acids (aspartate), purines and pyrimidines formation, and detoxification by conjugation. Citrate for fatty acid and amino acid synthesis. Alpha-ketoglutarate is used for glutamate formation, which is used for forming amino acids and neurotransmitters.
- The cycle uses the metabolites formed by Anaplerotic reactions. It means that the TCA cycle can use components synthesized from outside the cycle, such as oxaloacetate, fumarate, succinyl-CoA, and alpha-ketoglutarate.

8.5 Shuttle Mechanisms

The cytoplasmic organelle with a double membrane (inner and outer) is well known as a mitochondrion. The NADH produced in the cytosol can undergo oxidation and reduction reactions.

- When NADH loses electrons, or hydride gets converted to NAD+, the process is oxidation.

  Oxidation ----- NADH→NAD+
- When NAD+ gains, the electrons from a hydride get converted to NADH; the process is reduction.

  Reduction----- NAD+ → NADH

The NADH generated in glycolysis is extra-mitochondrial (cytoplasmic) and must be oxidized to NAD+ at the electrons transport chain located in the mitochondrion. However, NADH-generated electrons are impermeable to the inner mitochondrial membrane, so they cannot directly enter the mitochondria. This transfer of NADH into the mitochondria is made possible by the "shuttles" present on both sides of the mitochondrial membrane, which are the specific dehydrogenases. The shuttles help transfer NADH.

I. Types of shuttle systems

A. Malate-aspartate shuttle or malate shuttle

The malate-aspartate (M-A) shuttle is essential for its regulatory role in glycolysis and lactate metabolism. The shuttle is crucial for various organs (the kidney, heart, brain, and liver) to generate 3 ATP per mole of NADH.

The malate shuttle helps to maintain the balance of metabolites within the matrix and the cytosol, which is achieved with the help of two antiporters for the transport of metabolites. These electrons, when entering the electron transport chain, generate ATP.

The shuttle consists of four protein parts distributed in the inner mitochondrial membranes and the mitochondrial matrix (two proteins) & intermembrane space (two proteins). The four proteins of the shuttle system are Malate dehydrogenase, Aspartate aminotransferase, glutamate-aspartate antiporter (transporter I), and malate-alpha-ketoglutarate antiporter (transporter II).

Malate dehydrogenase and Aspartate aminotransferase are proteins in the mitochondrial matrix and intermembrane space.

The Aspartate aminotransferase and malate dehydrogenase found in the cytosol are different from those found in the mitochondrial matrix because of their difference in location and structure. They catalyze their reactions in opposite directions in this process. There are two sets of enzymatic reactions; one set of reactions occurs in the cytosol, and the other set of reactions takes place in the mitochondrial matrix.

The other two proteins found in the inner membrane of mitochondria are antiporters, namely glutamate-aspartate antiporter (transporter I) and malate-alpha-ketoglutarate antiporter (transporter II).

1, Glutamate-aspartate antiporter or transporter I . They import glutamate from the cytosol into the matrix and export aspartate from the matrix to the cytosol. Here, when there is an entry of aspartate from the mitochondrial matrix to the cytosol, there is an exit of glutamate and proton from the cytosol into the mitochondrial matrix. The aspartate/glutamate antiporter moves these amino groups.

Table 8.4: Table showing antiporters and transfer of molecules.

	Transfer	
Antiporters	**From Cytosol to Mitochondrial Matrix**	**From Mitochondrial Matrix to Cytosol**
Glutamate-aspartate antiporter or transporter I	Glutamate	Aspartate
Malate- alpha-ketoglutarate antiporter or transporter II	Malate	Alpha-ketoglutarate

2. The malate- alpha-ketoglutarate antiporter or transporter II. It imports malate from the cytosol into the matrix and, in exchange, exports alpha-ketoglutarate from the matrix to the cytosol. It transfers reducing equivalents from the cytosol into mitochondria.

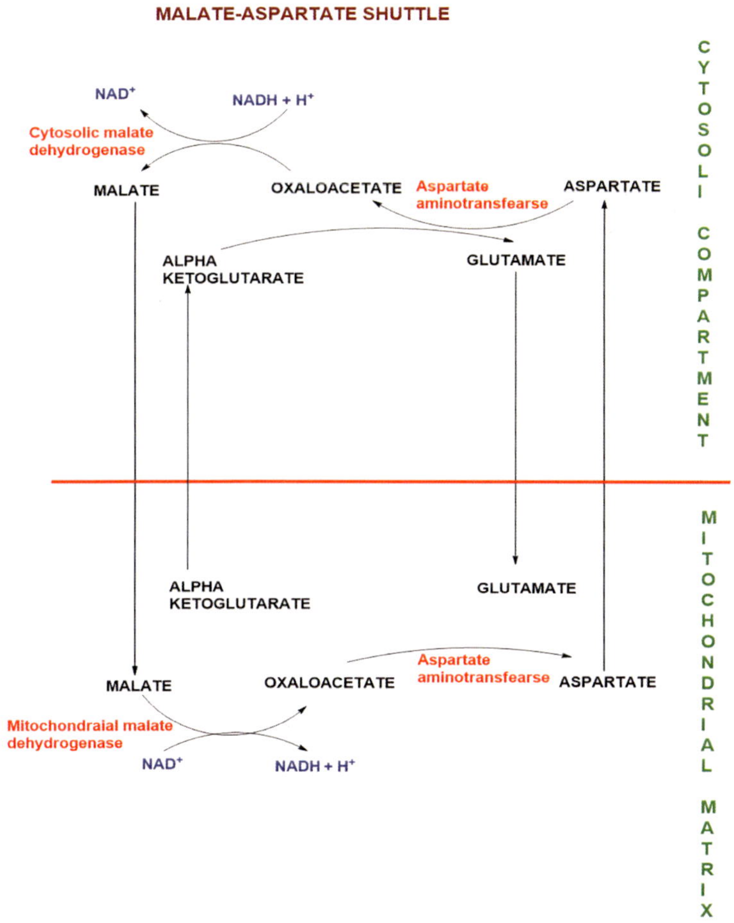

Fig 8.5: Malat-aspartate shuttle

• Cytosolic reactions

The types of reactions occurring on the cytoplasmic side include the conversion of aspartate to malate via oxaloacetate and the conversion of alpha-ketoglutarate to glutarate. These reactions are as follows:

a) An alpha-ketoglutarate's conversion into glutamate occurs when aspartate is converted to oxaloacetate. Conversion of cytosolic aspartate to oxaloacetate requires the removal of an amino group with the help of the cytosolic enzyme aminotransferase.

b) Oxaloacetate ($C_4H_4O_5$) H formed is further reduced to malate ($C_4H_6O_5$) by the cytosolic malate dehydrogenase. This leads to the oxidation of cytosolic NADH to NAD^+. The oxaloacetate cannot cross the mitochondrial membrane. Here, malate is imported to the matrix from the cytosol by malate-alpha-ketoglutarate. Simultaneously, it exports alpha-ketoglutarate from the matrix into the cytosol.

• Mitochondrial matrix

Two types of reactions occurring in the mitochondrial matrix side are

a) The conversion of malate (entered from the cytoplasm) to oxaloacetate, catalyzed by the mitochondrial malate dehydrogenase. Here, NAD is converted to NADH.

b) Oxaloacetate cannot cross the mitochondrial matrix to enter the cytosol. Therefore, oxaloacetate gets converted to aspartate and to alpha-ketoglutarate with the help of glutamate (which provides an amino radical) in the presence of the mitochondrial enzyme Aspartate aminotransferase. Aspartate can cross the mitochondrial membrane. Aspartate is also converted to oxaloacetate in the cytosol, completing the cycle of reaction

B. Glycerophosphate shuttle

This mechanism transports metabolites or chemical groups across the mitochondrial membrane. When the glycerophosphate is transported by electron shuttle into the mitochondria, the dihydroxyacetone phosphate moves out of the mitochondria . The inner mitochondrial membrane on its outer surface possesses an isozyme of Glycerol 3-phosphate dehydrogenase (G3PDH). This isozyme in the cytosol causes the oxidation of NADH and, in mitochondria, causes the reduction of FAD. The $FADH_2$ generates 2 ATP when it gets oxidized via ETC. In this shuttle mechanism, there is no involvement of membrane transport systems. Every time Dihydroxyacetone phosphate escapes into the cytosol, the shuttling continues. The reduction of dihydroxyacetone phosphate produces the glycerol phosphate. This shuttle system is in insect flight, white, skeletal, and brain muscle. It is also found in brown adipose tissue present in young ones for generating heat. The skeletal muscle and the brain utilize Glycerol-phosphate, where 2 ATP are generated from NADH.

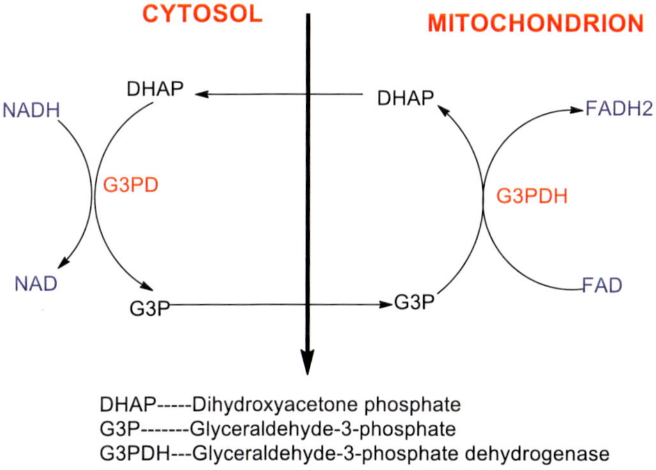

Fig. 8.6: Glycerophosphate shuttle

8.6. Plant Cycles

I. The glycolate cycle/ glyoxylate cycle

This cycle is a modification of the TCA cycle and is seen in the glyoxysome of plants and some microorganisms but not in animal cells. The cycle does not operate in animal cells due to the absence of two enzymes: isocitrate lyase and malate synthase. Isocitrate lyase acts on isocitrate to form succinate and glyoxylate. Both acetyl-CoA and oxaloacetic acid give citric acid. Citric acid is converted to isocitric acid. The isocitrate in the presence of isocitrate lyase gets converted to succinate and glyoxylate. Glyoxylate reacts with a second acetyl-CoA to form malate by the malate synthase, a glyoxysomal enzyme. Malate is further converted to oxaloacetate by the enzyme malate dehydrogenase. Animal cells utilize the acetyl-CoA by its oxidation to generate CO_2, H_2O, and ATP. Fats in the oily seeds are converted to glucose by this cycle during the seed's germination phase and the saplings' growth.

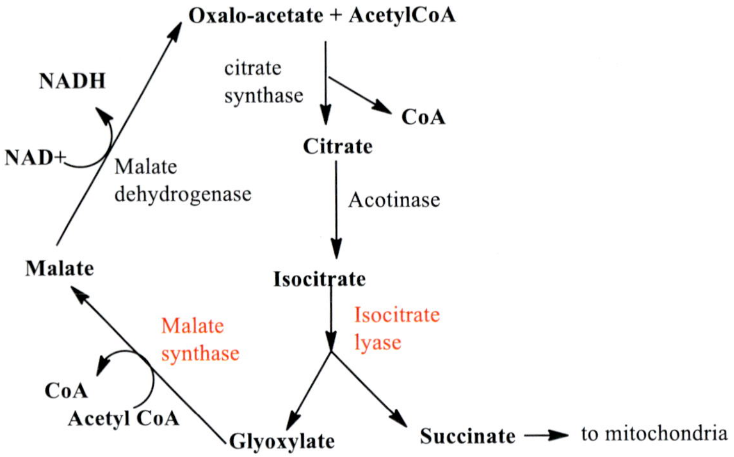

Fig. 8.7; The Glyoxylate cycle

The importance of the cycle

- This cycle is a modification of the TCA cycle. Bacteria and plants can produce glucose from fatty acids through this cycle.
- This cycle, like the TCA cycle, forms citrate, which is then converted to isocitrate. The reaction starts with the condensation of acetyl-CoA and oxaloacetate.
- It provides energy in the form of ATP.
- Acetyl-CoA here gives succinic acid, which is used for glucose formation. In animal cells, it is oxidized to generate CO2, H2O, and ATP.
- There is no formation of carbon dioxide, so all carbon is saved. In this cycle, two oxaloacetates are produced, and one net gain is achieved against the TCA cycle, with no net oxaloacetate production.
- This cycle enables the conversion of acetyl-CoA into glucose, which is not the case in animal cells.
- Every turn of the cycle produces one FADH and one NADH

II. The C2 oxidative photosynthetic carbon oxidation cycle/ photorespiration

This cycle metabolizes carbohydrates, converting two molecules of phosphoglycolate into one molecule each of serine and carbon dioxide. It is an anabolic pathway found in plants, bacteria, and fungi but not in animals. The cycle occurs stagewise in three compartments: the chloroplast, the peroxisome, and the mitochondria.

Glycolate (four carbons) biosynthesis from ribulose-1,5-biphosphate occurs in **chloroplasts**—oxygenation of ribulose-1,5-biphosphate forms 2-phosphoglycolate. The 2-phosphoglycolate is rapidly hydrolyzed to glycolate by chloroplast phosphatase.

Glycolate synthesized now enters the peroxisome, where it is oxidized by glycolate oxidase to generate glyoxylate and hydrogen peroxide. The hydrogen peroxide is destroyed by catalase. Glyoxylate formed undergoes transamination and is further converted to glycine by glutamate: glyoxylate aminotransferase present in the peroxisome. Glycine now enters mitochondria.

The third compartment is mitochondria, where the condensation of two glycine molecules, one NAD^{+}, gives rise to serine, NADH, and NH4+, and carbon dioxide is released during photorespiration. Ammonia diffuses from here to the chloroplasts to form amino acids by glutamine synthetase.

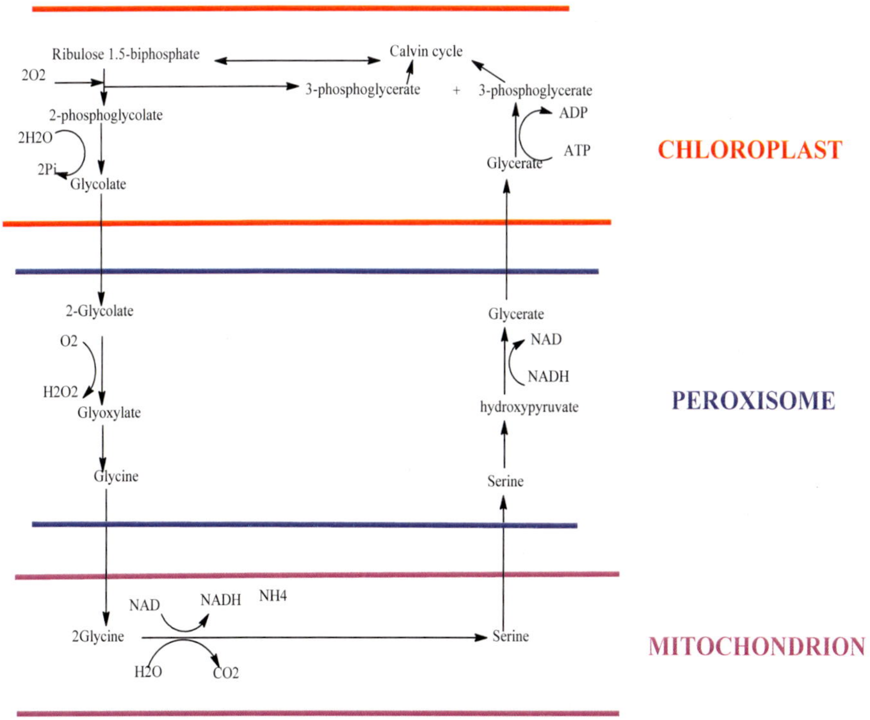

Fig. 8.8: The C2 oxidative photosynthetic cycle.

The serine from mitochondria enters the peroxisome, which is converted to glycerate via hydroxy pyruvate formation. A malate-oxaloacetate acetate mechanism provides NADH from the cytoplasm into the peroxisome. The glycerate again enters the chloroplast and gets phosphorylated, so the product formed is 3-phosphoglycerate, which can join the Calvin cycle. Ammonia released at the mitochondrial level is taken up by chloroplasts and incorporated into amino acids (glutamate). Photorespiration is a wasteful process as there is no fixation of CO, but it releases fixed carbon as CO2 with the use of ATP.

III. The C3 cycle or Calvin cycle, the CBB (Calvin-Benson-Bassham) cycle, the reductive pentose phosphate (RPP) cycle, the light-independent or dark reaction of photosynthesis

The Calvin pathway is found in all photosynthetic plants. The reactions of photosynthesis using carbon dioxide and water form glucose and other carbohydrates. The reaction uses the stored energy (ATP and NADPH) generated by the light-dependent reactions. It becomes an anabolic reaction. Photosynthesis occurs in the chloroplasts of plants. The reactions involved are carbon fixation, reduction, and regeneration. The carbon fixation and reduction reactions take place in the stroma. The Calvin cycle is called the C3 cycle because three-carbon products, 3-PGA, are formed. Most of the plants are C3 plants. Calvin cycle is a common cycle found in both C3 and C4 plants in forming sugars.

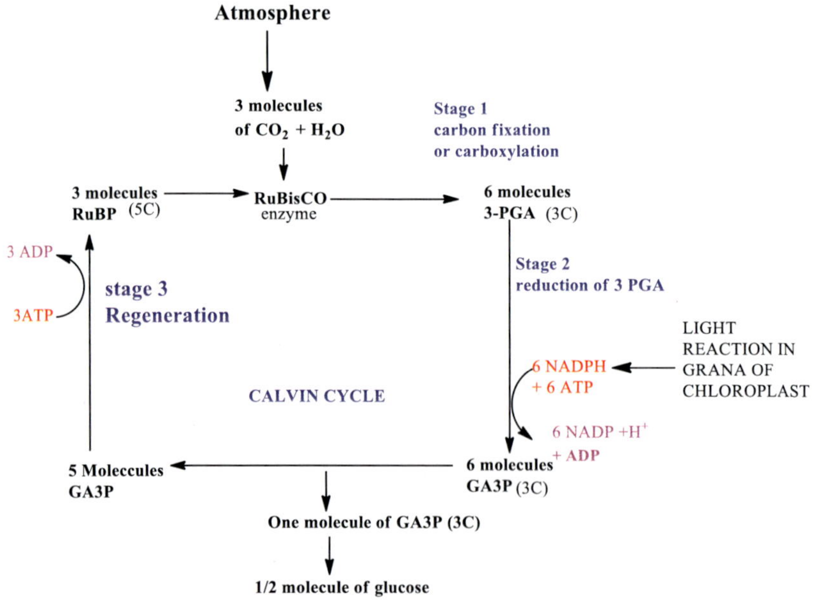

Fig. 8.9: The Calvin cycle

Every turn of the cycle uses one carbon dioxide molecule at a time. If 3 carbon molecules like GA3P are to form, the cycle runs three times, and to produce 6 carbon glucose molecules, it runs six times.

- Carboxylation or Carbon fixation -stage 1. Formation of 3-PGA: This is the first stage of the light-independent reaction occurring in the stroma of the chloroplast, where carbon fixation occurs, and the organic intermediates are formed. The carbon dioxide and water are obtained from the environment for this cycle to run. One carbon of carbon dioxide is fixed by the enzyme RuBisCO (ribulose biphosphate carboxylase), a chloroplast enzyme. When undergoing carboxylation, the reaction between one CO2 and one RuBP (5C) forms two molecules of 3PGA (three phosphoglyceric acids) each having 3C. Only one molecule of carbon dioxide and one molecule RuBP are used at every cycle turn to form two molecules of 3-PGA, each containing three carbons. Three atoms of CO2 and 3 RuBP (5C), when used, make 18 carbons to create six molecules of 3-PGA. Three ribulose bisphosphate (RuBP) each with five-carbon molecules are used here. As the CO_2 is "fixed" from an inorganic form into organic molecules, this process is called the carbon fixation stage or reaction.
- Reduction-stage 2. Formation of carbohydrate glyceraldehyde-3-phosphate (GA3P): In a reduction reaction, an atom or molecule is a receipt of an electron. This stage of the cycle is the reduction reaction stage because it involves the reduction of 6 PGA (three phosphoglyceric acids), gaining electrons. Six molecules of both NADPH and ATP are used while forming 6 GA3P (glyceraldehyde 3-phosphate) from 6 PGA. ATP releases energy when converted to ADP.NADPH releases energy and a hydrogen atom when converted to NADP. One glucose molecule is produced in the second stage through six turns of the cycle and the fixation of six molecules of CO_2.
- Regeneration-stage 3. Regeneration of ribulose 1,5-biphosphate: The third step is known as regeneration because the ribulose-bis phosphate (RuBP) that begins the cycle is regenerated from GA3P to continue the cycle. One of the G3P molecules from the Calvin cycle is sent to the cytoplasm. The remaining 5 GA3P in the Calvin cycle are used to regenerate RuBP. Regenerated RuBP molecules are used to fix more CO2 in the next turn of the cycle. Three ATPs are used in the regeneration reaction.

  Overall, the reactants for the Calvin cycle are CO2, ATP, and NADPH, and the products are G3P, ADP, and NADP+.

IV. The C4 cycle or Hatch and Slack pathway, the C4 dicarboxylic acid pathway, and the cooperative synthesis or β-carboxylation pathway

This is an alternative to the C3 cycle. It is used to fix carbon dioxide. Oxaloacetic acid is the first compound formed as a four-carbon compound. Thus, the name "C4 cycle" was given. The C4 plants' leaf anatomy differs from that of other plants.

Mesophyll cells and bundle-sheath cells are involved in this cycle. In mesophyll cells, one carbon from carbon dioxide is fixed by PEP carboxylase with three carbon phosphoenolpyruvates (PEP) to form four carbon oxaloacetate molecules, which are later converted to malate. Mesophyll cells lack the RuBisCO enzyme.

There are four steps

- Carboxylation: Phosphoenolpyruvate, a 3-carbon compound, reacts with carbon dioxide in the presence of water and PEP carboxylase to form 4-carbon oxaloacetate
- Breakdown: Oxaloacetate breaks down into malate in mesophyll cells. It then enters into bundle-sheath cells.
- Splitting: Carbon dioxide is released to form 3-carbon pyruvate. Released carbon dioxide enters the Calvin cycle in the chloroplasts of the bundle-sheath cells. Ribulose diphosphate (5C) accepts carbon dioxide in the presence of carboxy dismutase to form two 3-phosphoglyceric acids. Some are used to regenerate ribulose diphosphate, and some are used in the formation of sugars.
- Phosphorylation: The pyruvate molecules formed in bundle-sheath cells are transported to the chloroplasts of the mesophyll cells and, in the presence of phosphokinase, get phosphorylated using ATP to regenerate phosphoenolpyruvate (PEP).

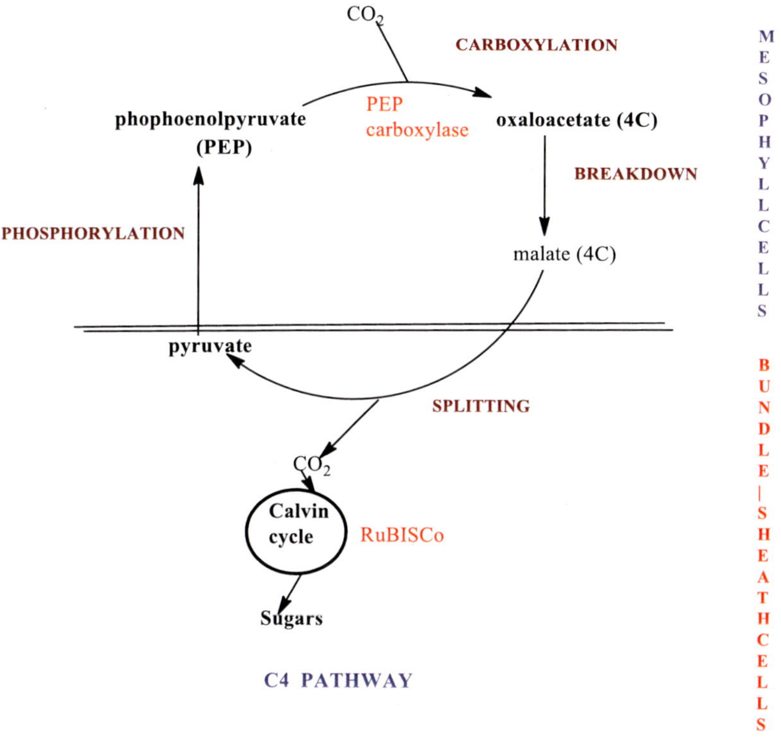

Fig. 8.10: C4 pathway

Table. 8.7: Difference between pants types & their cycles

S.N	Items	C3 plants & Cycle	C4 plants & Cycle
		C3 plants	**C4 plants**
1	Kranz anatomy in leaves	Not	Present
2	Bundle of sheath chloroplast	Not	Possess size large
3	Chloroplast	monomorphic	Dimorphic
4	Cycles present in the plant	Only C3 or Calvin cycle	Both C3 and C4 cycle
5	CO2 acceptor	One –RuBP	Two -RuBP and PEP
6	Efficiency of CO2 utilization	Efficient	Much more efficient
7	The first stable compound formed	Phospho-glyceric acid (3C)	Oxaloacetate (4C)
8	O2 effect on photosynthesis	Inhibitory	No such effect, but malic acid is formed
9	Photosynthetic yield due to photo-respiration	Reduced	No-photorespiration, so yield is increased
		C3 Cycle or Calvin Cycle	C4 cycle or Hatch and Slack Cycle

S.N	Items	C3 plants & Cycle	C4 plants & Cycle
10	The primary acceptor of CO2 in a given cycle	RuBP (5C)	PEP (3C)
11	The secondary acceptor of CO2	Not present	RuBP
12	The first stable compound formed	3-phosphoglyceric acid	Oxaloacetate
13	Release of fixed CO2 back	No	Yes
14	Rate of CO2 fixation	Slower	Faster
15	Net gain of the molecule	One hexose sugar from six CO2	No such gain but malic acid is formed

The C3 and C4 cycles are linked together in the Hatch and Slack pathways. The C3 plants do photosynthesis more efficiently compared to the C4 plants. In the case of the carboxylation reaction, the phosphoenolpyruvate carboxylase enzyme of the C4 cycle is more efficient than the ribulose diphosphate carboxylase enzyme of the C3 cycle.

Key Takeaway

- A monomer glucose molecule must be converted to acetyl-CoA to enter the TCA cycle. One glucose produces two acetyl-CoA molecules. A molecule of acetyl-CoA has two carbons.
- The pyruvic acid undergoes oxidative decarboxylation to form acetyl-CoA.
- In the TCA cycle, the anabolic and catabolic processes occur, so it is an amphibolic pathway.
- Oxidative decarboxylation of pyruvic acid needs a complex enzyme system known as a pyruvate dehydrogenase complex composed of three enzymes and five co-enzymes working in cooperation.
- The TCA cycle proceeds twice per glucose molecule because of the formation of two acetyl-CoA.
- The mitochondrial matrix is very important for the TCA cycle as it is the place where the required enzymes are available, except for the inner membrane SDH (succinate dehydrogenaseare), protein complex II in the electron transport chain.,
    - The products of the cycle are two molecules of **carbon dioxide, t**hree molecules of **NADH, t**hree hydrogen ions (**H+**), one molecule of **$FADH_2$, and** one molecule of **GTP.**
- Malate, upon oxidation, gets converted to oxaloacetate. This reaction regenerates the oxaloacetate molecule. The participating enzyme is malate dehydrogenase.

- Allosteric regulation by metabolites is observed in the regulation of the TCA cycle, which is primarily determined by product inhibition and substrate availability
- Citrate inhibits phosphofructokinase of glycolysis, and the availability of pyruvate and nd acetyl-CoA molecules gets reduced
- Four Water-soluble B complex vitamins participating are: 1. Riboflavin in the form of FAD, a cofactor in dehydrogenases (alpha-ketoglutarate dehydrogenase and succinate dehydrogenase). 2. Niacin- as the coenzyme in the form of NAD. 3. Thiamin as coenzyme TPP for decarboxylation reaction and 4. Pantothenic acid is a part of coenzyme A in the form of acetyl-CoA or succinyl-CoA.
- The malate-aspartate (M-A) shuttle used is an important mechanism to regulate glycolysis and lactate metabolism in various organs
- The malate-aspartate shuttle consists of four protein parts. The four proteins of the shuttle system are Malate dehydrogenase, Aspartate aminotransferase, glutamate-aspartate antiporter (transporter I), and malate-alpha-ketoglutarate antiporter (transporter II).
- The glutamatc-aspartate antiporter or transporter I does the function of import and export of different molecules. It imports glutamate from the cytosol into the matrix, and at the same time, it exports aspartate to the cytosol.
- The malate- alpha-ketoglutarate antiporter or transporter II. It imports malate from the cytosol into the matrix and exports alpha-ketoglutarate from the matrix to the cytosol.
- Glycerophosphate shuttle transports glycerophosphate into the mitochondrion and dihydroxyacetone phosphate out of the mitochondrion.
- The three critical steps of the Calvin cycle reactions do not depend on light; hence, it is known as the light-independent cycle. However, ATP and NADPH are required to be generated in light-dependent reactions.
- The first stage of the Calvin cycle occurs when light-independent reactions are initiated. In the Calvin cycle, CO_2 is fixed from an inorganic to an organic molecule.
- In the second stage, two 3-PGA are reduced into G3P, using ATP and NADPH. Therefore, ATP is converted to ADP, and NADPH is converted to NADP.
- In the last stage of the Calvin Cycle, more fixation of CO_2 is enabled as RuBP is regenerated.

- Calvin cycle reactions involved are carbon fixation, reduction, and regeneration.
- The reaction between one CO_2 and one RuBP (5C) forms two molecules of 3PGA (three phosphoglyceric acids) molecules, each having 3C.
- The Calvin cycle is called the C3 cycle because three-carbon products, 3-PGA, are formed. Most of the plants are C3 plants.
- The C4 cycle plants have PEP carboxylase to fix the carbon with PEP to form 4-C product oxaloacetate in mesophyll cells. Mesophyll cells lack the RuBisCO enzyme.

Intext Questions

1. Which reaction links pyruvic acid with the TCA cycle?
2. How much acetyl-CoA is produced from one glucose molecule? Why?
3. How often does a TCA cycle run for one glucose molecule? Why?
4. What is the location of the TCA cycle enzymes? What is the location of the substrate for the first reaction in the TCA cycle?
5. What is the substrate-level reaction? Give an example of it.
6. In which reaction of the TCA cycle does succinic dehydrogenase participate? Write the reaction in detail.
7. Which reaction in the TCA cycle regenerates the oxaloacetate molecule? Explain in detail.
8. Name the different high-energy bonds formed in the TCA cycle.
9. Which reactions convert the 6C molecule to the 4C molecule in the TCA cycle?
10. Which vitamins participate in the TCA cycle?
11. Which reaction produces GTP?
12. What are other names for the TCA cycle?
13. Conversion of pyruvic acid to acetyl-CoA needs coenzymes and enzymes. Enlist them.
14. What is the biological significance of the TCA cycle?
15. Enlist all enzymes involved in the TCA cycle.
16. How does the first reaction occur in the TCA cycle?
17. Explain the substrate-level phosphorylation reaction of the TCA cycle.

18. What is the amphibolic cycle and the Anaplerotic reaction?

19. What are the functions of Mesophyll cells and Bundle sheath cells?

20. Which cells produce CO_2 in case of C4 cycle?

21. What is the fate of CO_2 released in bundle sheath cells of the C4 cycle?

22. What is the fate of atmospheric CO_2 enters mesophyll cells and is then produced by decarboxylation in bundle sheath cells?

23. Which pathway is common in forming sugars for C3 and C4 plants?

24. Choose one correct option for the following from the options given below (The answer is given in **bold** letters):

a Which one of the following is tricarboxylic acid?

i. **Citric acid** ii. Alpha ketoglutarate

iii. Succinate

b How many acetyl-CoA are produced from one six-carbon glucose molecule?

i. **Two** ii. Three

iii. Four

c Glucose has six carbons, and acetyl-CoA has two carbons. How many acetyl-CoA are formed from a glucose molecule?

i. Three ii. **Two**

iii. Four

d Which of the following is oxidized form?

i. **NAD^+** ii. $FADH_2$

iii. $NADH_2$

e Which one of the following is a true statement?

i. **The enzymes required by the TCA cycle are present in the mitochondrial matrix except succinate dehydrogenase, located on the inner membrane of mitochondria.**

ii. The TCA cycle requires all enzymes to be present only in the mitochondrial matrix.

iii. The enzymes required by the TCA cycle are present in the mitochondrial matrix, except citrate synthase, located on the inner membrane of mitochondria.

f The number of carbon dioxide molecules formed in the TCA cycle is

i. Three ii. **Two**

iii. Five

g Which type of reaction converts fumarate to malate?

i. **Hydration reaction** ii. Decarboxylation

iii. Substrate level

h One glucose yields two pyruvates. When converted to acetyl-CoA, how much ATP do these pyruvates yield?

i. 4 ii. **6**

iii. 8

i The number of ATPs generated when one molecule of alpha-ketoglutarate is converted to succinyl- CoA.

i **3** ii 6

iii 8

j One acetyl-CoA is capable of generating.

i. **12 ATP** ii. 24ATP

iii. 6ATP

k Which vitamin is present in TPP?

i. Riboflavin ii. **Thiamin**

iii. Niacin

l Which vitamin is present in coenzyme A?

i. Niacin ii. **Pantothenic acid**

iii. Ascorbic acid

m The Calvin cycle is called the

i. **C3 cycle** ii. C4 cycle

iii. C2 cycle

Further Readings

A Textbook of Biochemistry. Berry, A. K. (2009), sixth edition, Emkay Publishers, Delhi.

A textbook on animal physiology. A, K, Berry. (1997). Sixth Edition. Emkay Publication, Delhi.

Berg JM, Tymoczko JL, Stryer L. Biochemistry. 5th edition. New York: W H Freeman; 2002. Chapter 17, The Citric Acid Cycle. https://www.ncbi.nlm.nih.gov/books/NBK21163/20.

Biochemistry. Naik, Pankaj (2009). Second Edition, Jaypee Brothers, New Delhi.

Cell Biology. E.D.P. De Robertis, Wiktor W. Nowinski, and Franciso A. Saez (1970). Fifth Edition, WB Saunders Company. Philadelphia, London, Toronto.

Cell Physiology. Arther C. Giese (1968), Third Edition, WB Saunders Company. Philadelphia, London, Toronto.

Donald Voet, Judith G. Voet, Charlotte W. Pratt. Fundamentals of Biochemistry, 4th ed.;2011.

Fundamentals of biochemistry. J.L.Jain, Sunjay Jain and Nitin Jain(2005). First edition., S.Chand & Company Ltd. New Delhi.

Harper's Illustrated Biochemistry, 30th Edition. Rodwell VW, Bender DA, Botham KM, Kennely PJ, Weil PA. Lange, 2015; Chapter 16&17

https://chem.libretexts.org/Bookshelves/Physical_and_Theoretical_Chemistry_Textbook_Maps/Supplemental_Modules_(Physical_and_Theoretical_Chemistry)/Acids_and_Bases/Acids_and_Bases_in_Aqueous_Solutions/The_pH_Scale.

https://energyeducation.ca/encyclopedia/The_pH_scale#:~:text=pH%20is%20a%20logarithmic%20scale,concentration)%20changes%20by%2010%20times.

https://vincerainstitute.com/why-vincera/news-articles/the-importance-of-ph-balance

https://www.numerade.com/ask/question/pure-water-at-25-c-the-concentration-of-water-is-555-m-grais-o-hzo-in-l-divided-by-its-gram-molecu-lar-w

Lehninger, A. L., Nelson, D. L., & Cox, M. M. (2000). Lehninger principles of biochemistry. New York: Mac Millan Worth Publishers.

Life Sciences.Fundamentals and practice I.Pranav Kumar and Usha Mina.2017.6th Edition, pathfinder Publication.New Delhi.India.

Lincoln Taiz and Eduardo Zeiger.plant physiology. Third edition. Panama publishing corporation. New Delhi/Banglore.

Lippincott's Illustrated Reviews Biochemistry, 5th Edition. Harvey RA, Ferrier DR. Lippincott Williams & Wilkins, 2011; Chapter 9.

Patel MS, Korotchkina LG. (2003). "The biochemistry of the pyruvate dehydrogenase complex." Biochem Mol Biol Educ. 31: 5.

Pretest Physiology. James P. Ryan and Ronald F. Tuma (1996), Eighth Edition, McGraw-Hill, New York, Milan Montreal New Delhi.

Principles of physical, organic, and biological chemistry. John R. Holum. (1969), John Wiley & Sons, Inc., New York, London, Sydney, Toronto.

Pyruvate Dehydrogenase Complex. https://pdb101.rcsb.org/motm/153 Molecule of the Month (September 2012). @Protein Data Bank.

Rodwell, V. W., Botham, K. M., Kennelly, P. J., Weil, P. A., & Bender, D. A. (2015). Harper's illustrated biochemistry (30th ed.). New York, N.Y.: McGraw-Hill Education LLC.

Silverman RB. The Organic Chemistry of Enzyme-Catalyzed Reactions (Revised ed., 2002, Academic Press).

Smith, C. M., Marks, A. D., Lieberman, M. A., Marks, D. B., & Marks, D. B. (2005). Marks' basic medical biochemistry: A clinical approach. Philadelphia: Lippincott Williams & Wilkins.

Staunton J. Primary Metabolism: A Mechanistic Approach (1978, Oxford University Press)

Textbook of Biochemistry for Medical Students. Vasudevan, D. M.; Sreekumar, S.; Vaidyanathan, Kanan (2019). Ninth Edition, Jaypee Bros. Medical Publication, New Delhi.

Textbook of Veterinary Biochemistry. Dhanotiya R.S. (2006). Reprint of the second edition. Jaypee Brothers, New Delhi.

Traaseth, N. et al. 2004. Role of calcium signaling in the activation of mitochondrial nitric oxide synthase and citric acid cycle, Biochimica et Biophysica Acta (BBA) - Bioenergetics, 1658, pp.64-[71].

Wan, B. et al. 1989. Regulation of citric acid cycle by calcium, Journal of Biological Chemistry, 264, pp. 13430-13439. https://www.ncbi.nlm.nih.gov/pubmed/2503501

9

Hexose Monophosphate (HMP) Pathway or Shunt (HMS)/ Pentose Phosphate Pathway (PPP)/ Pentose Cycle/Phosphogluconate Pathway/ Warburg-Dickens-Lipman Pathway

This alternative pathway for glucose oxidation has enzymes in the cytosol through which glucose is degraded to carbon dioxide. It is also called a shunt because carbon atoms from glucose 6-phosphate take a long route before they proceed down the EMP or glycolytic pathway. This pathway is anabolic in nature as it produces many intermediates. In this pathway, two NADPH are generated from one glucose 6-phosphate molecule. No ATP is generated or consumed.

Overall, this pathway needs three glucose-6-phosphate molecules to complete all the reactions. Therefore, one cycle run ***produces 2 mol of fructose-6-phosphate, 1 mol of glyceraldehyde-3-phosphate, and 3 mol of carbon dioxide from 3 mol of glucose-6-phosphate with the reduction of 6 mol of*** $NADP^+$ ***to NADPH.***

Overall reactions of the HMP pathway are as follows:

3 Glucose-6-P + 6 $NADP^+ \rightarrow$ 3 ribulose-5-P + 3 CO_2 + 6 NADPH

3 Ribulose-5-P $\rightarrow$ 2 xylulose-5-P + Ribose-5-P

2 Xylulose-5-P + Ribose-5-P $\rightarrow$ 2 fructose-6-P + Glyceraldehyde-3-P

The pentose phosphate pathway operates in two phases: 1. Irreversible or nonreversible ***oxidation phase,*** and 2. reversible non-***oxidation*** reactions.

9.1 Reactions of the Cycle

I Oxidation reactions phase (irreversible) or NADPH regeneration phase

There are three irreversible reactions in the oxidation phase of this shunt. The enzymes participating in this phase are a. glucose 6-phosphate dehydrogenase, b. gluconolactone hydrolase, and c. phosphogluconate dehydrogenase. Here, NADP acts as a hydrogen acceptor (whereas NAD is used as a hydrogen acceptor in glycolysis or EMP cycle) to form two molecules of NADPH after oxidation of one molecule of glucose 6 phosphate. Of the six carbons of glucose 6- 6-phosphate, 5 carbons form Ribulose 5-phosphate, and one carbon is used to form carbon dioxide. In the oxidative phase, glucose-6-phosphate can be further metabolized to convert to form nicotinamide adenine dinucleotide phosphate (NADPH) and ribulose-5-phosphate. The ribulose-5-phosphate was later converted to ribose-5-phosphate in non-oxidative phase reactions. This pathway metabolizes about 30% of glucose by the liver and RBCs.

Various enzymes involved in oxidative reactions include glucose-6-phosphate dehydrogenase gluconolactonase, and 6-phosphogluconate dehydrogenase.

The overall oxidative phase of HMP is summarized below:

Glucose 6-P+ 2NADP$^+$ +H_2O $\longrightarrow$ Ribulose-5-phosphate+2NADPH +2H$^+$ + CO_2

The cellular concentration of NADPH controls the entry of glucose 6-phosphate into HMP.

The above oxidative reactions are detailed stepwise as follows

A. Conversion of glucose to glucose 6-phosphate and 6-phosphogluconolactone.

1. **Formation of 6-phosphogluconolactone.** The first step of the reaction in the EMP cycle or glycolysis is to lock the free glucose molecule that enters the cell by forming glucose-6-phosphate. It is an endergonic reaction, and so requires the use of ATP. The enzyme that participates in this reaction is hexokinase. In a subsequent reaction, glucose-6-phosphate is oxidized to 6-phosphogluconolactone by an irreversible dehydrogenation. This reaction occurs in the presence of glucose-6-phosphate dehydrogenase (G6PD), coenzyme NADP, and Mg+2. NADP is converted to NADPH.

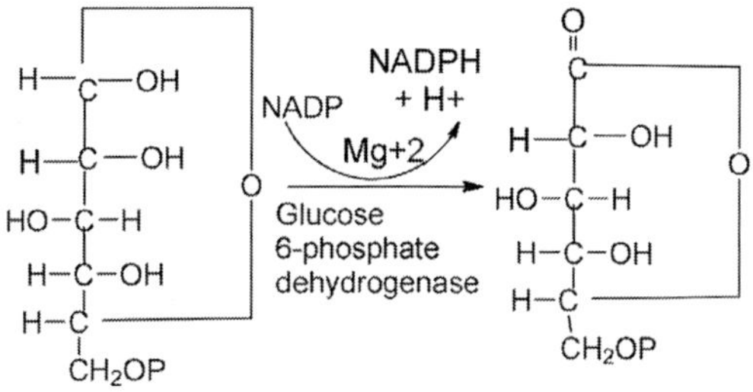

GLUCOSE 6-PHOSPHATE 6-PHOSPHOGLUCONOLACTONE

The rate-limiting enzyme is glucose-6-phosphate dehydrogenase, yielding NADPH. NADP+ stimulates this enzyme, and NADPH is a strong inhibitor of G6PD.

The cells use NADPH to overcome the oxidative stress. The reduced glutathione is formed with the help of NADPH and the enzyme glutathione reductase. The reduced glutathione converts reactive H_2O_2 into H_2O by glutathione peroxidase.

2. **Conversion of 6-phosphogluconolactone to 6-phosphogluconate**

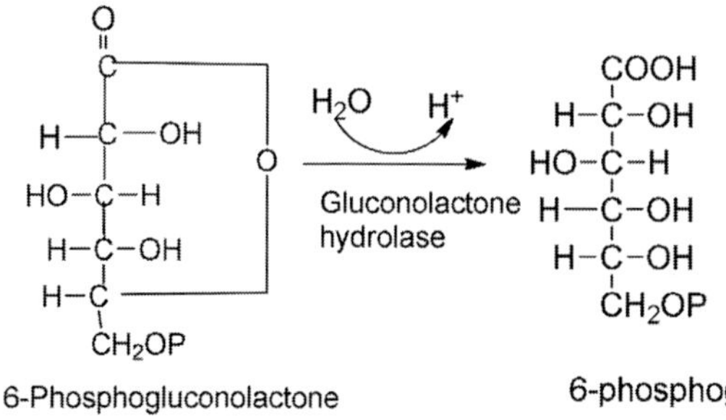

6-Phosphogluconolactone 6-phosphogluconate

A hydrolase enzyme catalyzes this reaction, and it is a hydrolysis reaction.

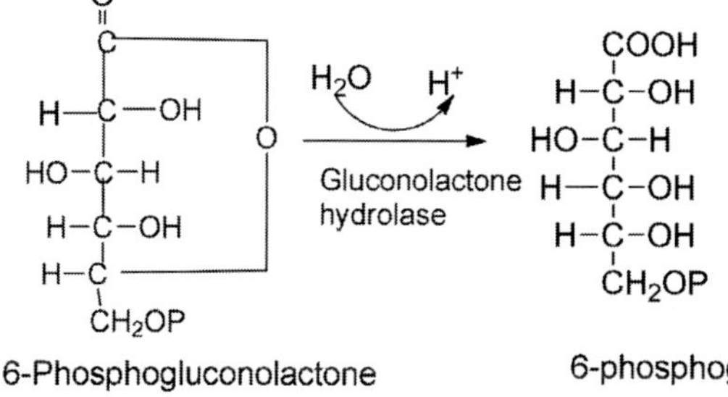

6-Phosphogluconolactone 6-phosphogluconate

3. **Conversion of 6-phosphogluconate to ribulose-5 phosphate:** The enzyme 6-phosphogluconate dehydrogenase reacts with the 6-phosphogluconate and catalyzes this reaction, forming 3-keto-6-phosphogluconate, which, after decarboxylation, forms ribulose-5 phosphate.

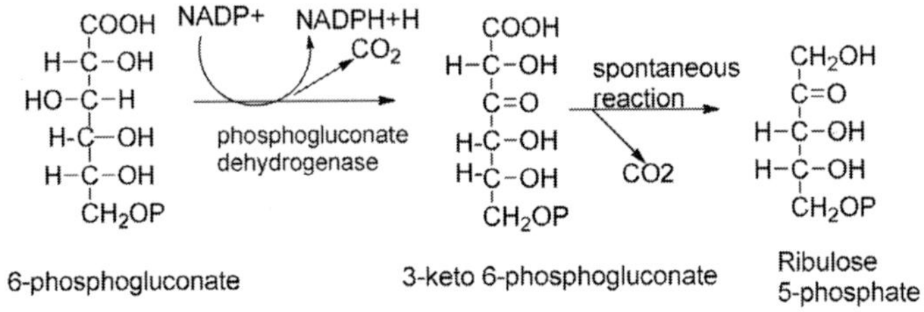

II. Non-oxidative reactions phase (Reversible or Regenerative Interconversion of sugars)

Four enzymes in the non-oxidative phase play an important role. They are a) ribulose-5-phosphate epimerase, b) ribulose-5-phosphate isomerase, c) transketolase, and d) transaldolase enzymes. Transketolase transfers 2-C units. Transaldolase transfers 3-C units. Both these enzymes transfer the groups from the ketone donor to the aldose acceptor.

Here, keto sugar, ribulose-5-phosphate, is converted to aldose sugar, ribose-5-phosphate, and there are a series of other reactions. The non-oxidative reaction phase is important due to the intermediates produced, e.g., ribose 5-phosphate and erythrose 4-phosphate. The ribose 5-phosphate is used in the biosynthesis of nucleotides and nucleic acids, whereas erythrose 4-phosphate is used to synthesize aromatic amino acids. NADPH molecules are not at all generated in the non-oxidative reactions phase.

The overall reactions are as follows:

A. Utilization of molecules of ribulose-5-phosphate in two ways

1. Conversion of ribulose 5-phosphate to xylulose-5-phosphate by ribulose-5-phosphate 3-epimerase.

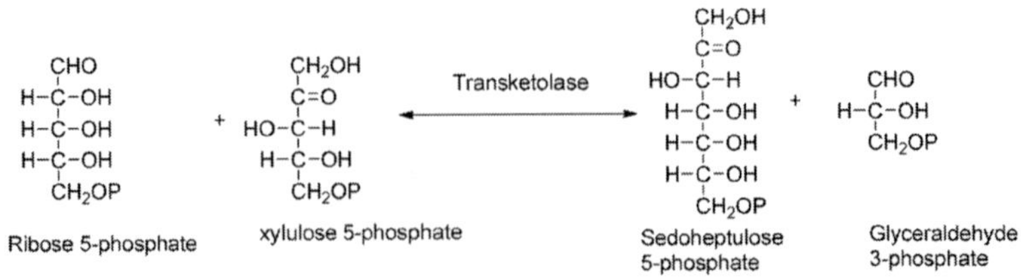

2 conversions of ribulose-5-phosphate to ribose-5-phosphate by ribulose-5-phosphate isomerase enzyme. The reaction converts a ketone sugar to an aldose sugar.

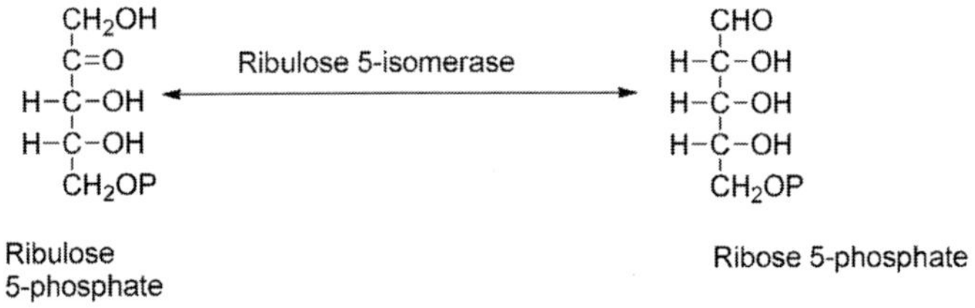

The remaining reactions will occur with the help of transketolase and transaldolase enzymes. These enzymes' peculiarities are that transketolase transfers two-carbon units, whereas transaldolase always transfers three-carbon units.

Ribose 5-phosphate is a precursor for the synthesis of nucleotides or nucleic acids.

B. Conversion of two 5C sugars to 7C and 3C sugars

This reaction makes use of two pentose sugars: one is an aldehyde, and the second is a ketone sugar. The enzyme transketolase rearranges ten carbons of these two pentose sugars. This enzyme transfers a 2C unit from ketone sugar xylulose 5-phosphate (5C) to the aldehyde sugar ribose5-phosphate (5C) to create two molecules with different numbers of carbon atoms. The 3C unit molecule is glyceraldehyde-3-phosphate, and the 7C unit molecule is sedoheptulose-7-phosphate. Transketolase activity is affected by thiamine deficiency. To evaluate Thiamin or vitamin B1's nutritional status, erythrocyte transketolase is measured.

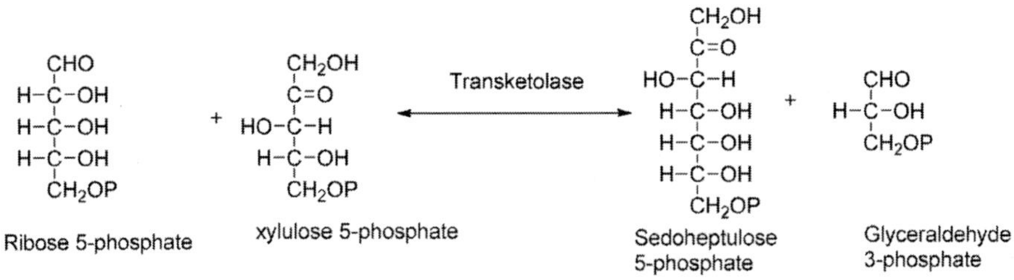

C. Transfer of a 3C unit from the 7C unit molecule sedoheptulose-7-phosphate to glyceraldehyde-3-phosphate by transaldolase.

This reaction forms a 6C unit of fructose-6-phosphate and a 4C unit of erythrose-4-phosphate. This enzyme transfers the 3-carbon unit from sedoheptulose keto sugar to erythrose aldehyde sugar. As a result, carbons are rearranged, leading to the formation of fructose-6-P and erythrose 4-P.

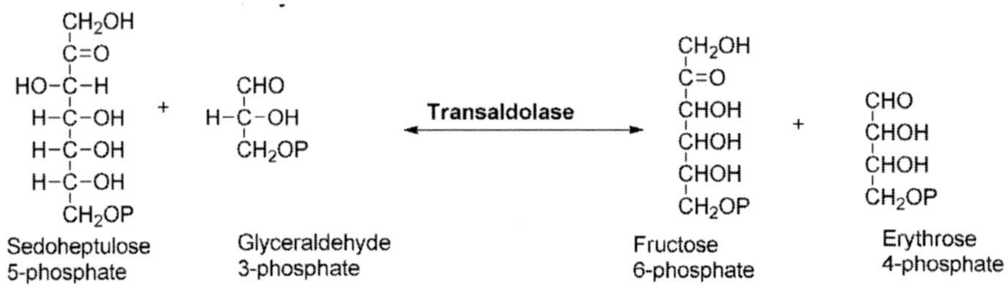

D. Transfer of a 2C unit from a second molecule of xylulose- 5-phosphate (5C) to a 4C unit molecule erythrose-4-phosphate by transketolase.

This reaction creates molecules having 3C and 6C molecules. The 3C molecule formed is glyceraldehyde-3-phosphate, and the 6C unit molecule formed is fructose-6-phosphate.

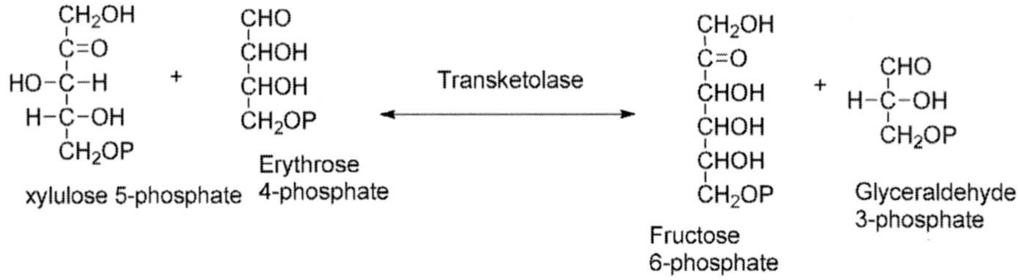

These fructose 6-phosphate and glyceraldehyde 3-phosphate molecules are used as per the body's needs.

Overall, the nonoxidative phase is characterized by several reversible reactions, and the intermediates formed can connect to several other pathways like nucleotide synthesis, aromatic amino acid synthesis, and glycolysis.

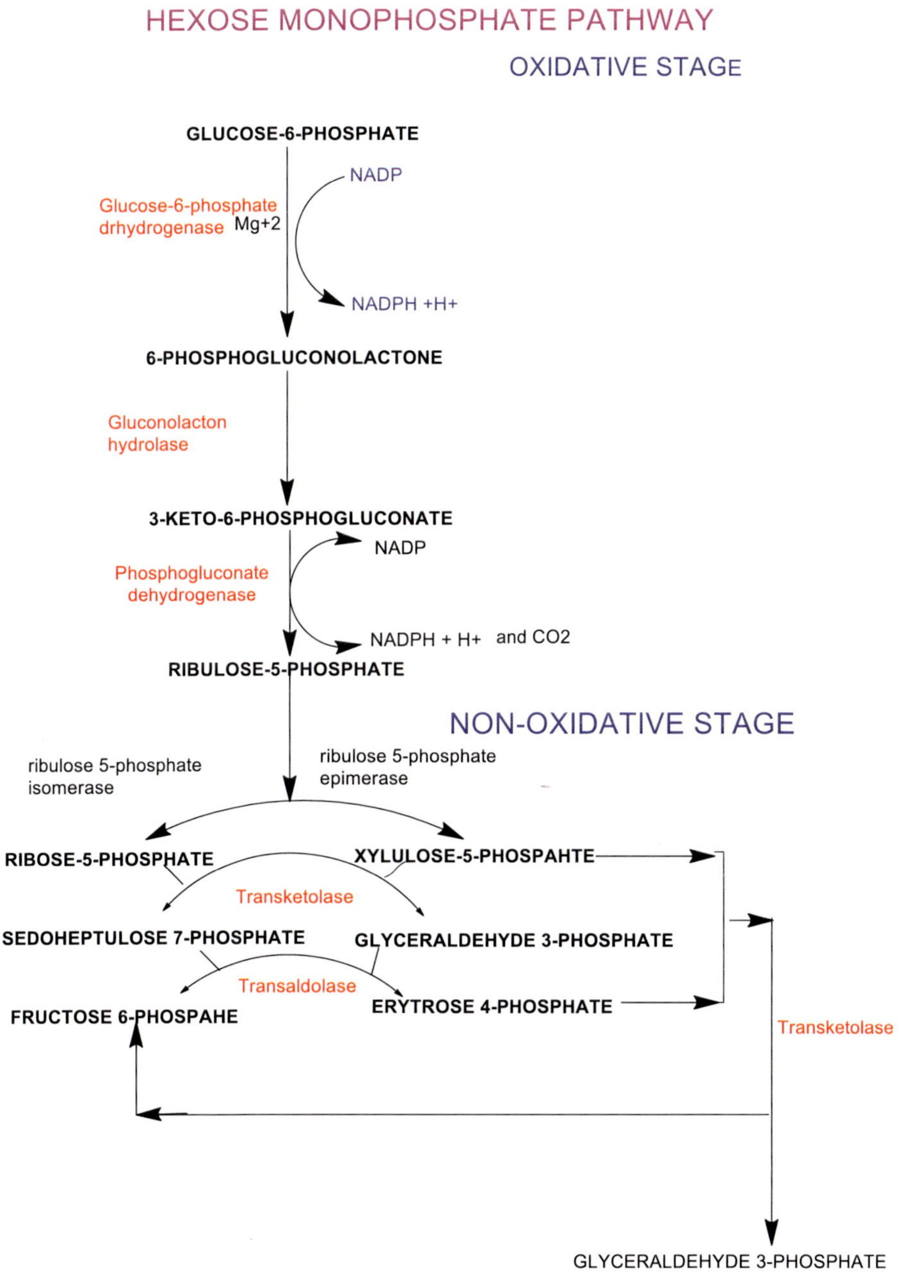

Fig. 9.1: HMP pathway with oxidative and non-oxidative phases.

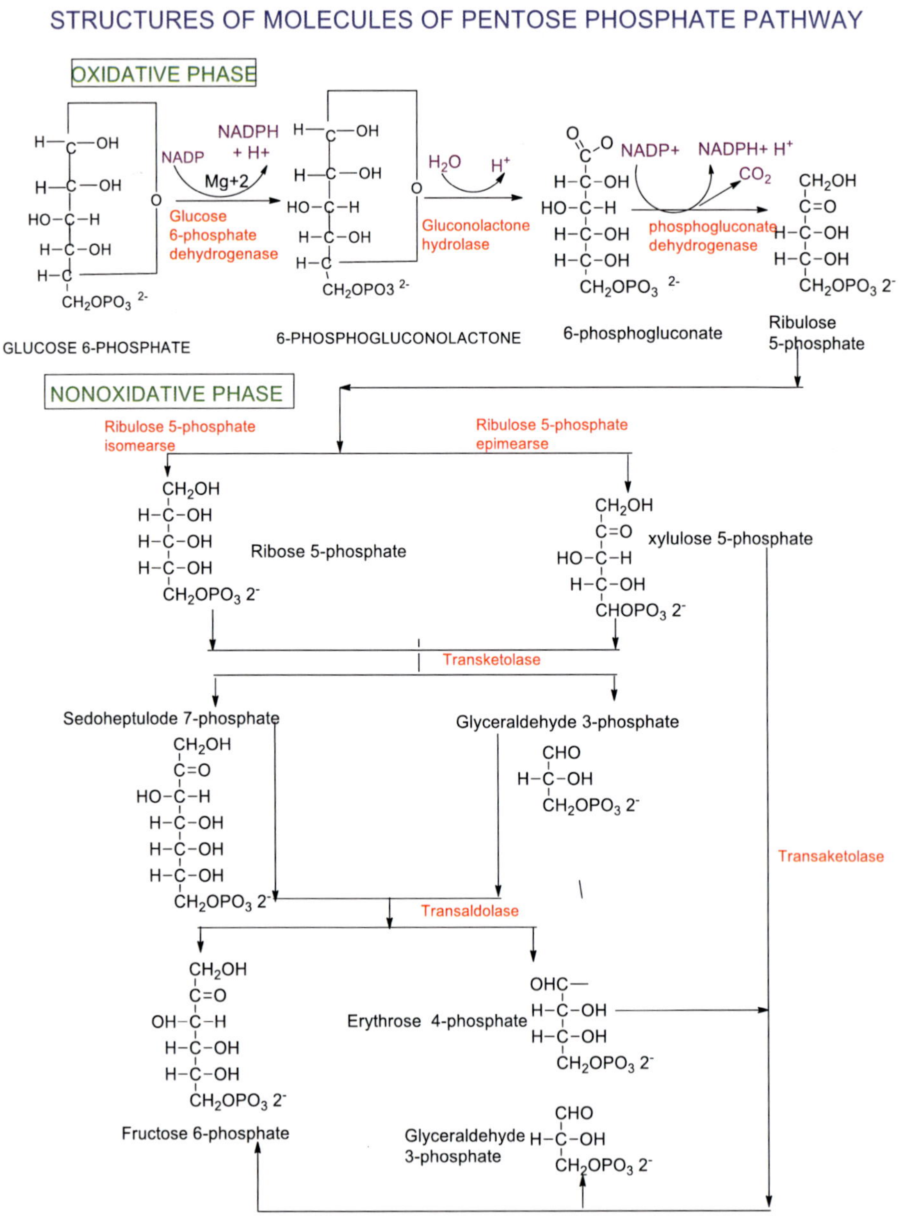

9.2 Important Points to Remember

- Overall, this pathway needs six glucose-6-phosphate molecules to produce six-carbon molecules, as glucose has six carbon molecules. When this pathway involves 6 molecules of glucose, 6-phosphate (36 carbons) yields four molecules of fructose 6-phosphate (24 carbons) and 2 molecules of glyceraldehyde 3-phosphate (6 carbons) in addition to 6 molecules of CO2. Twelve molecules of NADP+ are converted to NADPH.

- The pathway reactions occur in the cytosol of the cell. The rate and direction of reversible reactions will change depending on the needs of the cell.
- If NADPH generation is needed by the cell, fructose-6--phosphate and glyceraldehyde-3-phosphate are converted back to glucose by reverse glycolysis.
- If ATP generation is needed by the cell, then fructose-6-phosphate and glyceraldehyde-3-phosphate enter the TCA cycle through pyruvate formation.
- Formation of ribose-5-P can serve as a precursor of nucleotide and nucleic acids.
- Oxidized glutathione is converted into reduced form by glutathione reductase and NADPH. Reduced glutathione can detoxify free radicals and peroxides and protect the tissues from any free radical injury.
- Generated NADPH is required for reductive biosynthesis (fatty acid synthesis, steroids, and amino acids).
- This shunt operates in various specific tissues like the gonads and adrenal cortex (for steroid biosynthesis), the liver (fatty acid and cholesterol biosynthesis), adipose tissues, the eye lens (maximum concentration), red blood cells (reduced glutathione formation), and the lactating mammary gland (fatty acid biosynthesis).
- This pathway is inactive in the non-lactating mammary gland and has low activity in skeletal muscle. The production of erythrose-4-phosphate synthesizes aromatic amino acids, which are precursors of many biosynthetic pathways.

9.3 Clinical Aspects

1. Red blood cells lack mitochondria. They have an HMP pathway to generate and regenerate NADPH. Cellular NADPH reduces glutathione (via glutathione reductase), which converts reactive H_2O_2 into H_2O (by glutathione peroxidase), preventing RBCs from oxidative stress

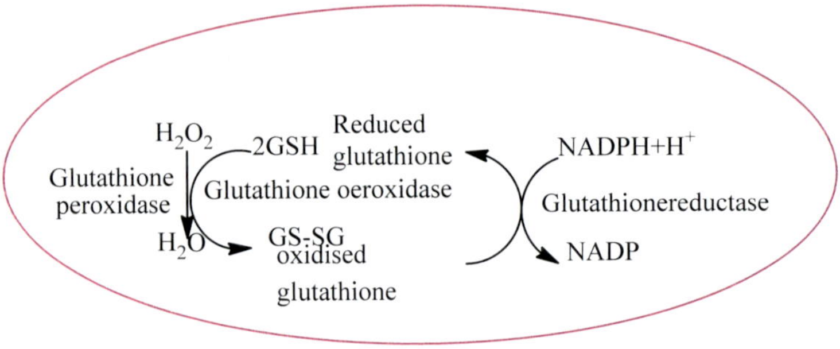

Fig. 9.2: NADPH regeneration and its importance in RBC

Without NADPH, free radicals generated from H_2O_2 rupture the membrane of erythrocytes, releasing hemoglobin (hemolysis), and the resulting condition is known as **hemolytic anemia**. The membrane rupture results from the peroxidation of lipids caused by free radicals in red blood cells deficient in the G6PD enzyme.

2. Hemolytic anemia in G6PD deficiency can also occur in response to many agents. They can be fava beans, nitrofurantoin medications, or anti-malarial drugs.
3. The aggregated denatured globin chains form the characteristic Heinz bodies seen in red blood cells of G6PD deficient patients.
4. A deficiency of G6PD in RBC results in the accumulation of methemoglobin.
5. Relative deficiency of G6PD in red blood cells is also, in some cases, beneficial as it protects individuals from falciparum malaria. (Malaria resistance). Malaria organisms, for their optimal growth, need reduced glutathione. Increased free radicals kill malarial parasites.
6. Maximum concentration of NADPH is in eye lenses, which helps to preserve their transparency.
7. Fava beans and certain medicines like quinine, Aspirin, Nonsteroidal anti-inflammatory drugs, Sulfa drugs, and antibiotics such as quinolones and nitrofurantoin can trigger red blood cell destruction.
8. In leucocytes, NADPH generates superoxide radicals to destroy the phagocytosed materials during phagocytosis.

9.4 Important Points to Remember

- Overall, this pathway needs multiple glucose molecules; three glucose-6-phosphate (18 carbons) yield two molecules of fructose 6-phosphate (12 carbons) and 1 molecule of glyceraldehyde 3-phosphate (3 carbons) in addition to 3 molecules of CO_2. Six molecules of NADP+ are converted to NADPH.

- If NADPH generation is needed in the cell, fructose-6--phosphate and glyceraldehyde-3-phosphate are converted back to glucose by reverse glycolysis.
- If ATP generation is needed by the cell, then fructose-6-phosphate and glyceraldehyde-3-phosphate enter the TCA cycle through pyruvate formation.
- Formation of ribose-5-P can serve as a precursor of nucleotides and nucleic acids.
- Glutathione reductase and NADPH are required to convert the oxidized glutathione into reduced glutathione. Reduced glutathione detoxifies free radicals and peroxides and prevents any cell injury that otherwise might occur due to free radicals.
- Generated NADPH is required for reductive biosynthesis of fatty acid, steroids, and amino acids.
- This shunt operates in various specific tissues like the gonads and adrenal cortex (for steroid biosynthesis), the liver (fatty acid and cholesterol biosynthesis), adipose tissues, the eye lens (maximum concentration), red blood cells (reduced glutathione formation), and the lactating mammary gland (fatty acid biosynthesis).
- This pathway is inactive in the non-lactating mammary gland but has low activity in skeletal muscle. The production of erythrose-4-phosphate can form aromatic amino acids, which are used as precursors of many biosynthetic pathways.

9.5 Regulation of the Pathway

- NADPH (a PPP product) inhibits G6PDH.
- NADPH and ATP inhibit 6-phosphogluconate.
- Insulin stimulates glucose availability for the HMP. It stimulates the G6PDH enzyme.
- Glucagon reduces insulin levels and inhibits this pathway.
- Higher quantities of glucose-6-phosphate can stimulate the process, while lower levels can shift glucose-6-phosphate to glycolysis.

Key Takeaway

- The HMP pathway is an alternative glucose oxidation pathway with enzymes in the cytosol through which glucose is degraded to carbon dioxide.

- It is also called a shunt because carbon atoms from glucose 6-phosphate take a long route before they proceed down the EMP or glycolytic pathway.
- This pathway is anabolic in nature as it produces many intermediates.
- In EMP, a single glucose molecule can be processed. The HMP pathway needs multiple glucose molecules to complete all the reactions.
- The pathway has two main functions. One is to provide NADPH. NADH is used for ATP synthesis, and NADPH is used for reductive biosynthesis. The second function is to provide ribose-5-phosphate for nucleotide and nucleic acid biosynthesis.
- There are two phases in the cycle. One is the oxidative phase, and the non-oxidative phase.
- A total of seven enzymes operates in the HMP pathway. Of these, three are for the oxidative phase, and the remaining four are for the non-oxidative phase.
- The oxidative phase is irreversible, and the non-oxidative phase is reversible.
- The first reaction of the HMP cycle is the oxidation of glucose-6-phosphate to form 6-phosphogluconolactone by an irreversible dehydrogenation. The enzyme is G6PD, an NADP-dependent enzyme. The first reaction forms NADPH and Mg^{2+}.
- The non-oxidative phase reaction ribose 5-phosphate is formed by isomerization of ribulose 5-phosphate.
- Ribose 5-phosphate is a precursor for synthesizing nucleotides or nucleic acids.
- The non-oxidative reaction requires various enzymes like a) ribulose-5-phosphate epimerase, b) ribulose-5-phosphate isomerase, c) transketolase, and d) transaldolase enzymes.
- Transketolase transfers 2-C units. Transaldolase transfers 3-C units. Both these enzymes transfer the groups from the ketone donor to the aldose acceptor.
- This reaction makes use of two pentose sugars: one is aldehyde Ribose 5-phosphate, and the second is ketone xylulose 5-phosphate.
- Both the enzymes transaldolase and transketolase create a reversible link to the glycolytic pathway.

- Oxidized glutathione is converted into its reduced form by glutathione reductase and NADPH. Reduced glutathione protects the tissues from any free radical injury.
- Generated NADPH is required for the biosynthesis of fatty acids, synthesis, steroids, and amino acids.
- This shunt operates in various specific tissues like the gonads, the adrenal cortex, the liver, the adipose tissues, the eye lens, the red blood cells, and the lactating mammary gland.
- A relative deficiency of G6PD in red blood cells may protect individuals from falciparum malaria and create malaria resistance.
- Transketolase activity is affected by Thiamine deficiency.

Intext questions

1. What are other names given to the HMP pathway?
2. Why is the HMP pathway known as the shunt pathway?
3. What are the two main objectives of the HMP pathway?
4. What are the two main phases of the HMP pathway?
5. What are the differences between the two phases of the HMP pathway?
6. How do EMP and HMP cycles differ from each other?
7. Which pentasaccharides are formed in the HMP pathway?
8. Enlist the enzymes of the oxidative and non-oxidative phases of the HMP pathway.
9. How is the HMP pathway important to protecting the tissues from any free radical injury and in the biosynthesis of other molecules?
10. What is the necessity of transaldolase enzyme in the HMP pathway?
11. What is the necessity of the transketolase enzyme in the HMP pathway?
12. Which enzyme and which phase in the HMP pathway produces glyceraldehyde 3-phosphate?
13. Which enzyme and which phase in the HMP pathway produces erythrose 4-phosphate?
14. Arrange all products of the HMP pathway in ascending number of their carbon atoms.

15. Which enzyme deficiency of the HMP pathway causes hemolytic anemia when treated for malaria? Why?

16. Which enzyme deficiency of the HMP pathway offers resistance to malaria? Why?

17. Why is the HMP pathway important for red blood cells and the lactating mammary gland?

18. Choose one correct option for the following from the options given below (The answer is given in **bold** letters):

a The HMP pathway is also known as

i. EMP pathway ii. **PPP pathway**

iii. TCA pathway

b How many phases are there in the HMP pathway?

i. Three ii. **Two**

iii. Three

c The enzyme glucose-6-phosphate dehydrogenase forms

i. Glucose 6-phosphate ii. 6-phosphogluconate

iii. **6-phopshogluconolactone**

d The occurrence of HMP pathway reactions is in

i. Mitochondria ii. **Cytosol**

iii. Nucleus

e Which one of the following is the product of the HMP pathway?

i. **NADPH** ii. FADH

iii. Lactic acid

f The precursor is required for the synthesis of nucleotides or nucleic acids.

i. Galactose phosphate ii. **Ribose 5-phosphate**

iii. Ribulose 5-phosphate

g The release of which molecule occurs when 3-keto 6-phosphogluconate is converted to ribulose 5-phosphate

i. Water ii. Hydrogen

iii. **Carbon dioxide**

h The enzyme ribulose 5-phosphate 3-epimerase is involved in

i. Conversion of xylulose 5-phosphate to ribulose 5-phosphate

ii. Conversion of ribulose 5-phosphate to xylulose 5-phosphate

iii. **Both i. and ii.**

i Formation of 7 C and 3 C sugars in the HMP pathway needs two pentasaccharides

I. Both are aldoses ii. Both are ketoses

iii One is aldose and second is ketose

j The HMP path generates NADPH. The generated NADPH is required for the biosynthesis of

i. Fatty acid synthesis ii. Steroids

iii. **Both i. and ii.**

k Generated NADPH has low activity in one of the following organs.

i. Adipose tissues ii. Liver

iii. **Muscles of the body**

l The HMP pathway is inactive in

i. Liver ii. Gonads

iii. **Non-lactating mammary gland.**

m Which condition occurs in the absence of NADPH in erythrocytes?

i. High PCV ii. **Hemolytic anemia**

iii. Erythroblastosis fetalis

n Which one of the following is correct for the G6PD enzyme

i. **Relative deficiency of G6PD in red blood cells may protect individuals from falciparum malaria and create Malaria resistance.**

ii. Relative deficiency of G6PD in white blood cells may protect individuals from falciparum malaria and create Malaria resistance.

iii. Relative deficiency of G6PD in platelet cells may protect individuals from falciparum malaria and create Malaria resistance.

o Higher levels of NADPH and ATP

   i. **Inhibit 6-phosphogluconate formation**

   ii. Stimulates 6-phosphogluconate formation

   iii. No effect on 6-phosphogluconate formation

p Insulin stimulates glucose availability for the HMP, and so

   i. **Stimulates the G6PDH enzyme**

   ii. Inhibits the G6PDH enzyme

   iii. No effect on G6PDH

q Which one of the following is the correct option for HMP?

   i. Generates NADH and pentose sugars

   ii. **Generates NADPH and pentose sugar**

   iii. It has a three-phase operation.

r Which of the following is the correct statement for the HMP pathway?

   i. **It does not use any ATP**   ii. It uses ATP in some reaction

   iii. It uses NADH in its reactions.

Further Readings

https://en.wikipedia.org/wiki/Pentose_phosphate_pathway

https://www.ebi.ac.uk/thornton-srv/m-csa/entry/148/

https://www.khanacademy.org/test-prep/mcat/biomolecules/carbohydrate-metabolism/a/pentose-phosphate-pathway

https://www.ncbi.nlm.nih.gov/books/NBK551687/

Lehninger, A. L., Nelson, D. L., & Cox, M. M. (2000). Lehninger principles of biochemistry. New York: Worth Publishers.

Life Sciences.Fundamentals and pr^ac^tice I.Pranav Kumar and Usha Mina.2017.6th Edition, pathfinder Publication.New Delhi.India.

Objective biochemistry. Sengur R.S. and Caudhary Rishu, (2014). New India publishing agency. New Delhi.

Rodwell, V. W., Botham, K. M., Kennelly, P. J., Weil, P. A., & Bender, D. A. (2015). Harper's illustrated biochemistry (30th ed.). New York, N.Y.: McGraw-Hill Education LLC.

Smith, C. M., Marks, A. D., Lieberman, M. A. (2005). Marks' basic medical biochemistry: A clinical approach.2nd edition. Philadelphia: Lippincott Williams & Wilkins.

Stincone, Anna, et al. "The return of metabolism: biochemistry and physiology of the pentose phosphate pathway." Biological Reviews 90.3 (2015): 927-963.

10

Uronic Acid Pathway or Glucuronic Acid Pathway

The uronic acid pathway is a minor and alternative pathway for glucose oxidation and for the production of uronic acid sugar. Uronic acids of sugars are formed when the terminal carbon's (C_6) hydroxyl group (CH_2OH) of sugar gets oxidized to a carboxylic acid. Uronic acid names depend on the name of the parent sugars. Uronic acid of glucose is glucuronic acid, hexose uronic acids are hexuronic acids, and pentose uronic acids are named penturonic acids. The uronic acid pathway is located in the cytosol of the liver and adipose tissue.

The glucuronic acid should not be confused with gluconic acid (aldonic acid) and glucaric acid (aldaric acid or dicarboxylic acid). Gluconic acid (aldonic acid or "onic" acid) is obtained from D-glucose by oxidizing its aldehyde group at C1 to a carboxyl group. Glucaric acid (aldaric or dicarboxylic acid) is formed upon oxidation of both the terminal hydroxyl groups (C1 and C6) and the aldehyde. They have two carboxylic acid groups. This pathway yields pentoses & ascorbic acid or vitamin C (not in primates & guinea pigs) from glucose.

10.1 Steps in Glucuronic Pathway

I. Formation of glucose 1-phosphate

Glucose, after its activation to Glucose-6-phosphate, gets converted to glucose-1-phosphate. This reaction needs the enzyme phosphoglucomutase.

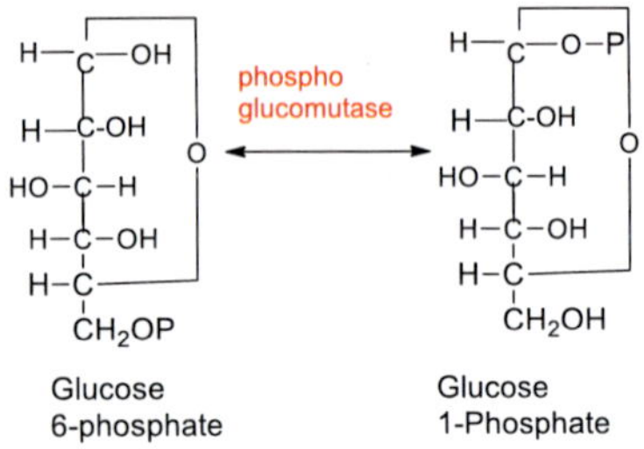

II. Formation of UDPG

This is a reversible reaction. Glucose 1-phosphate reaction with UTP (uridine-5'-triphosphate) results in the formation of UDPG (uridine diphosphate glucose). The reaction requires the enzyme UDPG pyrophosphorylase, and the forward reaction uses one UTP and liberates PPi.

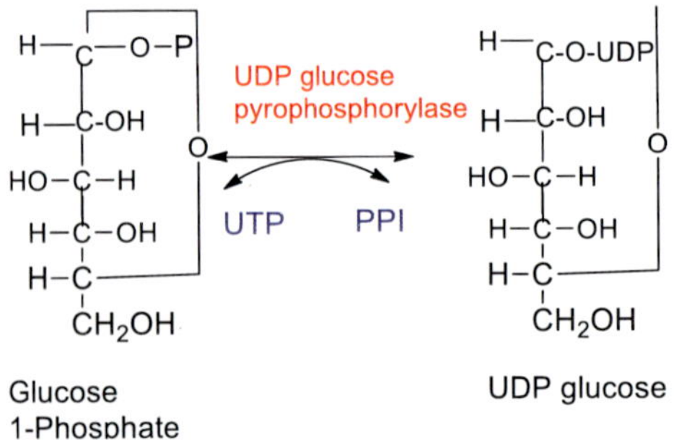

III. Formation of UDP-glucuronic acid

UDP glucose, in the presence of UDPG dehydrogenase and NAD^+, is oxidized to UDP-glucuronic acid. This UDP-glucuronic acid, in the presence of hydrolase enzyme, releases its UDP group to form D-glucuronate.

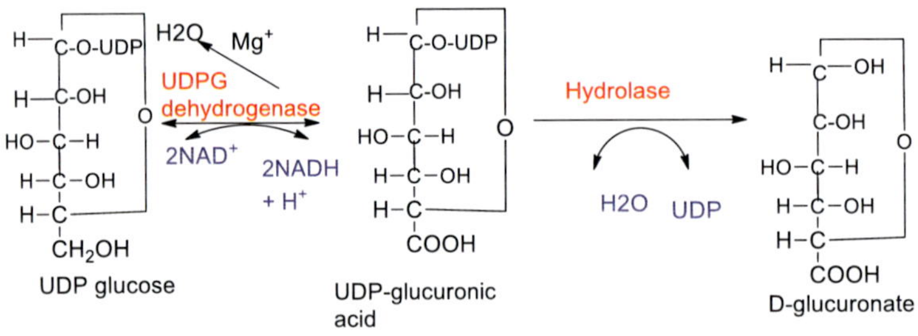

IV. Formation of D-glucuronate

L-gulonic acid is formed from D-glucuronate in presence of NADPH. L-gulonic acid can be converted to L-ascorbic acid if required (except in primates and guinea pigs), or it can be converted to a keto product known as a 2-keto-L-gulonolactone by spontaneous reaction. L-ascorbic acid can be converted to L-dehydroascorbic acid.

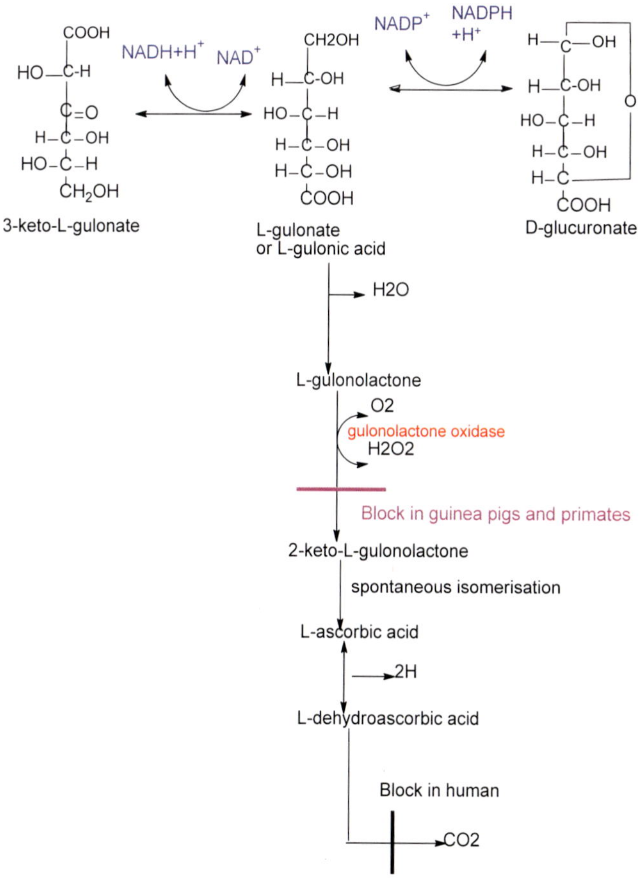

V. Formation of 3-keto-L-gulonate

One carbon is lost as carbon dioxide from 3-keto-L-gulonate to form five-carbon molecules called L-xylulose. Further reaction of L-xylulose depends upon the presence or absence of the enzyme L-xylitol dehydrogenase, which can cause a reduction of L- L-xylulose to xylitol.

In the presence of the enzyme L-xylitol dehydrogenase, L-xylulose is converted to D-xylulose in the presence of reduced NADPH+H to form NADP. The deficiency of this enzyme is a very rare hereditary condition. In the deficiency condition, D-xylulose is not formed from L-xylulose. As a result, L-xylitol gets excreted in urine as a pentose sugar, creating a condition known as "essential pentosuria."

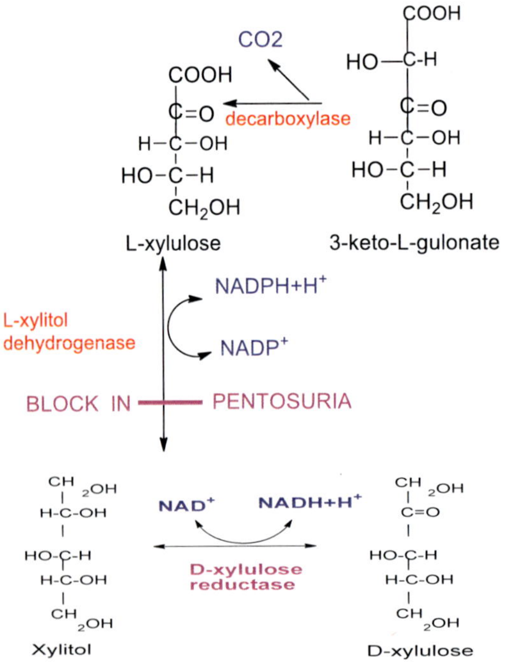

VI. Formation of L-xylulose

The 3-keto-L-gulonate decarboxylase acts on the 3-keto-L-gulonate to liberate carbon dioxide; thus, the 6c molecule is converted to a 5c molecule. If it is not metabolized, it further appears in urine, leading to an essential pentosuric condition.

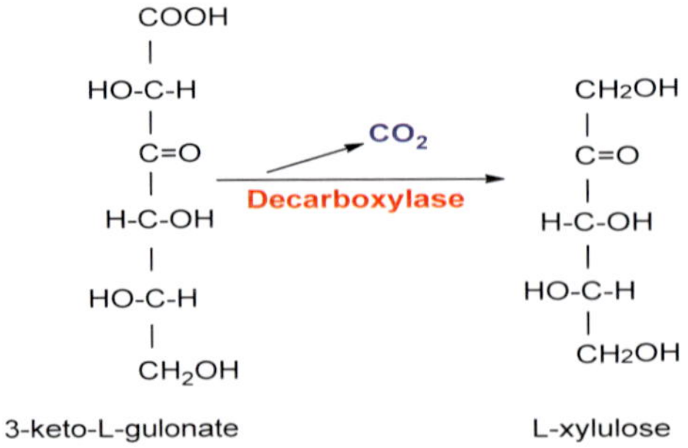

VII. Formation of xylitol

L-xylulose, in the presence of NADPH, gets converted to xylitol in the presence of dehydrogenase. This conversion is blocked in the pentosuria condition, so there is an accumulation of L-xylulose excreted in the urine.

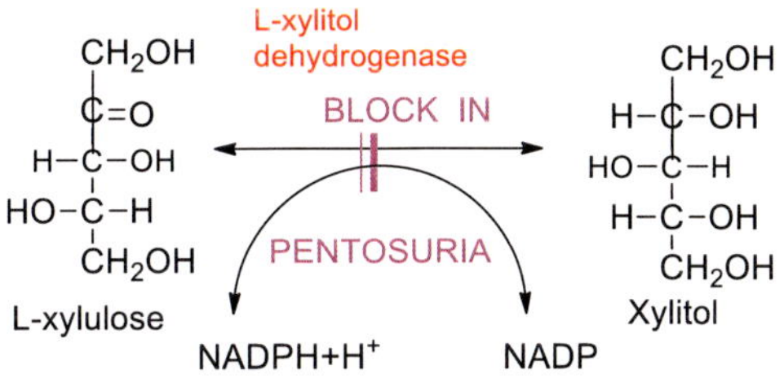

VIII. Conversion of xylitol to D-xylulose

Under the action of D-xylulose reductase and NAD+, Xylitol loses H^+ and is taken up by NAD to form $NADH+H^+$.

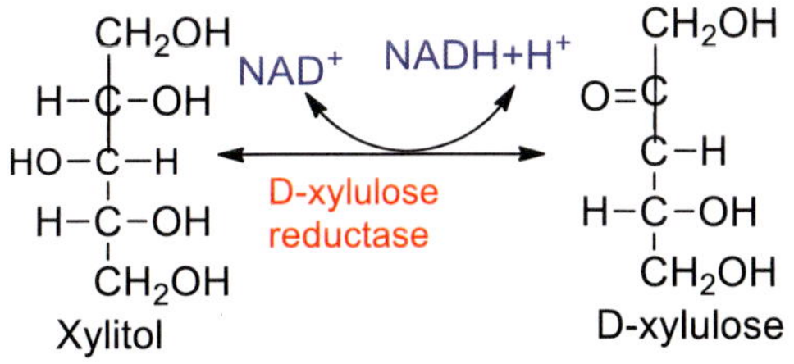

IX. Fate of D-xylulose

In the next step, D-xylulose is converted either to D-xylulose-5-phosphate or D-xylulose-1-phosphate by ATP.

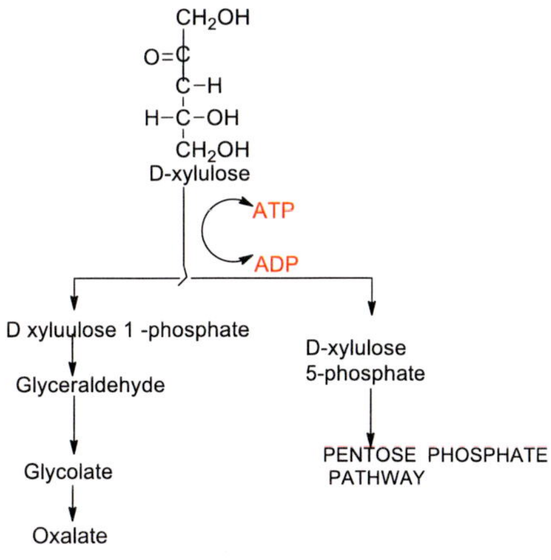

D-xylulose-5-phosphate is further metabolized in the HMP shunt. Whereas D-xylulose-1-phosphate is metabolized sequentially to glyceraldehyde, glycolate, and oxalate.

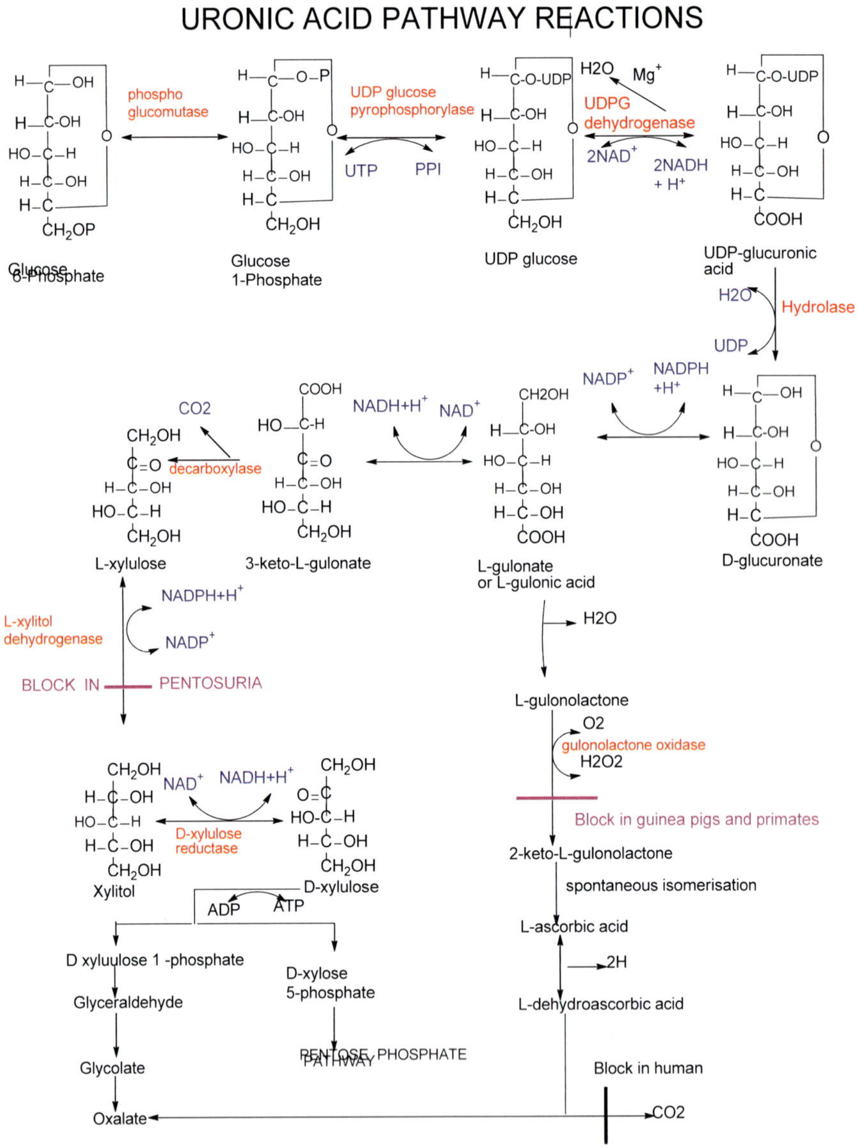

Fig.10.1: Glucuronic acid pathway.

10.2 Important Points of Glucuronic Pathway

1. It is an alternative minor oxidative pathway for glucose.
2. There is no energy production as ATP.
3. Vitamin C or ascorbic acid dietary supplementation is not required by animals other than primates and guinea pigs;
4. Glucuronic acid is helpful in various conjugation reactions in the liver (with bilirubin, many drugs, hormones, or other foreign compounds).
5. Conjugated products are excreted as glucuronate salts in bile or urine.
6. Glucuronic acid is found in heparin, hyaluronic acid, and chondroitin sulfate
7. Glucuronic acid is required in the cornea and cartilage to synthesize mucoprotein.
8. Essential pentosuria occurs due to NADP-dependent enzyme xylitol dehydrogenase and xylulose reductase deficiency. Under this situation, L-xylulose cannot be converted to xylitol, and L-xylulose is excreted in the urine.

Key takeaway

- The uronic acid pathway is a minor and alternative pathway for glucose oxidation
- The uronic acid pathway is for the production of uronic acid sugar.
- Uronic acid names are derived from the parent sugars. Pentose uronic acids are named penturonic acids.
- The uronic acid pathway is found in the cytosol and occurs in the liver and adipose tissue.
- The first reaction is the conversion of Glucose-6-phosphate to glucose-1-phosphate. The required enzyme is the phosphoglucomutase enzyme.
- Glucose 1-phosphate then reacts with UTP to form UDPG with the help of the enzyme UDPG pyrophosphorylase. One UTP is used, and PPi is liberated.
- Glucuronic acid is helpful in various conjugation reactions in the liver (with bilirubin, many drugs, hormones, or other foreign compounds).
- Xylulose 5-phosphate provides a link to the HMP shunt pathway.

- That 5-carbon containing different products are formed in the uronic acid pathway.
- There is no energy production as ATP. However, it is used by D-xylulose.
- Pentosuria condition is observed in deficiency of L-xylitol dehydrogenase, so L-xylulose cannot be converted to D-xylitol.

In-text questions

1. What is the other name for the uronic acid pathway?
2. The uronic acid pathway operates in which organs? Where does it occur within the cell?
3. Explain the first reaction, giving the name of the enzyme required for this reaction.
4. How is the UDP glucose formed?
5. What happens to dietary xylulose in the uronic acid pathway?
6. Explain the reactions in which is NAD used in forward reactions.
7. Name the reaction in which UTP is used.
8. Explain the reactions in which NADPP is used in forward reactions.
9. In which reaction UDP is liberated?
10. Which reaction products will be formed wherein D-xylulose, ATP, and D-xylulose reductase participate?
11. Which precursor undergoes spontaneous isomerization to form L-ascorbic acid?
12. What happens to L-xylulose in physiological conditions and when the conditions are not physiological with respect to the enzyme required by L-xylulose?
13. What is the pentosuria condition, and how does it occur?
14. Why can primates not synthesize ascorbic acid while ruminant animals can?
15. Which sugar is excreted in the essential pentosuria condition?
16. What is the fate of D-xylulose?
17. How are the L-ascorbic acid and L-dehydroascorbic acid formed?
18. Choose one correct option for the following from the options given below (The answer is given in **bold letters**):

a A location of the glucuronic acid pathway is

i. Golgi apparatus ii. **Cytoplasm**

iii. mitochondria

b D-glucuronic acid is required for

i. Detoxification

ii. Synthesis of mucopolysaccharides

iii. **Both i. and ii.**

c The vitamin synthesized in the uronic acid pathway is

i. Thiamin ii. **Ascorbic acid**

iii. Pantothenic acid

d Which one is the correct reaction in the glucuronic acid pathway

i. **Glucose 6-phosphate is converted to glucose 1-phosphate as the first step reaction**

ii. Glucose 1-phosphate is converted to glucose 6-phosphate as the first step of the reaction

iii. Glucose 1-phosphate is converted to fructose 1-phosphate in the first reaction.

e The enzyme required for the formation of uridine diphosphate glucose from glucose 1-phosphate

i. **UDP glucose pyrophosphorylase** ii. phosphoglucomutase

iii. Neither i. nor ii.

f Which of the following enzyme(s) is/are required for the conversion of uridine diphosphate glucose to uridine diphosphate glucuronate?

i. **UDPG dehydrogenase**

ii. UDP-glucose pyrophosphorylase

iii. Both i. and ii.

g What is released during the conversion of Uridine diphosphate glucuronate to glucuronate?

i. UTP ii. **UDP**

iii. PPi

h What is formed in a reaction converting glucuronate to gulonate?

i. H_2O ii. ATP

iii. **$NADP^+$**

i Formation of 3-keto-gulonate from Gulonate needs

i. NADP ii. **NAD**

iii. GTP

j Xylulose formation from 3-keto-gulonate releases

i. **Carbon dioxide** ii. Water

iii. Pyrophosphates

k Xylulose to xylitol formation requires

i. L-xylitol decarboxylase ii. **L-xylitol dehydrogenase**

iii. Both i. and ii.

l Organs having glucuronic acid pathways

i. Liver ii. Adipose tissues

iii. **Both i. and ii.**

m This pathway yields ascorbic acid from glucose, but not in

i. Guinea pigs ii. Primates

iii. **Both i. and ii.**

n What is the complete form of UTP in the glucuronic acid pathway?

i. Unshielded twisted pair ii. **Uridine triphosphate**

iii. Utridine triphosphate.

o Conversion of UDP glucose to UDP-glucuronic acid requires

i. NAD^+

ii. **UDPG dehydrogenase enzyme**

iii. Hydrolase enzyme

p Glucuronate is a

i. **Polar molecule** ii. Non-polar molecule

iii. Amphipathic molecule

q The condition pentosuria is due to a deficiency of

i. Hormones ii. Coenzymes

iii. **Enzyme**

r Sugar which appears in urine during pentosuria condition is

i. **L-xylulose** ii. D-xylulose

iii. Xylitol

s Water soluble vitamin obtained from L-gulonolactone in the uronic acid cycle.

i. Thiamine ii. **Ascorbic acid**

iii. Pyridoxine

Further Readings

Biochemistry by David E Metzler 2nd edition, ElsevierAcademic Press

Biochemistry by Thomas.M.Delvin, Edition no.7

Harper's illustrated Biochemistry, Edition no.26. Robert K. Murray, Daryl K, Granner, Peter A. Mayes, Victor W. Rodwell, lang MedicalBooks/McGraw-Hill Medical Publishing Division.

http:// biocheminfo.com/2020/05/14/uronic-acid-pathway-glucuronic pathway/

https://microbenotes.com/uronic-acid-pathway/

https://microbenotes.com/uronic-acid-pathway/

https://www.davuniversity.org/images/files/study-material/BCH5156-3.pdf

https://www.scribd.com/document/328465491/2-4-Biochemistry-Tca-Hmp-and-Uronic-Acid-Pathway

https://www.scribd.com/document/328465491/2-4-Biochemistry-Tca-Hmp-and-Uronic-Acid-Pathway

https://www.scribd.com/presentation/26355617/uronic-acid-pathway-fructose-metabolism

https://www.slideshare.net/sathi3366/presentation-29571791.

https://www.studocu.com/en-gb/document/university-of-salford/biology/summary-notes-the-uronic-acid-pathway/16079375

John W. Pelley, Edward F. Goljan (2011). Biochemistry. Third edition. Philadelphia: USA.

Medical Biochemistry By M.N.Chatterjea.Edition no.8

Rodwell, V. W., Botham, K. M., Kennelly, P. J., Weil, P. A., & Bender, D. A. (2015). Harper's illustrated biochemistry (30th ed.). New York, N.Y.: McGraw-Hill Education LLC.

Sumitra Ramachandran1 , Pierre Fontanille1 , Ashok Pandey2 and Christian Larroche1. Gluconic Acid: Properties, Applications and Microbial Production. Food Technol. Biotechnol. 44 (2) 185–195 (2006).

Textbook of Biochemistry by U Satyanarayana. 5th edition. Year 2019.

Textbook of Biochemistry by Vasudevan

11

Gluconeogenesis (Anabolic Pathway)

Plants, with the help of photosynthesis, convert simple molecules (like carbon dioxide and water) into more complex molecules (anabolism) through various chemical reactions. Animals synthesize more complex molecules (anabolism) from monomer units through multiple chemical reactions. Anabolism refers to a biosynthesis process involving the sequences of biochemical reactions by which complex molecules are formed in living cells from available monomer units. Anabolic processes include the biosynthesis of complex biomolecules like carbohydrates, proteins, and lipids. The anabolic processes require energy from high-energy bond compounds like ATP.

The anabolic processes related to the biosynthesis of carbohydrates include

(1) Gluconeogenesis and

(2) Glycogenesis

11.1 Gluconeogenesis

Gluconeogenesis is the synthesis of new glucose molecules from non-carbohydrate molecules. The non-carbohydrate molecules that are used for gluconeogenesis are derived from glycolysis, lipolysis, and proteolysis. Examples are pyruvate, lactate, glycerol, and glucogenic amino acids. The liver and kidneys are the preferred organs for the gluconeogenesis process. It also occurs to some extent in the brain. Blood glucose levels are maintained at physiological concentrations through the process of gluconeogenesis in humans and ruminants. The blood glucose drops below normal levels during various conditions like fasting, starvation, pregnancy, lactation, fever, and low carbohydrate diets.

Pyruvate is a common starting material for gluconeogenesis. Pyruvate is formed from lactate and amino acids like alanine, glycine (via serine), and cysteine. Lactate is also formed by RBC (lacking mitochondria) and active skeletal muscles. The gluconeogenesis process is very important as it provides a required amount of glucose as a source of energy to body cells.

Gluconeogenesis is much like glycolysis, except for some irreversible reactions in glycolysis, which uses different enzymes.

I Seps in gluconeogenesis

A. The first stage reaction in gluconeogenesis is the conversion of pyruvate to Phosphoenolpyruvic acid.

The pyruvate formed has two options. One is to enter the TCA cycle if oxygen is available. If oxygen is not available adequately, pyruvate gets converted to lactate.

1. **Lactate is converted to pyruvate** by the lactic dehydrogenase (LDH) enzyme, forming NADH from NAD. Pyruvate is formed in the cytoplasm of a cell but is oxidized in the mitochondria. Pyruvate entry into mitochondria requires a carrier known as the mitochondrial pyruvate carrier.
2. **conversion of pyruvate to oxaloacetate** . In the following steps, the irreversible **conversion of pyruvate to oxaloacetate** occurs with the action of pyruvate carboxylase found in the mitochondria. It requires biotin coenzyme. Transaminase enzymes can catabolize amino acids like alanine, aspartate, and asparagine into oxaloacetate.
3. **Conversion of Oxaloacetate into malate and again to oxaloacetate.** Oxaloacetate cannot cross the membranes of the mitochondria; hence, it gets converted into malate by malate-dehydrogenase (mitochondrial matrix). The converted malate can now cross the mitochondrial membrane to enter the cytoplasm. After the entry of malate into the cytoplasm, it gets converted back into oxaloacetate with cytoplasmic malate dehydrogenase, forming NADH from NAD.
4. **Oxaloacetate gets converted into Phosphoenol pyruvate.** Oxaloacetate, acted upon by phosphoenolpyruvate carboxykinase, gets converted into Phosphoenolpyruvate with the release of carbon dioxide and the formation of GDP from GTP. This reaction, being irreversible in nature, differs from that of glycolysis,

B. Formation of Phosphoenol pyruvate to 1,3-biphosphoglycerate (1,3 BPG)

It involves the conversion of phosphoenolpyruvate to 2-phosphoglycerate by the enzyme enolase.

- The 2-phosphoglycerate in the presence of the mutase enzyme gets transferred to 3-phosphoglycerate.

- This later gets phosphorylated to form 1,3-biphosphoglycerate (1,3 BPG) with the help of a kinase enzyme.

C. Conversion of 1,3, BPG to fructose -1-6-diphosphate

- Dehydrogenase acts on 1,3 BPG to form glyceraldehyde 3-phosphate.
- Further reaction will form 6C, fructose-1, 6-diphosphate if two molecules with three carbons are available, like glyceraldehyde 3-phosphate.
- Glycerol-3-phosphate gets converted to dihydroxyaceton phosphate with the help of an enzyme called Glycerol-3-phosphate dehydrogenase. Dihydroxyacetone phosphate from glycerol of triglycerides can also be converted to form a second molecule of glyceraldehyde 3-phosphate by the isomerase enzyme. Thus, fructose 1,6-diphosphate is formed by the aldolase from two C3 molecules of glyceraldehyde 3-phosphate.

D. Final stage for conversion of fructose-1,6-diphosphate to glucose.

1. This step for converting fructose-1,6-diphosphate to fructose-6-P is irreversible with the use of the enzyme fructose-1,6-phosphatase.
2. The same enzyme of glycolysis, phosphoglucoisomerase, converts fructose-6-P to glucose-6-P using the
3. The last irreversible step reaction differs from glycolysis in forming glucose from glucose-6-P with the glucose-6-phosphatase enzyme in the endoplasmic reticulum.

KEY ENZYMES DIFFERING BETWEEN GLUCONEOGENESIS AND GLYCOLYSIS

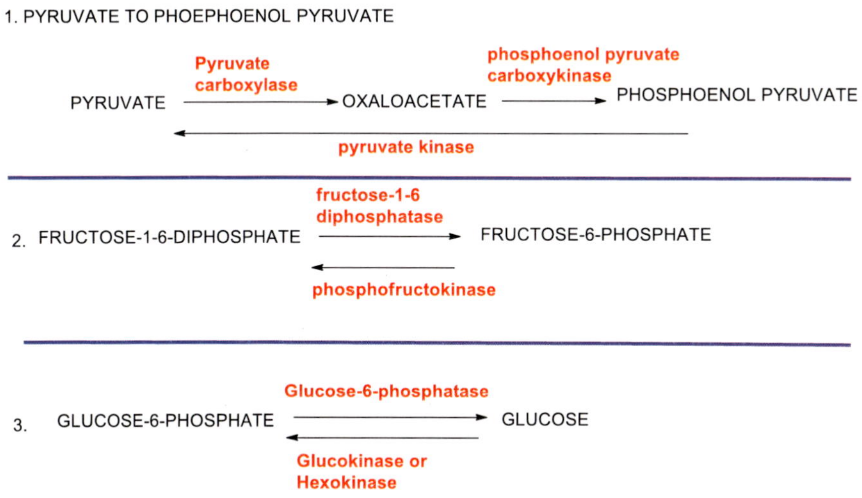

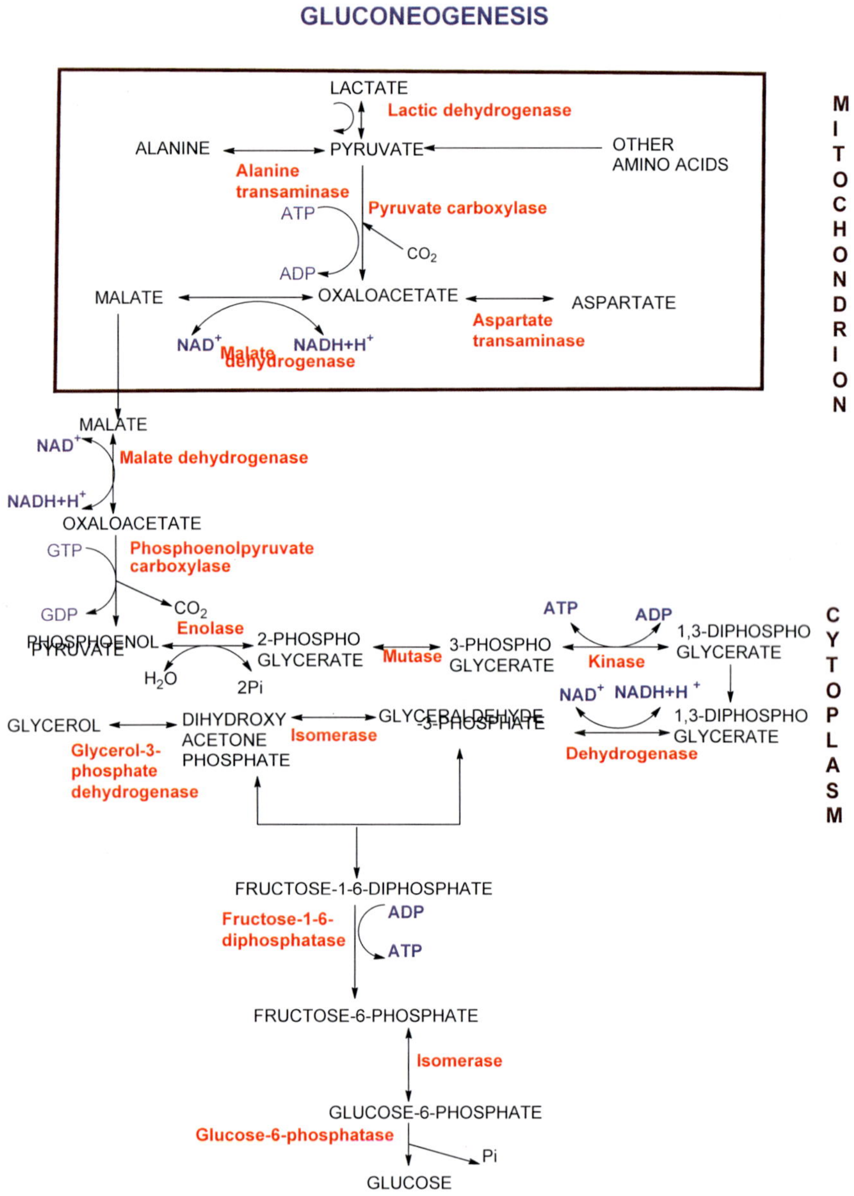

Fig. 11.1: gluconeogenesis cycle.

II Regulation of gluconeogenesis

A. The process of gluconeogenesis and glycolysis is regulated by the ATP/AMP ratio. The body increases gluconeogenesis and decreases glycolysis if surplus energy is needed (a large ATP to AMP ratio). In reverse conditions of lower energy than the body requires (low ATP to AMP ratio), the organism increases glycolysis and decreases gluconeogenesis.

B. The acetyl-CoA regulates the conversion of pyruvate to PEP because it (acetyl-CoA) activates the enzyme pyruvate carboxylase.

C. The conversion of fructose-1,6-BP to fructose-6-P occurs with the enzyme fructose-1,6-phosphatase (negatively regulated and inhibited by the molecules AMP and fructose-2,6-biphosphate).

D. Glucose is available from glucose -6-P by the action of glucose-6-phosphatase. Increased levels of glucose-6-P increase the glucose-6-phosphatase activity, and more glucose is produced, and glycolysis cannot proceed.

E. Various hormones regulate gluconeogenesis. Glucocorticoids, Thyroxine, Growth hormone, and Epinephrine stimulate it. Insulin inhibits it.

11.2 Cori Cycle or Glucose-Lactate Cycle

The Cori cycle is named after Carl and Gerty Cori (1947, shared Nobel Prize.). The steps of the cycle are as follows:

- In the Cori cycle, when metabolized under aerobic conditions, glycogen in muscle and other tissues is broken down to glucose (glycolysis). Further, these two pyruvate molecules are processed under anaerobic conditions to lactate molecules.
- The lactate molecules are carried to the liver via the blood circulation.
- The liver converts lactate to pyruvate molecules to form glucose (gluconeogenesis). The glucose released into the blood reaches back to muscles, other tissues, and cells that require it, such as red blood cells and neurons, thus closing the cycle.
- Under increased energy requirements of the muscles, there is a need to regenerate more glucose from lactate in the liver through the Cori cycle. If the lactate production is beyond the capacity of the Cori cycle, its excess amount will accumulate in the blood, leading to the condition known as lactic acidosis.

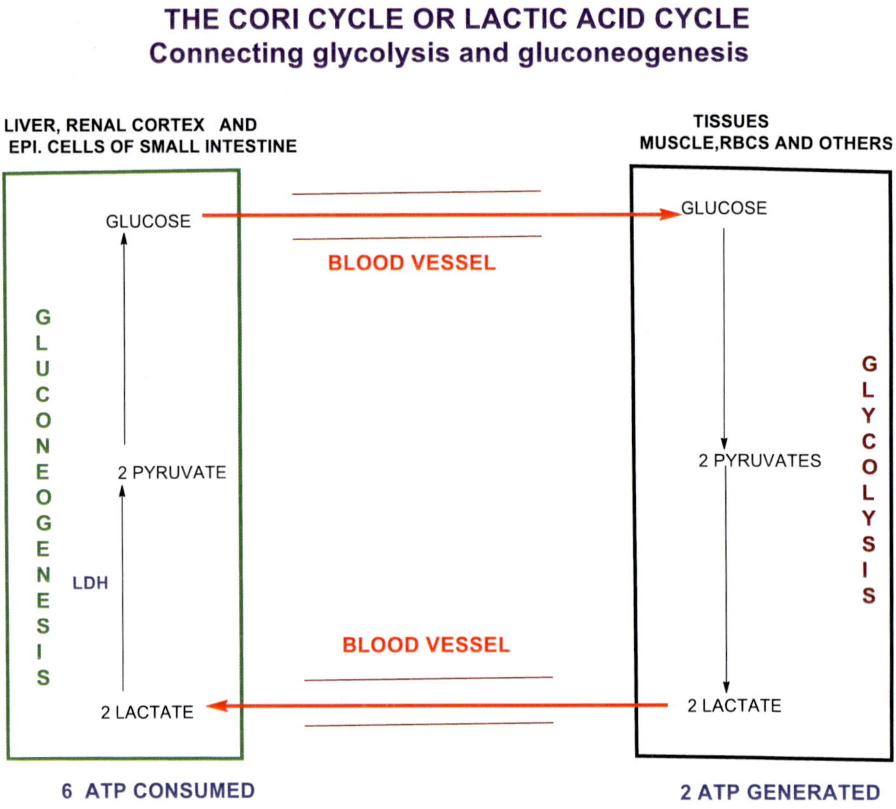

Fig. 11.2: The Cori cycle

I Energy cost of the Cori cycle

Glycolysis operating anaerobically produces two lactate molecules as the end product of glucose by generating 2 ATPs. Two lactate molecules convert to glucose in the liver and other tissues through gluconeogenesis, requiring 6 ATP. Thus, this cycle is inefficient for producing energy in the cell, as there is a net consumption of 4 ATPs.

II Importance of the Cori cycle

The Cori cycle allows the effective functioning of many extrahepatic cells when working under hypoxic or anaerobic conditions. The following are the examples:

- The Cori cycle is also essential during overnight fasting and starvation.

- Red blood cells are entirely dependent on anaerobic glycolysis for ATP production due to their lack of mitochondria. The lactate is partly disposed of by the Cori cycle.
- Anaerobic glycolysis, which occurs in skeletal muscles during intense exercise, produces energy and lactate. Lactate does not accumulate in muscles due to the Cori cycle.
- Greater lactate production occurs in hypoxic tissues under trauma, sepsis, burns, post-surgery time, cell proliferation during wound healing, and bone marrow. These conditions are associated with the increased energy demand of the patient's tissues. Nutritional supplements should be provided under such conditions. Progressive weight loss occurs in cancer patients.

Key Takeaway

- Lactate is converted to pyruvate by the lactic dehydrogenase (LDH) enzyme, forming NADH from NAD.
- Transaminase enzymes can catabolize amino acids like alanine, aspartate, and asparagine into oxaloacetate.
- The first stage reaction in gluconeogenesis is the conversion of pyruvate to Phosphoenol pyruvic acid.
- **Anabolism refers to a biosynthesis process involving** the sequences of biochemical reactions by which complex molecules are formed in living cells from available monomer units.
- **Gluconeogenesis** is the synthesis of new glucose molecules from non-carbohydrate molecules.
- The gluconeogenesis process occurs primarily in the liver and kidneys.
- The non-carbohydrate molecules that are used for gluconeogenesis are derived from glycolysis, lipolysis, and proteolysis.
- The blood glucose levels are maintained at their physiological concentrations by the process of gluconeogenesis in humans and ruminants.
- Lactate is also formed by RBC (lacking mitochondria) and active skeletal muscles.
- Gluconeogenesis is much like glycolysis, except that it uses different enzymes for some irreversible reactions in glycolysis.
- Lactate is converted to pyruvate by the lactic dehydrogenase (LDH) enzyme, forming NADH from NAD.

- Transaminase enzymes can catabolize amino acids like alanine, aspartate, and asparagine into oxaloacetate.
- Oxaloacetate cannot cross the membranes of the mitochondria; hence, it gets converted into malate-by-malate dehydrogenase (mitochondrial matrix). Malate crosses the mitochondrial membrane and enters the cytoplasm. The cytoplasm converts it back into oxaloacetate with malate dehydrogenase(cytosolic), forming NADH from NAD.
- Pyruvate is formed in the cytoplasm of a cell but is oxidized in the mitochondria.
- The conversion of fructose-1,6-diphosphate to fructose-6-P is irreversible with the use of the enzyme fructose-1,6-phosphatase.
- The conversion of fructose-6-P to glucose-6-P uses the same enzyme as glycolysis, phosphoglucoisomerase.
- The irreversible last step reaction differs from glycolysis. In forming glucose from glucose-6-P, the glucose-6-phosphatase enzyme is present in the endoplasmic reticulum.
- Two lactate molecules convert to glucose in the liver and other tissues through gluconeogenesis, requiring 6 ATP.
- Glycolysis operating anaerobically produces two lactate molecules as the end product of glucose by generating 2 ATPs., in the Cori cycle.
- In the Cori cycle, there is a net consumption of 4 ATPs.
- The Cori cycle operates between the muscles and the liver.

Intext questions

1. What is the first reaction of gluconeogenesis using pyruvate substrate?
2. Which organs are the place for gluconeogenesis?
3. Which processes contribute to the formation of glucose in the body?
4. What is the significance of gluconeogenesis in ruminant animals and humans?
5. What are the conditions associated with subnormal blood glucose levels in the body?
6. Lactate is also formed by RBC (lacking mitochondria) and active skeletal muscles.
7. Pyruvate is formed from lactate and amino acids like alanine, glycine (via serine), and cysteine.
8. Why is lactate formed in erythrocytes and skeletal muscles?

9. How does oxaloacetate cross the mitochondrial membrane?
10. How is cytoplasmic oxaloacetate converted to malate?
11. How is pyruvate converted to phosphoenolpyruvate?
12. Explain the interconversion of fructose 1,6-diphosphate to fructose 6-phosphate.
13. Describe the importance of the Cori cycle reactions.
14. Explain the energy calculation in the Cori cycle.
15. Choose one correct option for the following from the options given below (The answer is given in bold letters):

a A Pyruvate-to-oxalate conversion is catalyzed by an enzyme

i. **Pyruvate carboxylase** ii. Pyruvate kinase

i. Pyruvate dehydrogenase

b Coenzyme is required for the conversion of pyruvate to oxaloacetate

i. Thiamin ii. **Biotin**

iii. Nicotinamide

c Which of the following is different from the correct statement for gluconeogenesis?

i. Acetyl-CoA is used for gluconeogenesis as it is formed from pyruvic acid.
ii. Glycerol is used for gluconeogenesis.
iii. Glutamine is used for gluconeogenesis.

d Which of the following organs operates the Cori cycle?

i. **Liver** ii. Brain

iii. Muscle

e The precursor for gluconeogenesis is

i. Insulin ii. Galactose

iii. **Glycerol**

f Which one of the following is not used in the gluconeogenesis process?

i. Glycerol ii. **Glucose**

iii. Aminoacids

g Where is the location where gluconeogenesis reactions occur?

i. Golgi apparatus ii. Mitochondria

iii. **Cytoplasm**

h Glycolysis yields glycerol. Another source of glycerol to the body is from

i. Brain ii. **Adipose tissues**

iii. Skeletal muscles

i Which of the following is the correct statement for gluconeogenesis?

i. **Gluconeogenesis is much like glycolysis, except different enzymes are used for some irreversible reactions in glycolysis.**

ii. Gluconeogenesis is a glycogen synthesis.

iii. Gluconeogenesis is much like glycolysis, using the same enzymes for reactions in glycolysis.

j When lactate is converted to pyruvate, one of the HEB formed is

i. FADH ii. **NADH**

iii. NADPH

k Which one of the following three statements is true in sequentially arranged forward reactions which to occur in the gluconeogenesis process?

i. Lactate→oxaloacetate→pyruvate

ii. Pyruvate →lactate→oxaloacetate

iii. **Lactate→pyruvate →oxaloacetate**

l The conversion of pyruvate to Phosphoenol pyruvate involves the following

i. Pyruvate carboxylase.

ii. Phoshoenol pyruvate carboxykinase.

iii. **Both i. and ii.**

j Conversion of Phosphoenol pyruvate to pyruvate involves

i. **Pyruvate kinase** ii. Pyruvates carboxylase

iii. Both i. and ii.

k Interconversion of fructose 6-phosphate and fructose 1,6-diphosphate is an/a

i. Reversible reaction ii. **Irreversible reaction**

iii. Substrate level reaction

l The formation of fructose-6-P from fructose-1,6-diphosphate is irreversible. The enzyme required is

i. **Fructose-1,6-phosphatase.** ii. Fructose 6-phosphatase

iii. Fructose 1-phosphatesse

m The conversion of glucose-6-phosphate (G-6-P) to glucose occurs in the

i. Golgi apparatus of the liver

ii. **Endoplasmic reticulum (ER) of the liver**

iii. Mitochondria of the liver

n In the Cori cycle, the account of ATP is as follows**: There is a net consumption of 4 ATPs**.

i. There is a net gain of 6 ATP

ii. No ATP is used, nor is ATP produced.

o The Cori cycle has two processes: a. glycolysis and gluconeogenesis. Which process(es) is (are) endergonic in nature?

i, Glycolysis i. **Gluconeogenesis**

ii. Both i. and ii.

p Two parts in the Cori cycle are Gluconeogenesis and Glycolysis. The latter part occurs in which organ?

i. Kidney ii. **Muscles**

iii. Liver

Further Readings

Biochemistry. John W.Pelley and Edward F. Goljan. Third Edition. Publisher Elsevier.

By Sucheta Dandekar and Abbas Mahdi, second edition. Publisher Elsevier, year 2021.ISBN 9788131261507.

https://chem.libretexts.org/Bookshelves/Biological_Chemistry/Supplemental_Modules_(Biological_Chemistry)/Metabolism/Anabolism/Gluconeogenesis

https://microbeonline.com/cori-cycle-steps-regulation-and-importance/

https://www.sciencefacts.net/cori-cycle.html

Lehninger, A. L., Nelson, D. L., & Cox, M. M. (2000). Lehninger principles of biochemistry. New York: Worth Publishers.

Medical Biochemistry preparatory manual for undergraduate (2021)

Physiology, Gluconeogenesis – https://www.ncbi.nlm.nih.gov/books/NBK541119/

Shinde. Jaypee Brothers Medical Publishers (P) Ltd. New Delhi

Textbook of Medical Biochemistry. Dr (Brig) MN Chatterjea and Rana

Textbook of Veterinary Biochemistry (2006). Dhanotiya RS. Second edition.Jaypee brothers, New Delhi. India.

12

Glycogenesis

Glycogenesis helps to maintain blood glucose, as it is a primary source of energy for every cell. An inadequate supply of glucose to the body may lead to the failure of the vital organs. Glucose, when in excess, is synthesized into glycogen and stored (glycogenesis), mainly in the liver and muscles (including the uterus). It is also found in other tissues, such as the kidney, brain, heart, and adipose tissues. The stored liver glycogen is broken down (glycogenolysis) into blood glucose when needed. Still, the glycogen stored in muscle is broken down to glucose 1-phosphate and converted to blood glucose later. Either the synthesis or breakdown pathway is primarily active at a time in tissues. Glycogen synthesis and its breakdown are not reversible processes.

The percentage of glycogen stored in skeletal muscles is less (1% to 2%) than in the liver (10%). Still, the total quantity of glycogen is greater in muscles due to their greater mass than an animal's total liver mass. Muscle mass, on average, ranges from 40 to 45 % of body weight for the most mature mammals.

Glycogen synthesis (glycogenesis) is an anabolic process forming a branched homopolysaccharide made of several chains of alpha D-glucose units linked by α-1,4 glycosidic bonds in a linear chain and nearly every ten glucose residues; the chain branches off via α-1,6 glycosidic linkages. The α-glycosidic bonds give rise to a helical polymer structure. The number of glucose units in a single glycogen molecule varies (1,700 to -600,000 units). The glycogen granules are formed on the glycogenin protein in the center, so it acts as a primer, and their number determines the total number of glycogen particles.

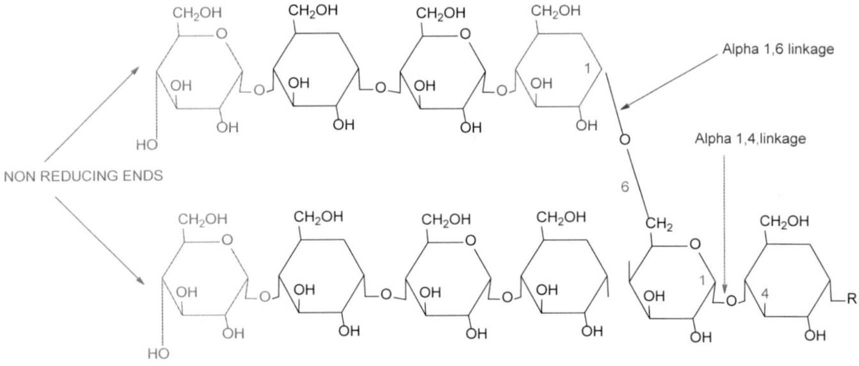

STRUCTURE OF GLYCOGEN

In the cells, glycogen exists as a complex with a glycosome protein. The glycogen in the glycosome can be in forms like alpha-granules, beta-granules, and gamma-particles. The glycogen beta-granules are a rapid source of glucose energy compared to the slower source of glucose energy from the alpha granules. The liver predominantly contains glycogen alpha-granules. The alpha-granules are formed from beta-granules.

12.1 Steps in Glycogenesis

The biosynthesis of glycogen from glucose or the polymerization of glucose into glycogen is called glycogenesis. This process requires a series of reactions, which are as follows:

I. Formation of glucose 6-phosphate, a rate-limiting reaction

Cell locks glucose by converting it to glucose 6-phosphate with the help of glucokinase (glucose substrate--liver)/hexokinase (six-carbon sugar as substrate --muscle, other tissues) enzyme and ATP, $Mg2^{+}$. Insulin secretion favors glycogen synthesis. Glucokinase is activated by higher blood glucose.

Glucose + ATP $\rightarrow$glucose 6-phospahte + ADP + H

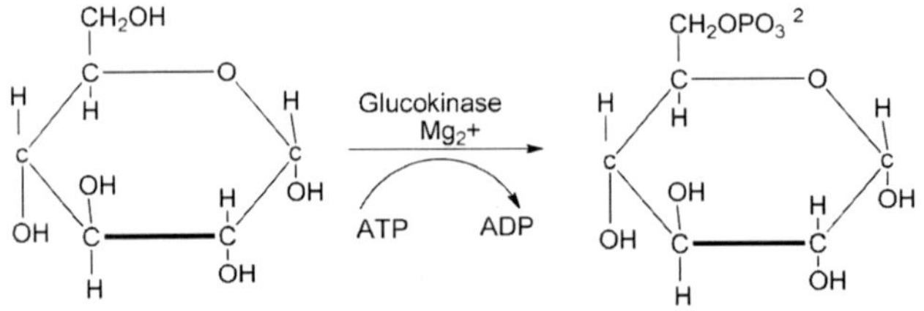

Glucose 6-phosphate

II. Isomerization of glucose 6-phosphate to glucose 1-phosphate.

Glucose 6-phosphate is reversibly converted to glucose 1-phosphate in the presence of phosphoglucomutase and $Mg2^{+}$. This enzyme involves the phosphate group transfer from the 6th carbon to the 1st carbon position.

III. Formation of UDPG (uridine diphosphoglucose), precursor for glycogen synthesis.

Glucose 1-phosphate cannot be incorporated into glycogen and must be converted to UDP glucose. This reaction needs glucose 1-phosphate, UTP, and the enzyme **UGPP (**UDP-Glucose Pyrophosphorylase). Glycogen synthesis requires energy supplied by uridine triphosphate (UTP).

G1P + UTP UDPG + PPi, PPi+H2O → 2Pi

Overall, it will be --G1P+UTP → UDP-glucose +2Pi.

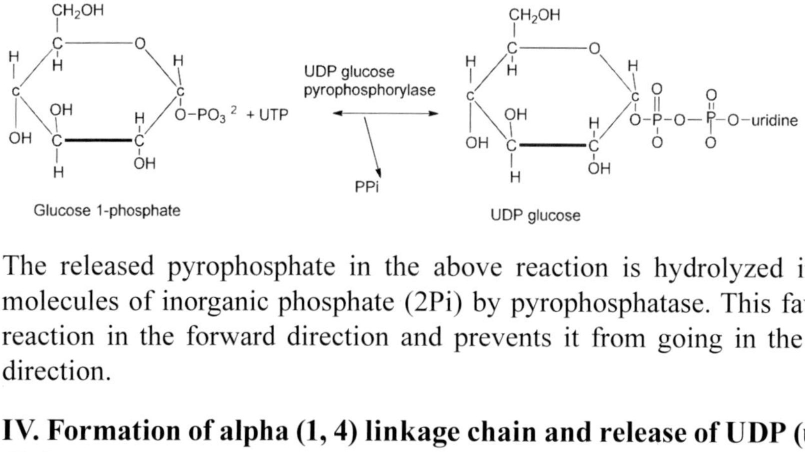

The released pyrophosphate in the above reaction is hydrolyzed into two molecules of inorganic phosphate (2Pi) by pyrophosphatase. This favors the reaction in the forward direction and prevents it from going in the reverse direction.

IV. Formation of alpha (1, 4) linkage chain and release of UDP (uridine diphosphate).

The UDPG formed has now dissociated to liberate UDP and glucose. This reaction is carried out by the enzyme glycogen synthetase or glucosyl transferase (tetrameric form). When UDP is dissociated from UDP glucose, the energy is released from the phosphor glycosyl bond of UDP glucose. Glycogen Synthase utilizes this energy to catalyze glucose incorporation into Glycogen. The UDP is subsequently released from the enzyme.

The glucose released from UDPG must be transferred to the pre-existing **"primer"** chain, with at least four glucose residues linked by 1-4 glycosidic linkage. This transfer occurs so that carbon number 1 link to carbon number 4 of an existing primer unit's non-reducing end of glucose. The glucose unit from UDPG is added to the non-reducing (outer) end of the glucose chain. Adding a glucose unit in an existing primer unit, an unbranched chain, leads to an elongation of the alpha (1, 4) glucose link chain. Glycogen synthase enzyme elongates the existing "primer" chain for glycogen synthesis.

Elongation of Glycogen chain from n residues to n+1 residues

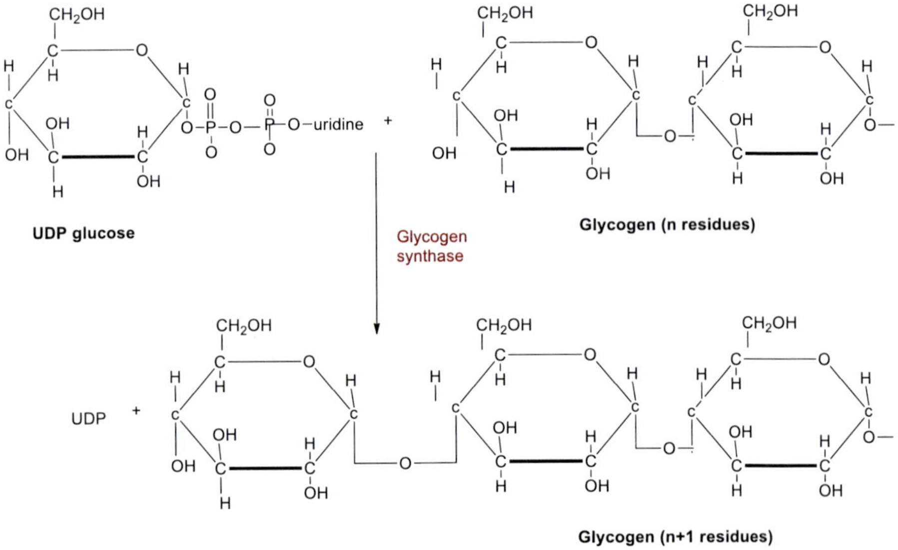

A glycogenin is a protein enzyme (a homodimer protein of 37-kDa subunits) that acts as a "primer" by polymerizing the first few glucose molecules, after which other enzymes take over. It is classified as a glycosyltransferase. Glycogenin is found to be associated with **glycogen particles** in the cytoplasm.

In *de novo* glycogen synthesis, the protein glycogenin catalyzes the attachment of UDP-glucose to itself. Glycogenin can accept glucose units from UDPG. Thus, glycogenin can undergo self-glucosylation to generate an oligosaccharide primer. The hydroxyl group of the amino acid tyrosine of Glycogenin is the site where the initial glucose attaches to catalyze the formation of linear chains of alpha 1 and 4 glucose linkages. Once the primer is formed, the glycogenin-bound glycogen synthase is released to help further lengthen the glucose polymer chain. Glycogen synthase 'a' is unphosphorylated and most active, whereas glycogen synthase 'b' is phosphorylated and less active.

Two isoforms of glycogenin are glycogenin-1, found in muscle, and glycogenin-2, found in liver and cardiac muscles. Glycogen phosphorylase can act on primed glycogenin, considering it as its substrate. It is also a substrate for glycogen phosphorylase.

V. Formation of branching through alpha (1,6) linkage

Once a glycogen synthetase makes a straight chain made up of 11 to 12 glucose molecules linked by alpha (1, 4) linkage, a branch point gets established by the branching enzyme at alpha 1, 6 linkages by transferring the minimum number (three) of detached alpha (1, 4) glucose residues from the terminal part of the straight chain.

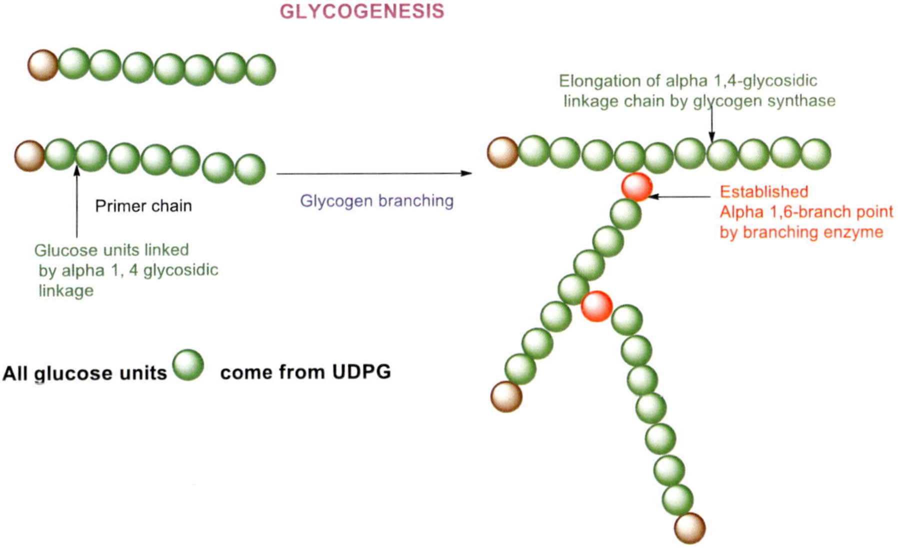

The branching enzyme or also termed as amylo-(1, 4-1, 6) transglycosidase, or glucosyl alpha -4-6 transferase performs two functions like

1. To break the terminal fragment of alpha (1, 4) glucose chains (minimum 4 glucose units) and release the energy.
2. Attach the detached alpha (1, 4) chain of glucose to the hydroxyl group of carbon number 6 of the neighboring pre-existing alpha (1, 4) linkage glucose chain using the released energy.

Both these actions, every time, will result in the formation of branch points at alpha (1, 6) linkage in glycogen molecule.

The action of both the glycogen synthetase and the branching enzyme will help to grow the glycogen molecule. Synthesis of a glycogen molecule is completed when the required polymer of glucose residues is linked by α-(1, 4) - and branched at α-(1, 6)-glycosidic bonds. This happens when glycogen synthase, acting at the non-reducing ends, separates from the glycogenin primer.

Summary of Steps of Glycogenesis

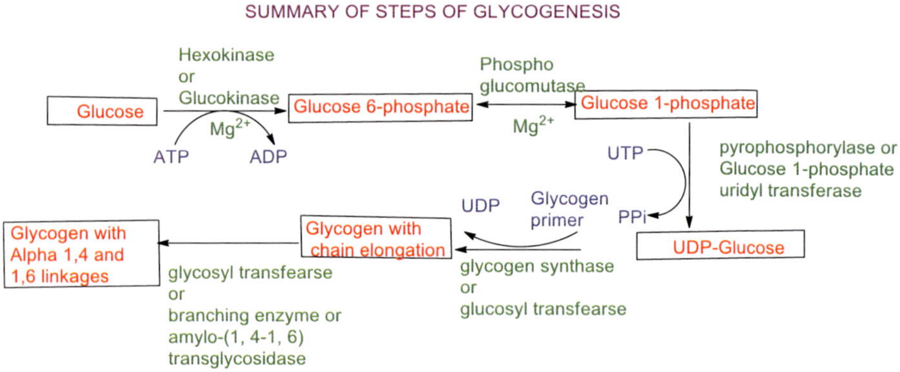

Overall, 2 is ATPs are utilized, one at the time of glucose phosphorylation, and a second is needed to convert UDP formed to regenerate UTP.

12.2 Regulations of Glycogenesis

The following are various ways to regulate glycogenesis:

I **Allosteric regulation**: It is regulated by Blood glucose level. High blood glucose concentration will lead to more intracellular entry of glucose. Glycogen synthase is activated allosterically by high glucose 6-phosphate intracellularly.

II **Covalent modification of glycogen synthase.** Active glycogen synthase **"a"** is dephosphorylated, and inactive glycogen synthase "**b**" is phosphorylated.

Covalent Modification of Glycogen Synthase

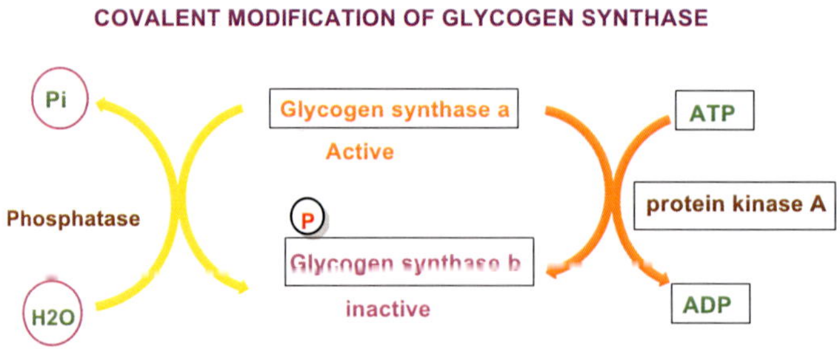

III. Hormones – insulin favors gluconeogenesis

The release of the Insulin hormone occurs in response to high levels of blood glucose (hyperglycemia). Insulin will reduce blood glucose by promoting the conversion of glucose to glycogen in various tissues, mainly the liver and skeletal muscles. To bring about glycogenesis, glycogen synthase is in an active form (dephosphorylated). Glycogen synthase activity is enhanced by insulin but decreased by adrenaline.

The hormone insulin increases the breakdown of cAMP to AMP through phosphodiesterase activity, thereby reducing the level of active PKA. This brings about a reduction in the level of phosphorylation of glycogen synthase.

The hormones glucagon and epinephrine are produced in response to low blood sugar (hypoglycemia). Muscle cells lack glucagon receptors.

IV. Regulation by calcium

Release of stored intracellular Ca^{2+} by various hormones and neurotransmitters leads to negative regulation of glycogen synthase activity.

12.3 Glycogen Storage Diseases

When various enzymes are deficient, glycogen is not properly mobilized, and abnormal forms of glycogen are deposited. These conditions are known as glycogen storage diseases. They are under

1. Type I Von Gierke's disease due to deficiency of the enzyme Glucose 6-phosphatase
2. Type II-Pompe's disease is due to a deficiency of alpah1,4 and 1,6 glucosidases.
3. Type III- Limit dextrinosis due to deficiency of debranching enzyme.
4. Type IV–Amylopectinosis due to deficiency of the enzyme branching enzyme.
5. Type V –McArdle's syndrome due to muscle phosphorylase deficiency.
6. Type VI-Her's disease due to deficiency of the enzyme liver phosphorylase.
7. Type VII deficiency of muscle phosphofructokinase

Key Takeaway

- Glycogenesis helps to maintain blood glucose.
- Glucose, when in excess, is synthesized into glycogen and stored (glycogenesis), mainly in the liver and muscles.

- The glycogen stored in muscle is broken down to glucose 1-phosphate and converted to blood glucose later
- The stored liver glycogen is broken down (glycogenolysis) into blood glucose.
- The glycogen synthesis process forms a branched structure of alpha D-glucose units linked by α-1,4 glycosidic bonds in a linear chain, and the chain branches off via α-1,6 glycosidic linkages.
- In the cells, glycogen is a complex with a glycosome protein. Glycosome is a membrane-enclosed organelle containing glycolytic enzymes. Glycogenin is an enzyme that can accept glucose units from UDPG.
- The glycosome can be in alpha-granules, beta-granules, and gamma-particles.
- Insulin secretion favors glycogen synthesis.
- Glucokinase is activated by higher blood glucose.
- Glycogen synthesis requires energy supplied by uridine triphosphate (UTP).
- Glycogen Synthase utilizes this energy to catalyze glucose incorporation into Glycogen.
- The glucose released from UDPG must be transferred to the pre-existing **"primer"** chain, with at least four glucose residues linked by a 1-4 glycosidic linkage.
- The glucose unit from UDPG is added to the glucose chain's non-reducing (outer) end.
- Thus, glycogenin can undergo self-glucosylation to generate an oligosaccharide primer.
- The hydroxyl group of the amino acid tyrosine of Glycogenin is the site where the initial glucose attaches to catalyze the formation of linear chains of alpha 1 and 4 glucose linkages.

Intext Questions

1. What is the distribution of glycogen in the liver and muscle, considering the percentage and volume of storage?
2. Why are glycogen synthesis and its breakdown not reversible processes?
3. What is the structure of glycogen?
4. What are glycogenesis and gluconeogenesis?

5. What is the complete expanded form of UDPG?
6. What is the source of energy for the formation of UDP-glucose?
7. Glycogen synthesis requires energy supplied by uridine triphosphate (UTP).
8. What is the primer chain?
9. How can glycogenin generate an oligosaccharide primer?
10. Choose one correct option for the following from the options given below (The answer is given in **bold letters**):

a The glycogen synthesis process forms a linear chain of

i. **Alpha D-glucose units linked by α-1,4 glycosidic bonds.**

ii. Beta D-glucose units linked by α-1,4 glycosidic bonds.

iii. Alpha D-glucose units linked by α-1,6 glycosidic bonds.

b The branch point is established in a linear chain of D-glucose units by glycosidic linkage. This glycosidic linkage, which forms branch points, is

i. β-1,6 glycosidic linkages. ii. **α-1,6 glycosidic linkages**.

iii. α-1,4-glycosidic linkages.

c In the cells, glycogen exists as a complex with

i. Lysosome proteins ii. Golgi proteins

iii. **Glycosome protein**

d The glycosome can be in the form of

i. **Granules** ii. Thread

iii. Mesh.

e The number of types of glycosome granules are

i. One ii. Two

iv. **Three**

f The glycosome granule types structurally are

i. alpha-granules ii. beta-granules

iii. **alpha-granules, beta-granules, and gamma-particles**

g The first step in glycogen synthesis is

i. Formation of glucose 1-phosphate

ii. **Formation of glucose 6-phosphate**

iii. Formation of glucose 1,6-diphosphate

h The first step enzyme in glycogen synthesis is

i. **Glucokinase** ii. Phosphorylase

iii. Hydrolysate

i The glycogenesis is favored by

i. Insulin hormone ii. Higher blood glucose

iii. **Both i. and ii.**

j The enzyme phosphoglucomutase in glycogen synthesis is involved in

i. Formation of glucose 6-phosphate

ii. **Formation of glucose 1-phosphate**

iii. Formation of fructose 1-phosphate

k What is the source of energy required for glycogen synthesis for the reaction using the UGPP enzyme?

i. ATP ii. **UTP**

iii. NADH

l Which one of the following is true?

i. The glucose released from UDPG must be transferred to synthesize the **"primer"** chain, with at least two glucose residues linked by 1-4 glycosidic linkage.

ii. **The glucose released from UDPG must be transferred to the pre-existing "primer" chain, with at least four glucose residues linked by 1-4 glycosidic linkage.**

iii. The glucose released from UDPG must be transferred to the pre-existing **"primer"** chain, with at least eight glucose residues linked by a 1-6 glycosidic linkage.

m Which of the following does not store glycogen?

i. **Red blood cells** ii. Liver

iii. Skeletal muscles.

Further Readings

Biochemistry, Glycogenesis. January 27, 2024. Sara C. Patino; Josephine A. Orrick. National library of medicine. National Center for Biotechnology Information.

Fundamentals of biochemistry. J.L.Jain, Sunjay Jain and Nitin Jain(2005). First edition., S.Chand & Company Ltd. New Delhi.

https://www.sciencedirect.com/topics/nursing-and-health-professions/glycogenin#:~:text=The%20glycogen%20granule%20itself%20is,as%20having%20potentially%2012%20tiers.

Life sciences. Fundamentals and practices I. sixth edition. Pranav Kumar and Usha Mina. (2017). Pathfinder publication. New Delhi

Objective Biochemistry. R.S.Senger and Reshu Chaudhary. (2014). New India publishing agency. New Delhi.

Textbook of Biochemistry for Medical Students. Vasudevan, D. M.; Sreekumar, S.; Vaidyanathan, Kanan.

The dynamic life of the glycogen granule. Clara Prats, Terry E. Graham, and Jane Shearer. J.Biol.Chem. ,2018, 293(19): 7089–7098..Published online 2018, doi:10.1074/jbc.R117.802843, PMID:PMC5949993, PMID: 29483195.https://www.ncbi.nlm.gov/pmc/articles/PMC594993/

13

Glycogenolysis

Glycogen breakdown is known as glycogenolysis. This breakdown occurs in the cytoplasm of cells of the organs, wherein glycogen is stored. The end product of glycolysis is glucose in the liver, which is glucose-1-phosphate in skeletal muscles.

Glycogen is a polysaccharide synthesized from glucose monomers. Glycogen is stored in the cell instead of glucose because glucose requires more water, whereas glycogen can be stored with much less water. Most of the stored glycogen is in the liver (5% of body weight) and muscle (1 to 2 % of body weight). Muscle and liver glycogen undergo glycolysis for different purposes. Muscle glycogen provides energy for muscular activity, whereas liver glycogen helps to maintain the blood glucose level.

Glycogen contains many glucose residues (up to 120,000) linked by alpha 1-4 glycosidic bonds, forming a straight chain. Glycogen has branches at alpha 1-6 glycosidic bonds at every ten glucose residues. At the reducing end, the carbon one of glucose is free, whereas it is not accessible in the case of non-reducing ends.

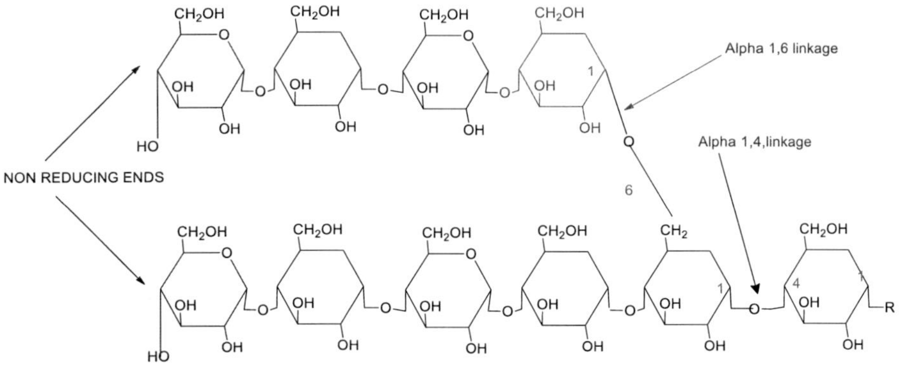

STRUCTURE OF GLYCOGEN

13.1 Steps in Glycolysis

I. Phosphorolysis/shortening of chains / Release of glucose 1-phosphate from glycogen

Hormones like glucagon and insulin activate the inactive enzyme glycogen phosphorylase. Epinephrine and glucagon are two hormones that activate glycolytic enzymes. Glucagon from the pancreas is very specific in activating only liver glycolytic enzymes. However, the epinephrine from the adrenal gland activates glycolytic enzymes in both the liver and kidneys. In other words, the liver glycolytic enzymes are activated by two different hormones independently (glucagon and epinephrine), and the kidneys are activated by a single hormone (epinephrine). The Glycogen phosphorylase breaks glucose units of glycogen linked by α-1, 4- glycosidic linkages and thus releases glucose-1-phosphate (G1P) one by one in a phosphorolysis reaction that occurs from the non-reducing end of a straight chain and does not require ATP, saving the cell energy. The phosphate used in the response is not from ATP but from the medium. Every time the release of glucose units from a non-reducing end shortens the alpha 1, 4-glycosidic glycogen chain till four glucose units are left near the branching point. This enzyme cannot act on alpha-1, 6-glycosidic linkages. The released G1P molecule, as such, cannot leave the cell.

Hydrolysis or phosphorolysis reactions can cleave the glycogen molecule into smaller molecules. Hydrolysis uses water to cleave the glycogen, whereas phosphorolysis reactions use phosphate for the same purpose. Glycogen phosphorylase employs the coenzyme pyridoxal-5'-phosphate(active form of vitamin B6).

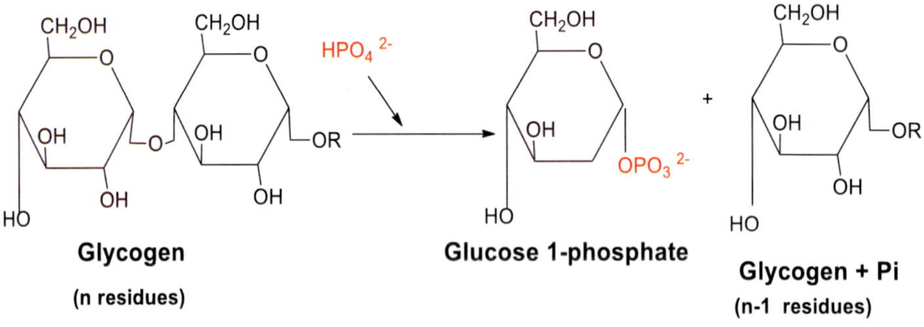

Fig. 13.1: Phosphorolysis reaction of Glycogen.

glycogen phosphorolysis

Glycogen + Pi ⟶ glucose 1-phosphate + glycogen

(n residues) **(n-1 residues)**

II. Debranching or removal of branches

When the glycogen phosphorylase stops its cleaving action, a separate enzyme called the debranching enzyme is required to remove the branches of the glycogen structure at alpha 1, 6-glycosidic linkage. The debranching enzyme has 2 independent active sites. It has two functions and is hence known as a bifunctional protein, as does the function of alpha-1-6-glucosidase, and can act as a transferase to continue the glycogenolysis process by phosphorylase. The transferase enzyme will transfer a block of 3 glycosyl units linked by alpha-1, 4- linkages from the remaining 4 glycosyl units present on one of the linear chains to another linear chain. A leftover single glucose molecule is linked by an alpha-1-6-glycosidic bond, which creates a branching point. This leftover glucose unit linked at the branch point will be acted upon by the debranching enzyme (alpha-1, 6-glucosidase) to cleave the alpha-1, 6 linkages and to release a glucose molecule. Glucose-6-phosphate (G6P) is formed from glucose either by hexokinase or glucokinase.

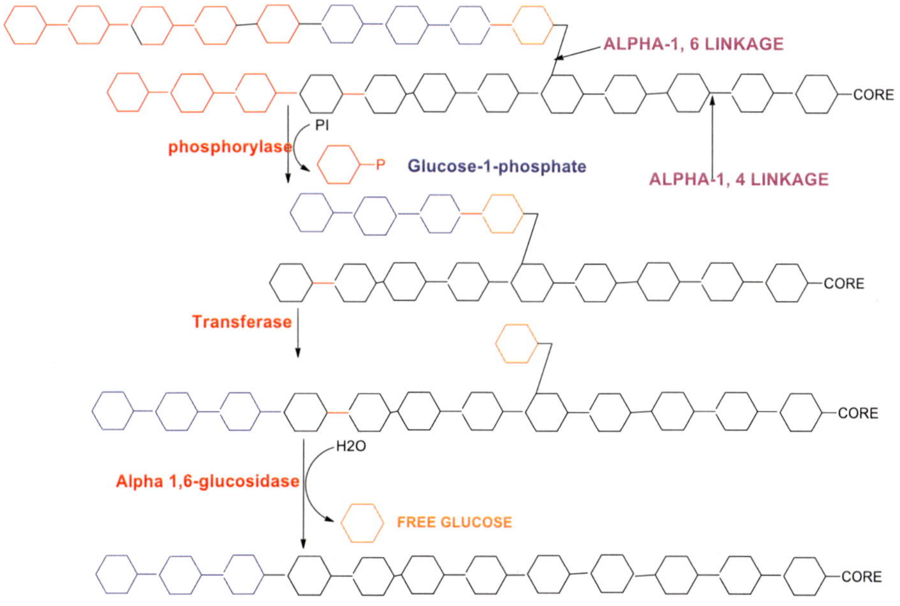

III. Recovery and Release

As a result of glycogenolysis, phosphorylated glucose is released as glucose 1-phosphate from a straight chain and a free glucose unit from a branch point. Glucose 1-phosphate is further converted to glucose 6-phosphate by the enzyme phosphoglucomutase.

The liver contains the enzyme glucose 6-phosphatase, which forms free glucose after it cleaves the phosphoryl group from glucose 6-phosphate. The free glucose enters the blood circulation. This enzyme is absent in muscle cells and will metabolize glucose 6-phosphate to either pyruvate or lactate, but not glucose.

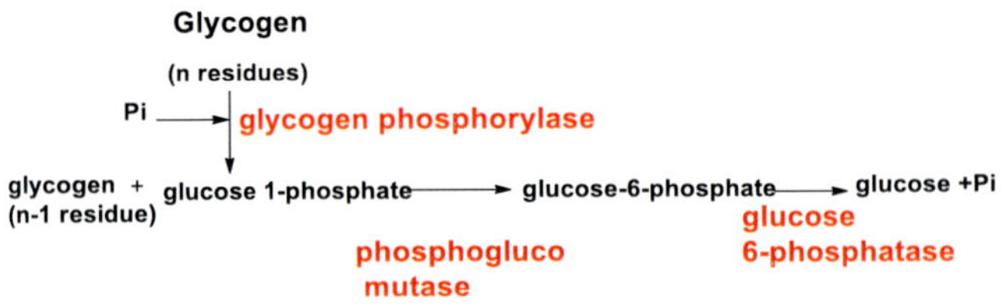

13.2 Significance of Glycogenolysis

- Glycogenolysis is a critical process to provide glucose for energy in case of the fight-or-flight response during fasting and exercise.
- Glycogenolysis in hepatocytes releases glucose into the bloodstream to deliver glucose to other cells. In other words, it contributes to regulating glucose levels in the blood and overcomes the hypoglycemic condition.
- In myocytes (muscle cells), glycogenolysis releases glucose-6-phosphate to provide energy for muscle contraction.

13.3 Regulations of Glycolysis

It depends on various factors regulating the turning on and off of glycolysis. Turning off glycolysis occurs in many ways:

- First is the exhaustion of the secretion of hormones, or when the hormones leave the receptors.
- The second step is turning off the activation of the alpha G-protein.
- Third is the breakdown of cAMP by the phosphodiesterase enzyme
- The fourth way is to inactivate phosphorylase kinase by removing phosphates and inactivating glycogen phosphorylase a, and converting it into an inactive glycogen phosphorylase b.

I. **Enzyme** Two mechanisms can activate the enzyme Phosphorylase kinase.

- When it gets phosphorylated by Protein Kinase A (PKA).
- Activation of phosphorylase kinase by calcium binding to the protein calmodulin. (a subunit of phosphorylase kinase). Calcium ion also stimulates muscular contraction and activate phosphorylase kinase. The phosphorylase kinase activates glycogen phosphorylase to release G1P from glycogen.

II. **Nerve signals**-- fight-or-flight response under stressful conditions through the sympathetic nervous system. The resultant effect is the release of the hormones --- epinephrine (adrenaline) and norepinephrine (noradrenaline), which will stimulate glycogenolysis.

III. **Calcium**. When a muscle gets stimulated, the sarcoplasm releases calcium, which binds to calmodulin (Calmodulin is a subunit of phosphorylase kinase) and activates phosphorylase kinase, which activates glycogen phosphorylase. Thus, calcium activates glycogenolysis and inhibits synthesis. It can be summarized as follows.

stimulus→Muscle →sarcoplasm→release calcium→binds to calmodulin→activate phosphorylase kinase→activate glycogen phosphorylase→-glycogenolysis occurs and inhibits glycogen synthesis.

IV. **Allosteric factors and covalent modification**. Glycogen Phosphorylase is inhibited by allosteric factors/high-energy substrates like ATP, Glucose 6-phosphate, and glucose. Glycogen Phosphorylase is activated by low-energy substrates, including AMP and others that activate. The enzyme uses a cofactor, pyridoxal phosphate (PLP).

V. **Substrate availability**: Glycogenolysis occurs when the body needs glucose and stops when there is enough glucose to fulfill its needs.

VI. **Hormones**--- Insulin, glucagon, and epinephrine/adrenaline are involved in regulating blood glucose levels.

- Insulin lowers blood glucose levels. It inhibits glycogenolysis via dephosphorylation and thus results in the inactivation of glycogen phosphorylase.
- Glucagon –increases blood glucose levels. Glucagon works preferentially on the liver.
- Epinephrine/adrenaline --- produced during stress, threat, fear, fight-or-flight response stimulates glycogenolysis and inhibits glycogenesis. Epinephrine exerts its most significant effects on muscle.

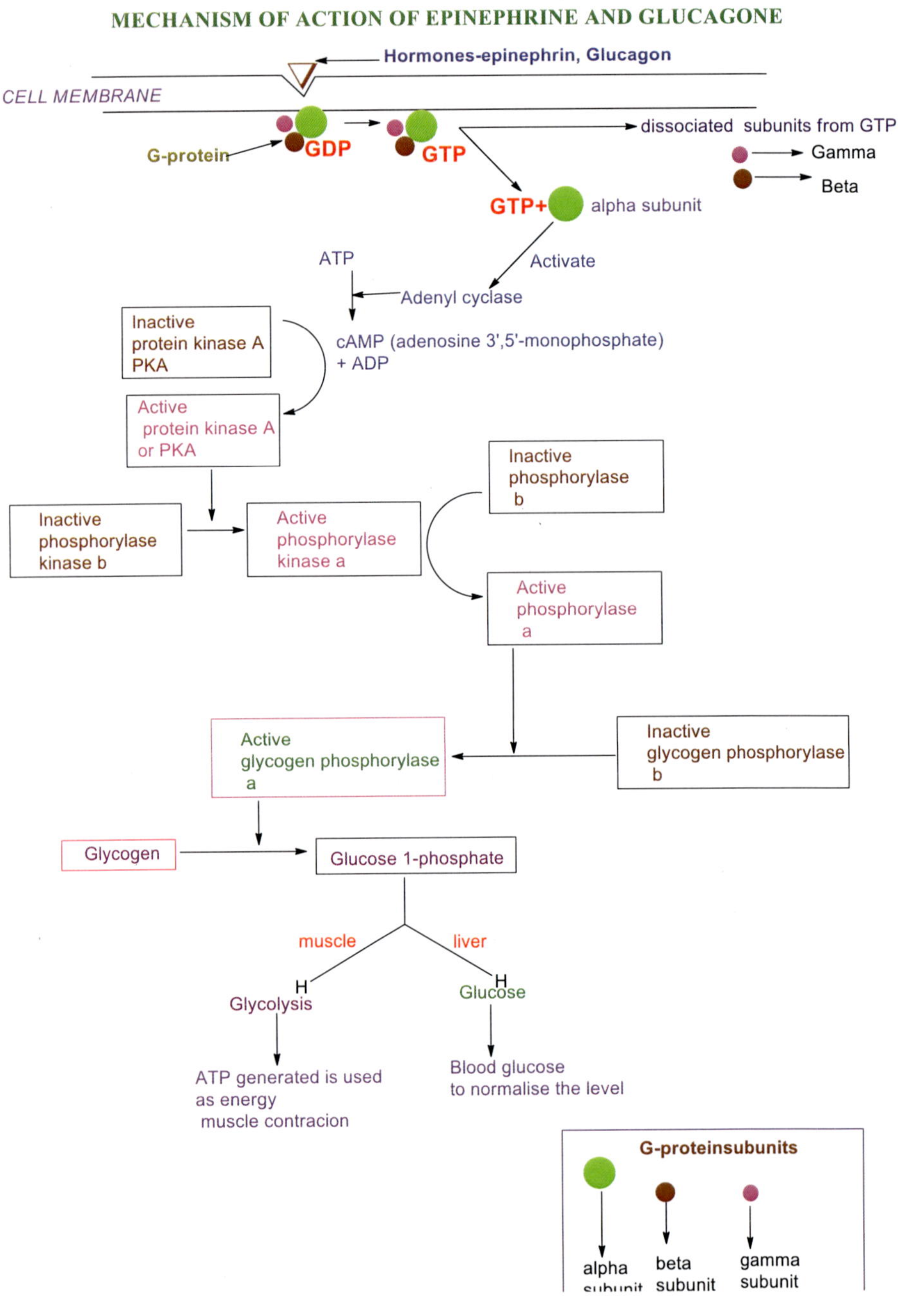

Fig. 13.3: Mechanism of action of epinephrin and glucagon

The above diagram explains that Epinephrine (or glucagon), released during 'fight or flight', when attached to the receptors, stimulates adenylate cyclase and cAMP formation so that it can activate protein kinase A. The protein kinase A will activate phosphorylase kinase from the inactive state to continue glycogenolysis to release G1P. The fate of released glucose 1-phosphate during glycolysis is different both in the liver and in the muscle. Liver enzyme glucose 6-phosphatase forms the free glucose to enter the blood circulation. The skeletal muscle will metabolize glucose 6-phosphate to either pyruvate or lactate, not glucose.

Key Takeaway

- Glycogen breakdown is known as glycogenolysis.
- This breakdown occurs in the cytoplasm of cells where glycogen is stored. The main organs of storage are the liver and muscle. To some extent, it can occur in the kidneys.
- The end product of glycolysis is glucose in the liver, which is glucose-1-phosphate in skeletal muscles.
- Glycogen is stored in the cell instead of glucose because glucose requires more water, whereas glycogen can be stored with much less water.
- Muscle glycogen provides energy for muscular activity, whereas liver glycogen helps to maintain the blood glucose level.
- At the reducing end, the carbon one of glucose is free, whereas it is not accessible in the case of non-reducing ends.
- Hormones like glucagon and insulin activate the inactive enzyme glycogen phosphorylase
- Glucagon from the pancreas is very specific in activating only the liver glycolytic enzymes.
- Epinephrine from the adrenal gland activates glycolytic enzymes both in the liver and kidneys.
- The Glycogen phosphorylase breaks glucose units of glycogen linked by α-1, 4- glycosidic linkages to release glucose-1-phosphate (G1P) one by one.
- A phosphorolysis reaction from the non-reducing end of a straight chain does not require ATP and saves the cell energy.
- Debranching enzyme removes the branches of glycogen structure at alpha 1, 6-glycosidic linkage.
- The debranching enzyme has two functions and is hence known as a bifunctional protein.

- The transferase enzyme will transfer a block of 3 glycosyl units linked by alpha-1, 4- linkages
- The leftover glucose unit at the branch point is acted upon by the debranching enzyme (alpha-1, 6-glucosidase) to release a glucose molecule.
- Glycogen phosphorylase employs the coenzyme pyridoxal phosphate.
- Due to glycogenolysis, phosphorylated glucose is released as a glucose 1-phosphate from a straight chain and a free glucose unit from a branch point.
- the enzyme glucose 6-phosphate, which cleaves the phosphoryl group from glucose 6-phosphate to get free glucose,
- This enzyme is absent in muscle cells and will metabolize glucose 6-phosphate to either pyruvate or lactate, but not glucose.
- Due to glycogenolysis, phosphorylated glucose is released as a glucose 1-phosphate from a straight chain and a free glucose unit from a branch point.
- Glucose 6-phosphatase is present in the liver but absent in muscles.
- Liver glucose 6-phosphate cleaves the phosphoryl group from glucose 6-phosphate to get free glucose.
- Glucose 6-phosphate is absent in muscle cells; it will metabolize glucose 6-phosphate to either pyruvate or lactate, not glucose.
- In muscle cells, activation of phosphorylase kinase by calcium occurs when it binds to the protein calmodulin.
- The fight-or-flight response releases the hormones namely epinephrine (adrenaline) and norepinephrine (noradrenaline).
- Allosteric factors/ high-energy substrates inhibit Glycogen Phosphorylase. The enzyme uses a cofactor, pyridoxal phosphate (PLP).
- Epinephrine stimulates adenylate cyclase and cAMP formation to activate protein kinase A.

Intext Questions

1. What is glycogenolysis? Which organs perform this process?
2. Why do the glycogenolysis end products differ in the liver and muscles?
3. Which hormones are released during the Fight-or-flight response? Why?
4. Write the actions of various hormones concerned with regulating glycogenolysis.

5. How does calcium regulate glycogenolysis?
6. Which allosteric factors regulate glycogenolysis?
7. What is the importance of glycogenolysis?
8. How is phosphorylase kinase activated?
9. Why is phosphorolysis preferable over hydrolysis?
10. Why does hydrolysis of glycogen not require hydrolysis of ATP?
11. Explain that "Glycogen phosphorylase alone cannot complete glycogenolysis.
12. Explain the functions of all enzymes engaged in glycogenolysis.
13. How does the glucose 1-phosphate get converted to glucose in glycolysis?
14. Why is the debranching enzyme considered a bifunctional protein?
15. Draw a sketch diagram explaining the mechanism of action of the epinephrine hormone.
16. Choose one correct option for the following from the options given below (The answer is given in **bold** letters):

a The enzyme that breaks α-1, 4- glycosidic linkages of glucose in the glycogen chain is

i. **Glycogen phosphorylase** ii. Glycogen phosphokinase

iii. Glycogen dephosphorylate

b The enzyme that breaks glucose units of glycogen linked by α-1, 4-glycosidic linkages release.

i. Glucose 6-phosphate ii. **Glucose-1-phosphate**

iii. Glucose 1,6-diphosphate

c The number of glucose units released during glycogenolysis by the action of the enzyme glycogen phosphorylase

i. Three units ii. Two units

iii. **One unit**

d The hormone that activates glycolytic enzymes is present both in the liver and kidneys.

i. Glucagon ii. **Epinephrine**

iii. Cortisol

e The hormone that activates glycolytic enzymes is present in the liver.

i. **Glucagon** ii. Epinephrine

iii. Cortisol

f The hormone Epinephrine activates glycolytic enzymes in

i. Liver ii. Kidneys

iv. **Both i. and ii.**

g The hormone Glucagon activates glycolytic enzymes in

i. **Liver** ii. Kidneys

iii. Erythrocytes

h The epinephrine hormone is released from

i. Liver ii. **Adrenal gland**

iii. Pituitary gland

i Glucagon hormone is released from

i. Liver ii. **Adrenal gland**

iii. Pancreas

j What is calmodulin?

i. Hormone

ii. **Binding protein for calcium** iii. Enzyme in glycogenolysis

k Which of the following statements is correct for glycogenolysis?

i. **In glycogenolysis, glucose 1-phosphate is released from a straight chain, and a free glucose unit from a branch point.**

ii. In glycogenolysis, glucose 6-phosphate is released from a straight chain, and glucose 1-phosphate is released from a branch point.

iii. In glycogenolysis, glucose 6-phosphate is released from a straight chain, and a free glucose unit from a branch point.

l The end product of glycolysis is glucose due to the action of the enzyme

i. **Glucose 6-phosphatase**

ii. Phosphoglucomutase

iii. Glucose epimerase

m How many glucose residues are away from a branching point when glycogen phosphorylase stops cleaving the glycogen?

i. **Four glucose residues.**

ii. Two glucose residues

iii. Six glucose residues.

n The primary purposes of glycogenolysis in the liver and muscles are

i. **To provide free glucose and energy for work.**

ii. To provide energy for work and to regulate blood glucose

iii. To reduce cholesterol and increase blood sugar.

o Phosphoglucomutase converts

i. **Glucose 1-phosphate to glucose 6-phosphate.**

ii. Glucose 1-phosphate to glucose.

iii. Glucose 6-phosphate to glucose 1-phosphate.

p Epinephrine forms cAMP with the help of adenylate to activate

i. **Protein kinase A.**

ii. Transferring of glucose units.

iii. Formation of glucose 6-phosphate

Further Readings

A textbook on animal physiology. A, K, Berry. (1997). Sixth Edition. Emkay Publication, Delhi.

Biochemistry, Glycogenolysis. Manuel A. Paredes-Flores; Nader Rahimi; Shamim S. Mohiuddin. National Library of Medicine. https://www.ncbi.nlm.nih.gov/books/NBK554417/

Biochemistry. Naik, Pankaj (2009). Second Edition, Jaypee Brothers, New Delhi.

Cell Biology. E.D.P. De Robertis, Wiktor W. Nowinski, and Franciso A. Saez (1970). Fifth Edition, WB Saunders Company. Philadelphia, London, Toronto.

Fundamentals of biochemistry. J.L.Jain, Sunjay Jain and Nitin Jain(2005). First edition., S.Chand & Company Ltd. New Delhi.

Glucose and glycogen in the diabetic kidney. Heroes or villains. Mitchell A., Sullivans. Josephine M. Forbes. (2019). https://www.thelancet.com/article/S2352-3964(19)30509-2/fulltext.

Glycogenolysis. https://www.britannica.com/science/glycogenolysis

Life sciences. Fundamentals and practices I. sixth edition. Pranav Kumar and Usha Mina. (2017). Pathfinder publication. New Delhi

Pretest Physiology. James P. Ryan and Ronald F. Tuma (1996), Eighth Edition, McGraw-Hill, New York, Milan Montreal New Delhi.

Textbook of Biochemistry for Medical Students. Vasudevan, D. M.; Sreekumar, S.; Vaidyanathan, Kanan (2019). Ninth Edition, Jaypee Bros. Medical Publication, New Delhi.

Textbook of Veterinary Biochemistry. Dhanotiya R.S.(2006).Reprint of the second edition. Jaypee Brothers, New Delhi.

14

Enzyme Deficiencies and Carbohydrate Metabolism

Carbohydrates (glucose) are the primary energy source for a body; when digested, glucose is the end product in human beings and simple stomach animals, but it yields volatile fatty acids in ruminants. Precise regulation of levels of blood sugars/glucose in the body is the net result of various processes like digestion, absorption, and metabolism operating under the interaction of various a) enzymes and b) hormones secreted by the different organs of the body. An abnormal secretion of any of the enzymes or hormones associated with the metabolism of carbohydrates disrupts the precise regulation of blood sugar, leading to various disorders.

14.1 Enzyme Deficiency and Disorders in Carbohydrate Metabolism

Usually, various enzymes are involved in the breakdown of carbohydrates, and if their secretion is insufficient, missing, or does not work correctly, it leads to different disorders.

I. Glycogen storage diseases (GSDs, glycogenosis, and dextrinosis)

Glucose is needed by the cells for the generation of energy and other cellular activities. If glucose happens to be in excess, it is stored as glycogen, mainly in the liver and muscles, for future needs. The stored glycogen can be converted back to glucose when the need arises. Many enzymes are related to this inter-conversion of glucose and glycogen. Deficiencies of one or more enzymes may lead to abnormal glycogen metabolism, leading to glycogen accumulation of normal or abnormal chemical structure in tissue, collectively called glycogen storage diseases (GSDs). Roman numerals identify different types of glycogen storage conditions. They are also classified under various categories like inherited metabolic diseases, autosomal recessive type, and X-linked type (VIII/IX type).

Table 14.1: Types of Glycogen Storage Diseases (GSDs)

GSD type Number	**Eponym**	**Deficient Enzyme (organs)**	**Role of the enzyme**
Type 0	(Lewis disease)	Glycogen synthase(GSY) (muscle—GSY1, liver—GSY2,). Autosomal recessive disorder.	Age group: infancy and early childhood. Moderately low glycogen stores in the Liver. Fasting hypoglycemia. Ketones in urine. Post-prandial hyperglycemia.
Type I-- (I A and IB)	Von Gierke's disease	Deficiency of Glucose-6-phosphatase occurs in the Liver and kidneys.	Autosomal recessive gene disorder. Deficiencies of these enzymes cause the storage of excess amounts of glycogen and fats in the body tissues. The body cannot break down liver glycogen or produce free glucose in this situation. This results in severe hypoglycemia and lactic acidosis.
Type II	Pompe's disease	Acid alpha-glucosidase (also called as Lysosomal alpha-glucosidase **α-1**, 4-glucosidase or acid maltase (all organs).	It is a recessive metabolic disorder/deficiency related to the enzyme. Glycogen is not degraded, so it accumulates within lysosomes and the cytoplasm, resulting in tissue destruction and muscle weakness.
Type III	Forbes-Cori disease	Amylo-1-6-glucosidase or Glycogen debranching enzyme	It is an inherited autosomal recessive trait. There is deposition of excess glycogen, which becomes toxic to the liver, the muscles, and the heart.
Type IV	Anderson's disease	Glycogen branching enzyme (liver, muscle, and most tissues)	It is an autosomal recessive gene trait. Here, glycogen is abnormally formed and such glycogen is deposed in the liver, muscle, and other tissues. The glycogen is structurally abnormal as it lacks the standard branched structure.
Type V	McArdle's disease	Muscle glycogen phosphorylase (Muscle)	Physiologically, muscle glycogen is broken down to provide energy during muscle contraction. Due to the deficiency of muscle glycogen phosphorylase. muscle glycogen is not broken down.
Type VI	Hers' disease	Liver glycogen phosphorylase (Liver)	Liver glycogen phosphorylase degrades glycogen to glucose, which contributes to regulating blood glucose levels. Failure of this enzyme impairs the breakdown of glycogen in the liver and glucose regulation in the blood. This results in hypoglycemia.

GSD type Number	Eponym	Deficient Enzyme (organs)	Role of the enzyme
Type VII	Tarui's disease	phosphofructokinase-PFK (Muscle, RBC)	It is needed for glycolysis. The enzyme deficiency accumulates glycogen in the muscle tissues.
Type IX	-	Phosphorylase kinase The gene involved is PHKA1 (muscle).PHKA2, PHKB, PHKG1, PHKG2 (liver), CALM1, CALM2, CALM3.	The enzyme in normal individuals helps in glycogen breakdown. The deficiency of enzymes affects the excessive amount of glycogen that is deposited. The deficient cells cannot break glycogen stored in the organs.
Type XI	Fanconi-Bickel syndrome(FBS)	Glucose transporter, GLUT2	It is an autosomal recessive disorder. GLUT2 is a glucose transporter from blood to tissues. Its deficiency affects glucose transport.
Type XII	Red cell aldolase A deficiency or ALDOA deficiency	Aldolase A or fructose-bisphosphate aldolase,	It is found in red blood cells and, if deficient, causes hemolytic anemia. It is also found in muscle tissue and, if insufficient, causes myopathy and hypoglycemia.
Type XIII	β enolase deficiency	Beta enolase or phosphopyruvate hydratase	It appears as an inherited autosomal recessive pattern. It is involved in the catalysis of converting 2-phosphoglycerate (2-PG) to phosphoenolpyruvate (PEP). It causes myalgia in deficiency conditions.

II Galactosemia

Galactose (a monosaccharide sugar) is one of the components of lactose sugar found in milk and dairy products. Lactose, when digested, releases its component two monosaccharides - galactose and glucose.

Galactosemia is characterized by inadequate utilization, resulting in a higher level of galactose in the blood. This condition has three types due to a deficiency or lack of one of the specific enzymes necessary for metabolizing galactose. Type 1 or Classic galactosemia is the most common and most severe form. It causes liver damage, kidney problems, and underdevelopment of the brain. It occurs when there is a deficiency (due to a genetic condition) of galactose-1-phosphate uridyl transferase.

Type 2 and Type 3 galactosemia are due to galactokinase deficiency and UDP galactose epimerase, deficiency, respectively. Type 2 is a milder form, and Type 3 is rare. The galactosemic condition is concerned with cataracts and damage to the eye lens.

III. Lactose Intolerance

Lactose sugar in milk and dairy products is digested physiologically by an enzyme called 'lactase' secreted into the small intestine. Lactose, when digested by lactase, releases two simple sugars, glucose and galactose. These sugars, after absorption from the intestine, enter the bloodstream. The lactose-tolerant patient cannot digest lactose due to a lactase enzyme deficiency. Undigested lactose draws water into the small intestine and causes diarrhea. Undigested lactose can travel to the large intestine, fermented by the bacteria to release acids and gases. It is safe to avoid foods containing lactose.

Lactose intolerance can be distressing and may be present right at birth (congenital/inherited) or develop at any time following enteritis or chemotherapy. Lactose intolerance, which occurs after gastroenteritis, is temporary and often improves as the gut lining heals, which takes several weeks. In lactase deficiency, undigested lactose from the small intestine passes into the colon, where lactose is fermented by bacteria-producing gas.

The Hydrogen breath test confirms this condition in humans, wherein more Hydrogen gas is produced.

IV. Hereditary fructose intolerance (HFI) or fructosemia or fructose aldolase B-deficiency or fructose 1,6-biphosphate aldolase deficiency

Fructose is a ketone monosaccharide sugar mainly found in fruits and table sugar. The fructose intolerance may be 1. dietary fructose intolerance or fructose malabsorption 2. Hereditary fructose intolerance.

In the case of dietary fructose intolerance, fructose is not absorbed from the intestine due to a deficiency of fructose carriers in the small intestine's enterocytes. This results in higher fructose in the intestine. Hereditary fructose intolerance is quite different from dietary fructose intolerance. Hereditary fructose intolerance is a genetic and fatal condition with an enzyme deficiency that metabolizes fructose.

There are three Aldolase isozymes: Aldolase A, B, and C. Aldolase A is found in muscle, Aldolase B in the liver, and Aldolase C in the brain. Undigested fructose, then accumulated in the body, may damage the liver (hepatomegaly, cirrhosis) and kidneys.

Aldolase B (ALDOB) plays a vital role in glycolysis and gluconeogenesis. Aldolase B breaks six-carbon fructose-1-phosphate into two molecules, each having three carbons: glyceraldehyde and dihydroxyacetone phosphate. Deficiency of ALDOB leads to the accumulation of fructose-1-phosphate in the liver, kidney, and small intestine, leading to fructose intolerance. As a result, phosphorus is not available to the body, leading to low ATP levels. This results in the liver inhibiting glycolysis and gluconeogenesis, resulting in hypoglycemia. Avoiding foods containing sugars like fructose, sucrose, and sorbitol is better in this condition.

V. Congenital sucrose intolerance or genetic/congenital sucrase-isomaltase deficiency (GSID/CSID) or SI deficiency or disaccharides intolerance I

This condition is a genetic disorder related to the inability to digest sugars such as sucrose and maltose. Sucrose and maltose are disaccharides (made of two simple sugars or monosaccharides). Usually, they get digested in the small intestine by enzymes, broken down into their monosaccharide components, which are absorbed and added into the bloodstream for body use.

Sucrose or table sugar is found in fruits, sugarcane, sugar beet, palms, and sugar maple. Sucrose is converted into glucose and fructose, and the sucrase enzyme releases its simple sugar components upon digestion.

Maltose, or malt sugar found in grains, is made up of two glucose molecules as its monosaccharide components, which are released by the action of iso-maltase.

A reduction or absence in activities of these two digestive enzymes (sucrase and iso-maltase) can occur due to an autosomal-recessive minor intestinal disorder resulting in congenital sucrose intolerance or SI deficiency. Under this condition, it is impossible to digest sucrose and maltose sugars.

VI. Disorders of pyruvate metabolism or pyruvate dehydrogenase complex deficiency (PDCD)

Typically, pyruvic acid, formed in the metabolism of carbohydrates and proteins after entering the mitochondrial matrix, is oxidized to acetyl-CoA with the help of the pyruvate dehydrogenase (PDH) multi-enzyme complex.

Overall aerobic conversion reaction will be as follows:

Pyruvate + HS-CoA + NAD —>Acetyl CoA +NADH +CO_2

If pyruvate is not utilized, it undergoes anaerobic conversion to lactic acid.

The PDH-enzyme complex is constituted of three enzymes and five cofactors. The three enzymes present in the enzyme complex are namely

1. Pyruvate dehydrogenase (decarboxylate or removes carbon dioxide from pyruvate)
2. Dihydrolipoyl transacetylase (remaining substrate is transferred to coenzyme A)
3. Dihydrolipoyl dehydrogenase (transfers the H2 to NAD^+).

A deficiency of one of the above three enzymes can cause pyruvate dehydrogenase complex deficiency. The deficiency condition in most individuals is related to an abnormality in the PDHA1 gene located on the X chromosome. An autosomal recessive inheritance may occur in a smaller percentage of affected individuals. In metabolic PDH deficiency, lactic acidosis occurs. Various neurological problems and structural abnormalities in the central nervous system characterize other cases. Individuals usually die in early childhood if PDCD symptom onset occurs in the prenatal period or infancy.

VII. Pyruvate carboxylase deficiency

This biotin-dependent enzyme forms oxaloacetate, which is required as an intermediate to run the citric acid cycle. Oxaloacetate is produced from pyruvate through the action of pyruvate carboxylase.

It is rare to notice the occurrences of inborn errors of metabolism related to pyruvate carboxylase deficiency. The PC deficiency is inherited as an autosomal recessive genetic condition. It is of different types: Type A (infantile form - moderately severe), Type B (severe neonatal form), and Type C (affecting the nervous system).

VIII. Pyruvate kinase (PK) deficiency

Pyruvate kinase has four isozymes, e.g., L (liver), R (erythrocytes), M1 (muscles, heart, and brain), and M2 (detected in tissues). Its deficiency is an inherited (autosomal recessive) disorder characterized by premature red cell hemolytic anemia.

This enzyme is essential in glycolysis. Pyruvate kinase directly transfers phosphate from Phosphoenolpyruvate (PEP) to ADP and forms ATP and pyruvate. The red blood cells deficient in Pyruvate kinase remain deficient in ATP. As a result, red blood cells undergo hemolysis (hemolytic anemia), affecting the oxygen supply to the body's tissues. The hemolytic anemia is of the hereditary non-spherocytic type.

IX. Hexokinase deficiency

Hexokinase is responsible for the rate-limiting first step of glycolysis in erythrocytes, which involves the phosphorylation of glucose to glucose-6-phosphate. Enzyme deficiency is a genetic, autosomal recessive disease (HK1 gene). Hemolytic anemia is a nonspherocytic (NSHA) type that is more severe than pyruvate kinase deficiency. The NSHA is caused by a shortage of 2,3 DPG, which usually facilitates tissue oxygenation.

X. Glucose 6-phosphate isomerase (GPI) deficiency

Mutations in the GPI gene are rare to cause GPI deficiency. The GPI enzyme plays a role in glycolysis for the reversible isomerization of D-glucose 6-phosphate to D-fructose 6-phosphate or the interconversion of glucose 6-phosphate and fructose 6-phosphate. This condition causes premature hemolysis of red blood cells, which causes chronic hemolytic anemia. Here, red blood cells' ability to maintain their structure is impaired, causing hemolysis.

XI. Transketolase deficiency

Transketolase deficiency is a genetic condition inherited from both parents in an autosomal recessive manner.

It is an enzyme involved in the pentose phosphate pathway of carbohydrate metabolism and connects the pentose phosphate pathway to glycolysis. A deficiency of thiamine causes transketolase deficiency, as the enzyme requires. Thiamine (vitamin B1) is Thiamine pyrophosphate, which is a cofactor. This enzyme transfers a 2-C fragment from xylulose-5-phosphate to either ribose-5-phosphate or erythrose-4-phosphate.

Key Takeaway

- Carbohydrates, when digested, glucose is the end product in human beings and simple stomach animals, but it yields volatile fatty acids in ruminants.
- Glucose in excess is stored as glycogen, mainly in the liver and muscles, for future needs.
- Deficiencies of one or more enzymes may lead to abnormal glycogen metabolism, leading to glycogen accumulation of normal or abnormal chemical structure in tissue, collectively called glycogen storage diseases (GSDs).
- Roman numerals identify different types of glycogen storage conditions.
- Glycogen storage diseases are classified under various categories, like inherited metabolic diseases, autosomal recessive type, and X-linked type.
- The enzyme glycogen synthase deficiency occurs in Glycogen storage disease type 0 (GSD 0).
- GSD1 is also known as von Gierke's disease. The enzyme deficient is Glucose-6-phosphatase.
- GSD1 condition results in storing excess glycogen and fats in the body tissues.
- GSD2 is also known as Pompe Disease. Glycogen deposits inside lysosomes.
- Glycogen debranching enzyme deficiency is the cause of GSD III. It is also known as Forbes-Cori disease.
- Glycogen branching enzyme deficiency is related to the GSD IV condition. Here, glycogen synthesized is abnormally formed.
- McArdle's disease is a GSD V condition due to a deficiency of muscle glycogen phosphorylase. Muscle glycogen is not broken down.
- Liver glycogen phosphorylase deficiency is found in the GSD VI condition. It is also Hers' disease.

- Phosphofructokinase deficiency is type GSD VII. It is also known as Tarui's disease.
- Type GSD IX is due to Phosphorylase kinase deficiency.
- Fanconi-Bickel syndrome (FBS) is another name for GSD type XI. There is a deficiency of glucose transporter GLUT2.
- Type GSD XII is due to Aldolase A deficiency or ALDOA gene defect in red blood cells. It causes hemolytic anemia.
- Type GSD XIII is due to beta enolase deficiency.
- Galactosemia results in a higher level of galactose in the blood.
- Type 1 or Classic galactosemia is the most common and severe form.
- Deficiency (due to a genetic condition) of galactose-1-phosphate uridyl transferase is seen in type I galactosemia.
- The lactose-tolerant patient cannot digest lactose due to a lactase enzyme deficiency. It is safe to avoid foods containing lactose.
- There are three Aldolase isozymes: Aldolase A, B, and C. Aldolase A is found in muscle, Aldolase B in the liver, and Aldolase C in the brain.
- Sucrose is found in fruits, and Maltose is found in grains.
- The PDH-enzyme complex is constituted of three enzymes and five cofactors.
- Pyruvate carboxylase is a biotin-dependent enzyme that forms oxaloacetate, an intermediate to run the citric acid cycle.
- Pyruvate kinase has four isozymes. Its deficiency in the red blood cells causes hemolytic anemia.

In Text Questions

1. How do glycogen storage diseases occur?
2. How are glycogen storage diseases classified?
3. Name the enzyme deficient in glycogen storage disease type 0.
4. Why is hyperglycemia seen during post-prandial time?
5. What does GST1 stand for?
6. What is the other name of GSD1? Which enzyme is deficient?
7. What happens in GSD II disease?
8. Which condition is caused by glycogen deficiency of the debranching enzyme?

9. What will happen if the glycogen branching enzyme deficiency prevails?
10. What is the GSD V condition?
11. Name the enzyme deficient in Hers' disease.
12. What happens in the GSD VII condition? Why?
13. Phosphorylase kinase deficiency is seen in which type of GSD?
14. Which type of GSD is Fanconi-Bickel syndrome? Why is glucose transport affected in this disease?
15. Why does the hemolytic anemia condition occur in GSD type XII?
16. Which type of GSD is beta enolase or phosphopyruvate hydratase deficiency found in?
17. Define the term galactosemia. Explain its types with the deficient enzymes.
18. Define the term fructose intolerance. How do its dietary and hereditary types differ?
19. How does pyruvate kinase deficiency in erythrocytes affect the oxygen supply to the tissues?
20. Nonspherocytic hemolytic anemia is caused by a deficiency of hexokinase, which creates a shortage of 2,3 DPG, which usually facilitates tissue oxygenation.
21. Choose one correct option for the following from the options given below (The answer is given in **bold** letters):

a Which enzyme deficiency causes type 0 GSD?

i. Glucose 6-phosphatase ii. **Glycogen synthase**

iii. α-1, 4-glucosidase

b Which numeral system identifies different types of glycogen storage conditions?

i. **Roman numerals** ii. Binary numerals

iii. Decimal numerals

c GSD1 is also known as

i. Pompe's disease ii. Lewis disease

iii. **Von Gierke's disease**

d In GSD II, glycogen accumulates in

i. **Lysosomes.** ii. Nucleus

iii. Ribosomes

e Glycogen debranching enzyme deficiency is the cause of the condition

i. GSD II. ii. GSD 0.

iii. **GSD III.**

f Enzyme deficiency in **Von Gierke's disease is**

i. Glycogen synthase ii. **Glucose 6-phosphatase**

iii. α-1, 4-glucosidase

g The enzyme deficiency is related to the GSD IV condition.

i. Glycogen transferring enzyme. ii. **Glycogen branching enzyme**

iii. Glycogen debranching enzyme.

h A deficiency of muscle glycogen phosphorylase occurs in

i. **McArdle's disease** ii. GSD IV condition

iii. Pome's disease

i The GSD VI condition is due to a deficiency of enzyme

i. **Glycogen phosphorylase** ii. Glycogen phosphatase

iii. Glucose phosphorylase

j Type GSD VII occurs due to a deficiency of enzyme

i. Glycogen phosphorylase ii. **Phosphofructokinase**

iii. Glucose 6-phosphatase

k Type GSD IX has a deficiency of which enzyme?

i. Phosphofructokinase ii. **Phosphorylase kinase**

iii. Glycogen phosphorylase

l Glucose transporter GLUT2 deficiency occurs in which type of GSD?

i. GSD III ii. GSD VII

iii. **GSD XI**

m Which type of GSD does hemolytic anemia occur in?

i. **GSD XII** ii. GSD IX

iii. GSD II

n Beta-enolase deficiency is observed in which type of GSD?

i. **XIII** ii. XI

iii. III

o Galactose levels, when higher in the blood, cause the condition

i. Galactosuria ii. **Galactosemia**

iii. Galactoinsemnia

p The enzyme deficient in type I galactosemia is

i. **Galactose-1-phosphate uridylyl transferase**

ii. Glucotransferase

iii. Galactose epimerase

q Which type of GSD is Tarui's disease?

i. GSD IX ii. **GSD VII**

iii. GSD XIII

r Which is the most severe form of galactosemia?

i. **Type 1** ii. Type 2

iii. Type 3

s How many Aldose isozymes are present

i. Five ii. Two

iii. **Three**

t Hexokinase deficiency-related Nonspherocytic hemolytic anemia causes a shortage of a product in R.B.Cs. that usually facilitates tissue oxygenation.

i. **2,3 Diphosphoglycerate.** ii. Fructose 1,6-diphosphate

iii. UDP-galactose

Further Readings

A Textbook of Biochemistry. Berry, A.K. (2009), sixth edition, Emkay Publishers, Delhi.

Biochemistry. Naik, Pankaj (2009). Second Edition, Jaypee Brothers, New Delhi.

Fundamentals of biochemistry. J.L.Jain, Sunjay Jain and Nitin Jain (2005). First edition., S.Chand & Company Ltd. New Delhi.

https://health.clevelandclinic.org/what-is-fructose-intolerance/

https://my.clevelandclinic.org/health/diseases/15553-glycogen-storage-disease-gsd

https://rarediseases.info.nih.gov/diseases/2125/glycogen-storage-disease-type-13

https://rarediseases.info.nih.gov/diseases/600/glycogen-storage-disease-type-12

https://rarediseases.org/gard-rare-disease/nonspherocytic-hemolytic-anemia-due-to-hexokinase-deficiency

https://rarediseases.org/rare-diseases/glycogen-storage-disease-type-i/

https://rarediseases.org/rare-diseases/pyruvate-dehydrogenase-complex-deficiency/

https://tktdeficiency.com/

https://www.agsdus.org/type-ix.php

https://www.betterhealth.vic.gov.au/health/conditionsandtreatments/lactose-intolerance

Life sciences. Fundamentals and Practice I. Pranav Kumar and Usha Mina (2006). Sixth Edition. Pathfinder publication. New Delhi.

Objective Biochemistry. Sengar, R.S. and Chaudhary Rishu (2014). New India Publishing Agency. New Delhi.

Textbook of Veterinary Biochemistry. Dhanotiya R.S.(2006). Second edition.Jaypee Brothers, New Delhi.

15

Hormonal Disturbances and Carbohydrate Metabolism

The number of hormones secreted by the body exhibits control over the physiological regulation of glucose sugar level (homeostasis) in the manner needed by the body. Depending upon the situation, hormones may increase or decrease the level of glucose sugar present in the blood or tissues. The glucose is oxidized to yield energy, carbon dioxide, and water.

Glucose, more than the body's needs, is stored or converted mainly to glycogen and, to some extent, protein and fat. If glucose is less than the need of the body, its denovo synthesis takes place from amino acids or fatty acids, or is formed from the breakdown of stored glycogen. Various hormones are secreted by multiple endocrine glands to control the blood glucose level. If it fails, it leads to hypoglycemia or hyperglycemia conditions. Gluconeogenesis is not developed at birth in the case of a baby pig. Therefore, hypoglycemia in baby pigs is observed within 2 days of age if starved, leading to hypoglycemia, weakness, confusion, and death.

15.1 The Hormones Involved in Homeostatic Regulation of Blood Glucose Level

The hormones influencing carbohydrate metabolism are explained below:

I. Insulin

The endocrine part of the pancreatic beta cells of the islets of Langerhans secretes it. It is secreted to lower blood glucose through various mechanisms when there are increased blood glucose levels. The mechanism involves increasing uptake by target cells (for ATP generation) and increasing the conversion of glucose to glycogen (glycogenesis in the liver) for its future needs. If insulin secretion is disturbed, its level may be lower (hypoinsulinism) or higher (Hyperinsulinism) than normal.

The biological functions of insulin require a typical chemical structure and disulfide interchain and intra-chain linkages in its polypeptide chains. Insulin is a polypeptide hormone composed of 51 amino acids, of which 21 are in peptide chain A (Glycine chain) and 30 amino acids in chain B (phenylalanine chain). Two interchain disulfide linkages link these two chains, A and B. Chain A and B are linked by the first disulfide linkage between the 7th amino acids of each chain and the second disulfide linkage between the 20th

amino acid of chain A and the 19th amino acid of chain B. Chain A only has intrachain disulfide linkage between the 6th and 11th amino acids. The sequence of amino acids in insulin chains differs in different animal species. Chain amino acids differ between species, are 8th, 9th, and 10th, whereas they differ when amino acids are positioned at the 30th position.

A. **Hypoinsulinism ---Insulin deficiency below the normal levels in the blood.** This condition can arise due to a) defective production and b) decreased production. Any one of these conditions can result in **diabetes mellitus (DM).** Normally, insulin stimulates glucose uptake by target cells to burn the glucose and lower blood glucose levels within physiological limits. Under DM conditions, the starvation of cells leads to weakness and unconsciousness and, if unattended, can terminate into death.

B. **Hyperinsulinism--- Insulin over-secretion above the normal levels in the blood.** This condition can arise due to excess internal insulin secretion in the body or external administration of high insulin doses. Hyperinsulinism can be associated with a) hyperglycemia due to reduced sensitivity to insulin and b) hypoglycemia due to excessive insulin secretion or external administration of insulin.

1. **Hyperinsulinemic hyperglycemia or reduced sensitivity to insulin and hyperglycemia:**

   This is found when the insulin target cells become less sensitive or more resistant to insulin. In this situation, the blood glucose is not used by target cells resistant to insulin, creating the condition of hyperglycemia. Hyperglycemia ultimately signals the secretion of more and more insulin and favors the hyperinsulinism condition.

2. **Hyperinsulinemic hypoglycemia or excessive insulin secretion with hypoglycemia.**

A standard regulatory mechanism is when insulin and blood glucose levels are disturbed, and insulin secretion continues beyond physiological levels (hyperinsulinemia) even when the blood glucose levels fall below the physiological range (hypoglycemia). Even under the prevailing hypoglycemic condition, glucose uptake into the target cells continues due to unregulated insulin secretion. Further, glucose formation is inhibited through various processes (gluconeogenesis, glycogenolysis, lipolysis, and ketogenesis) and disturbances in glucose regulation by cortisol and glucagon.

Unregulated insulin secretion can occur due to defective genes (congenital) regulating insulin secretion. Other recognized causes include perinatal stress, maternal diabetes mellitus, insulinoma, and injected insulin.

II. Glucagon

The endocrine part of the pancreatic alpha cells of the islets of Langerhans secretes it. Even though the effects of insulin and glucagon are opposite, both regulate blood glucose levels in the physiological range. Glucagon, released in response to hypoglycemia (during exercise), will elevate blood glucose levels to the normal range. The required glucose is released by the glucagon hormone from liver glycogen, and glucagon hormone also stimulates gluconeogenesis. Under hypoglycemic conditions, cells use free fatty acids as a source of energy. Under the condition of glucagonomas (tumors of the pancreas), secret excess glucagon can cause diabetes mellitus: while low secretion or deficiency of glucagon secretion results in hypoglycemia.

III. Amylin

It is another secretion from the pancreatic beta cells. It suppresses glucagon secretion and lowers blood glucose levels. It inhibits food intake, gastric emptying, and postprandial hepatic glucose production.

IV Somatostatin

It is secreted from the delta cells of the pancreas. Fasting enhances the release of glucose from the liver and raises blood glucose levels.

V Thyroid hormones

The hormones secreted from thyroid glands are T3 (tetraiodothyronine) and T4 (triiodothyronine) under the influence of thyroid-stimulating hormone (TSH) of the anterior pituitary. These hormones act on the mitochondria to metabolize nutrients, including glucose, to produce energy as ATP. Normal thyroid function is very important in regulating metabolism and controlling blood glucose. Any deviation in thyroid function can affect the course of diabetes. Thyroid hormone enhances intestinal absorption of glucose. Enhances gluconeogenesis in the liver. Stimulates glycogenolysis in the liver. It may cause the breakdown of insulin.

A. **Hyperthyroidism** is a condition of increased thyroid hormone levels and a hypermetabolic state compared to the euthyroid condition. In Graves' disease (autoimmune thyroid disease), antibodies stimulate thyroid enlargement and overproduction of thyroid hormone. Thyroid nodules and thyroiditis are other conditions associated with hyperthyroidism. There is increased gluconeogenesis and glycogenolysis to liberate free glucose, which is met with increased glucose oxidation. This is associated with the overall depletion of glycogen stored in the liver. It can also stimulate lipolysis and release free fatty acids in plasma. It promotes tissue fatty acid oxidation.

B. **Hypothyroidism** is a condition of the underactive thyroid gland. It can occur when functional acini are reduced in number in case of ablation of the thyroid gland in part or in complete form, or due to damage to the thyroid tissues upon radiation exposure. It is due to a deficiency of iodine or an inherited autoimmune condition called Hashimoto's thyroiditis. The basal metabolic rate is reduced. Hypothyroidism is observed among people with Type 2 diabetes.

VI. Adrenal gland

The adrenal gland has two parts that have endocrine functions: the adrenal cortex and the adrenal medulla. The hormones from the medulla are epinephrine or adrenaline and norepinephrine or noradrenaline, which are structurally related and called catecholamines. Adrenal cortex hormones related to carbohydrate metabolism are glucocorticoids.

A. **Epinephrine or Adrenaline:** It is an adrenal medullary hormone released into the bloodstream in response to excitement, emotion, anger, fear, or a "fight or flight Response." It provides blood glucose for energy production through gluconeogenesis from muscle lactic acid and the release of glucose through glycogenolysis in the liver. It affects sugar metabolism and exerts effects on blood pressure and heart rate. The binding of epinephrine/adrenaline to alpha-adrenergic receptors causes inhibition of insulin secretion by the pancreas and stimulation of both glycogenolysis and glycolytic pathways. Binding with beta-adrenergic receptors triggers glucagon secretion in the pancreas and increases lipolysis.

B. **Norepinephrine or Noradrenalin:** This hormone triggers the release of glucose and normalizes the blood glucose level by stimulating glycogenolysis in the liver and gluconeogenesis from blood lactic acid in muscles.

C. Glucocorticoids: They are one of the major groups of hormones of steroid nature secreted by the cells of the Zone fasciculate of the adrenal cortex. The glucocorticoids are hydrocortisone (cortisol) and corticosterone. Cortisol is the most important glucocorticoid. They have two side effects. On one side, they elevate the blood glucose concentration through gluconeogenesis. On the other hand, they inhibit glucose uptake by the peripheral tissue due to the anti-insulin effect. Cortisol aldosterone) The level is lower when the adrenal cortex fails to produce it and is higher due to its overproduction in a Cushing's syndrome condition, which is caused by a tumor in the adrenal gland or overstimulation of the ACTH hormone by the anterior pituitary gland.

VII. Hormones of Adenohypophysis

Three hormones, ACTH, TSH, and STH, or Growth hormone (GH), regulate carbohydrate metabolism. STH antagonizes insulin action and elevates blood glucose levels by enhancing liver glycogenolysis. It promotes an increase in alpha cells in the islets of the pancreas. It reduces the uptake of glucose by muscles. Cortisol release ACTH stimulates the release of cortisol from the adrenal gland, and TSH, acting on the thyroid gland, releases thyroid hormones influencing basal metabolic rate.

VIII. GLP-1(Glucagon-like peptide 1)

It is a peptide containing 30 amino acids, secreted by intestinal L cells and the colon, containing the proglucagon gene. Interestingly, GLP-1 receptors are in islet cells, the central nervous system, the heart, stomach, and adipose tissues. It is secreted upon the ingestion of food. It slows gastric emptying. It boosts insulin secretion but suppresses the secretion of glucagon; both of these results lower the levels of blood glucose. GLP-1 levels are low when fasting but elevate 15 -30 minutes after eating. It is an incretin hormone (increases insulin secretion in response to a meal) and gets metabolized by the enzyme known as dipeptidyl peptidase-4 (DPP-4). Inhibitors of DPP-4 can increase endogenous GLP-1. It has a short half-life, which is 1 to 2 minutes.

IX. Gastric inhibitory polypeptide—GIP or glucose-dependent insulinotropic polypeptide

It is an incretin (the hormone that increases insulin secretion in response to a meal) produced in the K cells of the small intestine. It stimulates the release of insulin following the intake of food. Somatostatin prevents the release of GIP. It is a 42-amino acid peptide derived from a 153-amino acid proprotein encoded by the GIP gene.

X. Asprosin or A Fasting-Induced Glucogenic hormone

It was recently discovered that white adipose, a glucogenic adipokine, are its source during fasting. Fasting enhances the release of glucose from the liver G protein-cAMP-PKA pathway and raises blood glucose levels. It is a C-terminal cleavage product of profilin (encoded by gene FBN1).

15.2 Diabetes Mellitus

Diabetes mellitus is also known as diabetes. Initially, there are higher blood sugar levels (hyperglycemia), then their excretion in urine (glucosuria). Blood glucose is under the control of various endocrine secretions or hormones. The main symptoms of diabetes are three P's, which are the polyuria (frequent urination), polydipsia (great thirst), and polyphagia (more appetite). Generally, blood glucose levels (fasting and postprandial) and HbA1c tests are done to diagnose diabetes mellitus.

I. Types of Diabetes Mellitus

There are different types of diabetes mellitus.

A. Most common types

1. Type 1 diabetes mellitus or Juvenile diabetes mellitus.

In this condition, the pancreas cannot produce enough insulin as required by the body. The body cells, without insulin, cannot use blood glucose to generate energy for their use. This leads to a hyperglycemia condition. To control hyperglycemia, the body depends on the external insulin administered (Insulin Dependent Diabetes Mellitus-IDDM). It is a cell autoimmune disease wherein the body's immune system generates secretion of substances that attack the beta cells of the pancreas, leading to little or no insulin secretion. It is common in young individuals (juveniles), and the insulin hormone treats it.

Glutamic acid decarboxylase antibodies, islet cell antibodies, and insulin antibodies are employed as markers for diagnosing Diabetes mellitus type I. The IDDM can be: 1. Autoimmune destruction type 2. Idiopathic. In the autoimmune destruction type, there is autoimmune destruction of beta-cells. Such patients may also have autoimmune destruction types, such as Graves' disease, Hashimoto's thyroiditis, and Addison's disease. Idiopathic forms of Type I diabetes etiology are unknown, but have no evidence of autoimmunity.

2. Type 2 diabetes mellitus or adult-onset diabetes mellitus

It is often a milder form of diabetes as compared to type 1. It is most common and is increasing all over the world. There are two ways to look at this condition. In one condition, insulin is produced, but it is insufficient for the cells. The second problem is that the cells present in muscle, fat, and the liver become resistant to the circulating insulin. Since the body cells and tissues resist insulin, they do not absorb glucose and remain in the bloodstream.

In either case, glucose is not effectively used by the body cells despite Insulin's presence; as such, it does not depend upon an external source of Insulin administration to survive; therefore, it is non-insulin-dependent diabetes mellitus (NIDDM) and adult-type diabetes mellitus.

Diabetes is sometimes associated with elevated blood ketones. Ketone bodies with abundance produced are acetoacetate and 3-β-hydroxy butyrate, while the third is acetone, the least abundant ketone body. Ketone bodies are physiologically produced by the liver but used by extrahepatic tissues as a source of energy. In acute diabetic ketoacidosis, the ratio of ketone bodies, which is 1:1 for 3-β-hydroxybutyrate: acetoacetate, rises to as high as 10:1. Acetone provides a sweet odor to the breath. Acetone is produced upon decarboxylation of acetoacetate—ketone bodies used by the brain as an energy source.

3. Gestational diabetes mellitus (GDM)

The women before, during, and after pregnancy have different blood glucose levels. Before pregnancy, they have normal blood glucose; during pregnancy, they develop diabetes, and it disappears after the termination of pregnancy. Such women are likely to have type 2 diabetes mellitus. It means possibilities for the development of type 2 diabetes mellitus in the time to come. Different hormones produced during pregnancy make the insulin effect ineffective. If the condition is not corrected, the growth and development of young ones are affected.

B. Other specific types of diabetes mellitus

1. **Neonatal diabetes mellitus (NDM)**. It is due to a single gene mutation or is in monogenic form. It occurs in the first 6 to 12 months of life. They do not produce enough insulin. It may be permanent neonatal or transient neonatal diabetes. The young ones may be smaller than their gestational age.

2. **Cystic fibrosis-related diabetes mellitus (CFRD).** It is due to cystic fibrosis. In this condition, there is relative insulin insufficiency due to destruction or dysfunction or decrease in islet cell mass and number of beta cells in pancreatic islets.
3. **Maturity-onset diabetes of the young (MODY).** It is due to a mutation of the gene. It runs in families. There are different types like MODY1 (chromosome 20-HNF4alpha, HNF complete form is Hepatocyte nuclear factor), MODY2(chromosome 7 glucokinase gene or GCK), MODY3,(chromosome 12, HNF1alpha), MODY4 (chromosome 13, IPF-1) and MODY5(chromosome 17, HNF1 beta). They are caused by genetic defects of beta-cell function.
4. **Steroid-induced diabetes.** It is due to chronic corticosteroid (glucocorticoid) overexposure. They are used widely as potent anti-inflammatory and immunosuppressive drugs, and their two modes of action cause hyperglycemia. Blood glucose can increase either by increasing its production or decreasing cell uptake. Corticosteroids stimulate gluconeogenesis in the liver and also decrease insulin-stimulated glycogen synthesis. They also decrease glucose uptake in skeletal muscle and white adipose tissue by antagonizing the insulin response.
5. **Latent Autoimmune Diabetes in Adults (LADA).** It is a slow-progressing form of autoimmune diabetes. Here, the pancreas does not produce adequate insulin. It is a bit different from both type 1 and type 2 diabetes.
6. **Types 3c diabetes, pancreatogenic diabetes, or secondary diabetes.** It is a type of diabetes in which enough insulin is not produced due to damage to the pancreas caused by another disease (pancreatic cancer, pancreatitis, or removal of the pancreas).
7. **Type 3 diabetes.** It is believed that insulin resistance might cause Alzheimer's disease.
8. **Type 4 diabetes.** It is a proposed type and is found in older non-obese people.
9. **Wolfram syndrome.** It is an inherited, rare genetic disorder. It is associated with childhood-onset insulin-dependent diabetes mellitus and optic atrophy. It has four most common features: diabetes insipidus, diabetes mellitus, optic atrophy, and deafness.
10. **Alstrom syndrome.** It is a rare, genetically inherited autosomal recessive trait (defects in ALMS 1 gene). There is a progressive loss of vision and hearing.

II. Complications of Diabetes Mellitus

The complications generally are hyperglycemia, ketoacidosis, hyperlipidemia, and polyuria: thirst, lack of energy, irritability, blurred vision, and weight loss. Type 1 diabetes typically has more complications than type 2 diabetes. They develop acute and chronic complications.

A. Acute complications

1. Diabetic ketoacidosis: It is characterized by a. hyperglycemia, b and increased lipolysis, and ketonemia, c. Increased proteolysis and decreased protein synthesis. In acute conditions, it can cause a diabetic coma.
2. Hyperglycemia hyperosmolar state: High blood glucose levels increase blood osmolality, drawing water out of cells into the blood.
3. Hypoglycemia: It is due to many factors like more insulin, over-exercise, or oral intake of more hypoglycemic drugs. Hypoglycemia is characterized by low blood sugar compared to the normal level. The patient needs to increase blood glucose to improve brain function.
4. Diabetic coma: A state of unconsciousness is known as a coma. This coma condition occurs due to severe diabetic hypoglycemia, hyperglycemia, ketoacidosis, severe hyperglycemia, hyperosmolar state, and dehydration, which can result in coma conditions.
5. Atherosclerosis: Atherosclerosis occurs due to the development of plaque made up mainly of fats and cholesterol on the artery walls, resulting in narrowing and reducing or blocking blood flow.

   In hyperglycemic conditions, high glucose levels react with immune proteins, leading to inflammatory changes that favor plaque formation in arteries. The plaque will narrow the blood vessel lumen, creating difficulty in blood circulation to the target organs.
6. Hypertension: Hypertension in diabetic conditions may be due to plaque formation in blood vessels, as one of the reasons.
7. Other infectious conditions may occur, including respiratory infection, periodontal disease, Foot damage, and disease affecting skin and mouth conditions.

B. Chronic complications

1. Microangiopathy. It occurs due to damage to small blood vessels due to diabetic conditions and creates conditions like diabetic cardiomyopathy, diabetic retinopathy, and Diabetic nephropathy.

2. Macroangiopathy. It occurs due to damage to the arteries. The conditions are coronary artery disease, myocardial infarction, muscle wasting, peripheral vascular disease, and stroke.

II. Laboratory tests related to diabetes mellitus

A. Quantitative tests

The blood sugars are measured as mg/100ml, mg/ dL, or millimoles per liter (mmol/L). One can convert mg/100ml or dL to mmo/L by multiplying the mg/dL value by a factor of 0.0555. One can convert mmol/L to mg/dL by multiplying the value with factor 18.018.

- Fasting and postprandial blood sugar estimation. The fasting blood sugar levels are measured after at least eight hours of the last food or drink intake. A blood sample is collected two hours after eating to measure postprandial blood sugar levels.
- Oral Glucose tolerance test (GTT). Here, a person is given 75 grams of sugar or glucose dissolved in water. Blood samples are taken 1 hour and 2 hours after glucose intake to measure sugar levels. Ordinary people will have glucose levels lower than 140mg/dL or 7.8 mmol/L. Anything above is indicative of a prediabetic or diabetic condition.

  Oral GTT is preferred in humans and dogs.
- Intravenous glucose tolerance test. It is done in humans and animals.
- Random blood sugar test. It measures blood sugar at any time of day, without any time restriction on when you last ate.
- HbA1C (hemoglobin A1C) estimation. It is estimated that the glycosylated hemoglobin levels are over 60 to 90 days. High HbA1C levels give a prediction of diabetic status. These are signs of high blood glucose from diabetes. A value below 5.7% is normal, 5.7% to 6.4% is prediabetic status, and more than 6.5% is diabetic status.

Glucose tolerance is affected by various factors. The factors in which the glucose tolerance curve appears to be a diabetic curve are starvation, a high-fat diet, liver disease, severe thiamine deficiency, and hypoinsulinism. The factors that are expected to influence the tolerance curve are exercise and hyperinsulinism.

B. Qualitative tests for sugar in a solution

As discussed in earlier chapters, carbohydrate sugars may be reducing or non-reducing, monosaccharides, disaccharides, or polysaccharides.

1. Common test

Any sugar gives a common test. It includes two tests:

- Molisch's test. In Molisch's test, the acid reacts with sugar and forms reddish violet furfural derivatives.
- Anthrone reaction. In this test, furfural formation further reacts with anthrone, giving a blue-green complex color.

2 Reduction tests

a. Methylene blue reduction test. When sugar is reduced when it reacts with methylene blue, the solution gets decolorized because methylene blue gets reduced to colorless leucomethylene blue.

b. Picric acid reduction test. Reducing sugar will react with picric acid so that picric acid is reduced, giving Mahogany a red color.

c. Nylander test or Bismuth reduction test. Reducing sugar upon reaction, a black color appears due to the precipitation of metallic bismuth.

d. Alkaline copper sulfate reduction tests.

   - Trommer's test. Any sugar having a free group acts a weak acid and, in an alkaline solution, forms unstable but reactive enediols. The reaction of sugar with copper forms yellowish-to-red precipitates of cuprous oxide.
   - Fehling's test. The reducing sugar will react with copper and form yellow or red precipitates due to the formation of cuprous oxides.
   - Benedict's qualitative reagent test. A reducing sugar will also form green, red, or yellow Cu_2O precipitates.
   - Barfoed's test. The reducing sugars react with $CuSO_4$ and convert to Cu_2O by monosaccharide.

e. Only for ketone monosaccharide sugars.

   - Rapid furfural test. The ketone sugar will react to give a red precipitate.
   - Seliwanoff's test. Like the rapid furfural test, when this positive test gives a red color, indicating the presence of a ketone sugar like fructose.

f. Osazone tests. This test identifies the presence of glucose, fructose, maltose, or lactose sugar. The test consists of a sugar reaction with phenylhydrazine hydrochloride and sodium acetate. It gives a yellow mass of crystalline sugar of phenylosazone. The crystals are observed under a microscope for their shape.

Needle-shaped crystals are formed by glucose and fructose. Sunflower-shaped crystals are given by maltose sugar, and Hedgehog crystals are given by lactose.

g. Color with Iodine (0.2N solution). Mix the given sugar solution with the iodine solution. The color produced indicates the presence of polysaccharides. If the color shown is Navy blue, then starch is present. If the given color is red, then glycogen is present.

15.3 Bovine Ketosis/Acetonemia/Sweet Breath/ Hypoglycemia

Blood glucose is the key source of energy for animals. In non-ruminants, glucose is available through carbohydrate digestion in the digestive tract. In ruminants, it is produced mainly through gluconeogenesis, as the end product of carbohydrate digestion is VFA, not glucose. Therefore, cows or ruminants cannot be fed glucose directly due to their complex digestive processes. Propionate, produced in a high-grain diet, is used in gluconeogenesis in the liver and kidneys , and some glucose is absorbed from the small intestine. Overall, ruminants' blood glucose levels are lower than those of non-ruminants. Hence, ruminants are more prone to ketosis than non-ruminant animals. When blood glucose is inadequate, the body will metabolize fats to get energy; thus, ketone bodies are produced in excess than required, leading to ketosis.

Many factors contribute to low blood glucose levels. Among them are 1. Underfeeding or the availability of inadequate feed, 2. Low quality of feed or spoiled silage with more butyric acid 3. Lactating status 4. Starvation 6. Pregnancy, 7 other diseases or clinical conditions like hypocalcemia, mastitis, or metritis reduce the appetite and feed intake.

Bovine ketosis occurs in high milk-producing lactating cows (or buffalo) due to the drainage of carbohydrates in milk as lactose, resulting in more energy demand than energy intake, creating a negative energy balance. Lactose is synthesized and secreted by the lactating mammary gland. The negative energy balance of energy generally occurs during the latter part of pregnancy or during the stage of highest milk production in the early stage of lactation. During negative energy demand, ketone bodies, namely acetone, acetic acid, and beta-hydroxybutyric acid, appear in the blood, and when their levels exceed, they appear in urine and milk. Breathed-out air contains the sweet smell of acetone. In non-ruminants, ketone bodies are produced in the liver, whereas in ruminants, they are produced by the rumen epithelium and mammary gland.

Clinical signs

Acetonemia or ketosis has two primary clinical forms of manifestation-the wasting form and the nervous form wasting form

There is a slow decrease or loss in appetite and milk yield in 2 to 5 days. There is weight loss. These occur during the first 60 days of lactation, but the prevalence is higher at the 4th calving. Expired air or breathed-out air, and sometimes milk contains the sweet smell of ketone, especially acetone.

Acetonemia can sometimes be a secondary condition to other disorders in the animal. It includes displacement of the fourth stomach, mastitis, metritis, peritonitis, milk fever, and other conditions reducing the appetite. Finally, there are hypoglycemia, ketonemia, ketonuria, and nervous disturbances.

Nervous form

This form is characterized by walking in circles, straddling or crossing the legs, aimless wandering, staggering, vigorous licking, and nervousness. The animal is very sensitive when touched, and its limbs' movement seems stiff and uncoordinated.

Treatment

Restoration of the condition to normal is possible if blood glucose is restored to normal with treatment by adopting a route other than oral. The oral route can worsen the acetonemia or ketosis condition.

Prevention

This can be done by adequate feeding and management during late pregnancy. High-energy feed must be given after calving. Fresh and lush pastures have high moisture and may not be suitable for supplying adequate energy to cope with the high producers' high demand. Feeding good-quality silage, hay, and cereal grains early in lactation may meet the energy demand. Drenching of propylene glycol immediately after calving may avert ketosis in cows.

15.4 Pregnancy Toxaemia

- Pregnancy toxaemia or ketosis occurs in cattle, ewes, and does. It is the most common metabolic disorder of small ruminants like ewes and does, and occurs during the final stage of pregnancy when the energy demand is not met due to the disturbed carbohydrate and fat metabolism. It affects extremely fat or excessively thin ewes and does. In cattle pregnancy, toxaemia is known as a fatty liver syndrome and a form of ketosis. Normal blood glucose, around 40 to 60 mg%, is reduced to 10 mg%. On the other side, blood ketone, which is around 1 to 4 mg % in blood, increases to 15 to 80 mg %, depending on the severity of the condition. High-energy concentrates and grain feeding during the last month of pregnancy are some of the preventive measures.

Causes

- Overfed animals deposit abdominal fat, reducing the capacity of the rumen, and fat deposition in the liver can cause liver dysfunction.
- Pregnancy. When the energy demand is increased and not met with the increased energy demand by the foetus in late gestation, the mobilization of fat stores is increased to provide energy to cope with the required level. But sometimes, it goes beyond the capacity of the liver, resulting in hepatic lipidosis and accumulation of ketone bodies, leading to pregnancy toxaemia.
- Multiple foetuses. The animals supplied with inadequate feed and feed of poor quality, besides reduced rumen capacity due to fetal growth in multiparous (twins or triplets) pregnancy, can lead to pregnancy toxaemia or hypoglycemia condition.
- Ketone bodies are produced as a source of energy, but become toxic and cause the death of animals over a prolonged period.
- Subclinical ketosis can quickly convert to pregnancy toxaemia if feed intake is reduced under various conditions like adverse weather, transportation stress, shearing, or other diseases like foot rot and pneumonia, etc. The hypoglycemia condition deprives the brain of glucose energy, creating nervous system symptoms.
- Lack of exercise
- Underfeeding
- Stress due to cold, wet, windy weather, and heavy infestation.

Preventive measures

- The best preventive measure is to increase feeding of high-energy concentrates and grains during the last month of pregnancy.
- Ewes bearing twins or triplets should be separated and offer a higher plane of nutrition.

Treatment

- Treatment in ewes/does is feeding concentrates and oral administration of propylene glycol or corn syrup as quick energy sources. The electrolyte solution is given during dehydration conditions in animals.

Key Takeaway

- The endocrine part of the pancreatic beta cells of the islets of Langerhans secretes insulin.
- It increases the conversion of glucose to glycogen (glycogenesis in the liver).
- Hypoinsulinism is an Insulin deficiency below the expected levels in the blood. This can result in diabetes mellitus (DM).
- Hyperinsulinism is a condition wherein insulin levels are above the normal levels in the blood.
- Hyperinsulinemia, hyperglycemia is due to reduced sensitivity to insulin and hyperglycemia
- This is found when the insulin target cells become less sensitive or more resistant to insulin.
- Hyperinsulinemia hypoglycemia, or excessive insulin secretion is found with hypoglycemia.
- The endocrine part of the pancreatic alpha cells of the islets of Langerhans secretes glucagon. The effects of insulin and glucagon are opposite. Stimulates gluconeogenesis.
- Amylin is a secretion from the pancreatic beta cells. It suppresses glucagon secretion.
- Somatostatin is released from the pancreatic delta cells.
- A normal thyroid regulates metabolism and controls blood glucose. Stimulates glycogenolysis and causes the breakdown of insulin.
- In Graves disease (autoimmune thyroid disease), antibodies stimulate the thyroid enlargement and overproduction of thyroid hormone.
- There is increased gluconeogenesis and glycogenolysis
- Hypothyroidism is due to a deficiency of iodine or an inherited autoimmune condition called Hashimoto's thyroiditis.
- The adrenal gland secretes Epinephrine or Adrenaline, norepinephrine or noradrenaline, and corticoids.
- Epinephrine or Adrenaline is an adrenal medullary hormone released in response to a "fight or flight response"
- Epinephrine helps in energy production through gluconeogenesis from muscle lactic acid and the release of glucose through glycogenolysis in the liver.

- The epinephrine binding to alpha-adrenergic receptors and beta-adrenergic receptors trigger different actions.
- Binding with alpha-adrenergic receptors causes inhibition of insulin secretion by the pancreas and stimulation of both the glycogenolysis and glycolytic pathways.
- Binding with beta-adrenergic receptors triggers glucagon secretion in the pancreas and increases lipolysis.
- Glucocorticoids are secreted by the cells of the zona fasciculata of the adrenal cortex.
- The level of glucocorticoids is higher in a Cushing's syndrome condition, resulting in the overstimulation of the ACTH hormone by the anterior pituitary gland.
- Three hormones, ACTH, TSH, and STH, or Growth hormone (GH), regulate carbohydrate metabolism.
- GLP-1 in its complete form is called Glucagon-like peptide 1. It is a peptide containing 30 amino acids, secreted by intestinal L cells and the colon.
- GLP-1 is an incretin hormone secreted upon ingesting food, and gastric emptying takes longer.
- Gastric inhibitory polypeptide—GIP. It is an incretin hormone. It is a 42-amino acid peptide.
- Type I and type II are two types of diabetes mellitus.
- The pancreas does not produce enough insulin. The body has to depend on the external insulin administered (Insulin Dependent Diabetes Mellitus-IDDM) to control hyperglycemia.
- Type 2 diabetes is non-insulin-dependent diabetes mellitus (NIDDM) and adult-type diabetes mellitus.
- The blood sugars are measured either as mg/100ml or mg/dL or as millimoles per liter (mmol/L).
- The fasting blood sugar levels are measured after at least eight hours of the last food or drink intake.
- HbA1c (hemoglobin A1c) estimation. It gives an idea of the mean sugar levels in the last two to three months. A value below 5.7% is average, 5.7% to 6.4% is prediabetic status, and more than 6.5% is diabetic status.
- Ruminants are more prone to ketosis than non-ruminant animals
- The negative energy balance of energy generally occurs during the latter part of pregnancy or during the stage of highest milk production in the early stage of lactation.

- Pregnancy toxaemia affects extremely fat or excessively thin ewes and does.

In-text questions

1. Give the complete form of GIP and GLP-1 secretions and explain their role in carbohydrate metabolism.
2. How do thyroid hormones regulate glucose metabolism?
3. How does the body produce optimal blood glucose levels through various processes?
4. Explain the abbreviations IDDM AND NIDDM used concerning diabetes mellitus.
5. How does hyperinsulinism cause hyperglycemia or hypoglycemia condition?
6. What is gestational diabetes mellitus, and its significance?
7. Write about the consequences that occur if the diabetes mellitus condition is unattended.
8. Which cells of the pancreas secrete insulin glucagon, amylin, and somatostatin, respectively?
9. The adrenal gland play its role in glucose metabolism. Explain this statement.
10. How does the insulin secretion get unregulated?
11. What is the benefit of measuring HbA1C?
12. What are the different ranges of HbA1C for normal, prediabetic, and diabetic status?
13. Name the hormones of the thyroid gland with their role in carbohydrate metabolism.
14. How do different hormones of the adrenal glands play a vital role in blood glucose level regulation?
15. What are the target organs of hormones, ACTH, TSH, and STH?
16. Why are GLP-1 and GIP increasing hormones?
17. What are the similarities and differences between type I and type II diabetes?

18. Steroid hormones can induce diabetes mellitus. Explain the statement with scientific reasoning.
19. What are the conditions for estimating fasting and PP2BS sugar levels?
20. What information do we get from the HbA1c estimate?
21. Choose one correct option for the following from the options given below (The answer is given in **bold** letters):

a Which type of cells of the pancreas secrete insulin?

i. Alpha cells ii. **Beta cells**

iii. Delta cells.

b Insulin increases the conversion of

i. **Glucose to glycogen** ii. Glycogen to glycogen.

iii. Glucose to galactose

c Glycogen is stored in one of the following organs, which can release free glucose upon glycolysis.

i. Muscle ii. Brain

iii. **Liver**

d Is it possible to get Hyperinsulinemic hyperglycemia?

i. **Yes** ii. No

iii. Cannot say

e The secretion source for the hormone glucagon is

i. Beta cells ii. **Alpha cells**

iii. Delta cells.

f Insulin and glucagon have actions that are contradictory to each other, but they regulate the blood levels of

i. Glycerol ii. Gentianose

iii. **Glucose**

g Amylin suppresses glucagon secretion. It is a secretion from the pancreatic,

i. **Beta cells** ii. Alpha cells

iii. Delta cells

h Name the secretion released from the delta cells of the pancreas.

i. Insulin ii. **Somatostatin**

iii. Glucagon

i Graves' disease and Hashimoto thyroiditis are conditions related to organ

i. Pancreas ii. **Thyroid**

iii. Adrenal

j The hormone secreted from the adrenal gland is

i. **Epinephrine** ii. STH

iii. Insulin

k The binding of which receptors triggers glucagon secretion in the pancreas.

i. **with beta-adrenergic receptors**

ii. with alpha-adrenergic receptors

iii. With delta-adrenergic receptors.

l Which hormone is secreted by the fasciculate zone of the adrenal cortex?

i. Glucagon ii. **Glucocorticoids**

iii. Epinephrine

m GLP-1 is a peptide secreted by

i. Pancreatic cells ii. Thyroid cells.

iii. **Intestinal cells.**

n The binding of epinephrine to which receptors inhibit insulin secretion from the pancreas

i. **Alpha-adrenergic receptors.** ii. Beta-adrenergic receptors.

iii. Delta-adrenergic receptors.

o Which type of Diabetes Mellitus is insulin Dependent Diabetes Mellitus?

i. Type 2. **ii. Type 1.**

iii. Type 4.

p Which type of Diabetes Mellitus is non-insulin-dependent diabetes?

i. Type 1 ii. **Type 2**

iii. type 3

q Which one is a false unit of expression of blood glucose concentration?

i. mg/100ml. ii. millimoles per liter.

iii. **millimoles per dL**

r HbA1c estimation is done at

i. Fasting level ii. Postprandial level

iii. **Random level.**

s The range level of HbA1c in an average person is

i. **below 5.7%** ii. between 5.7% to 6.4%

iii. more than 6.5%

Further Readings

Amori RE, Lau J, Pittas AG (July 2007). «Efficacy and safety of incretin therapy in type 2 diabetes: systematic review and meta-analysis». JAMA. 298 (2): 194–206. doi:10.1001/jama.298.2.194. PMID 17622601.

Functional carbohydrates. Zhang, Guoqiang, and Qinhong Wang. (2017), pp.269-298. (https://www.researchgate.net/publication/345618278). https://www.naturalproductsinsider.com/formulation/understanding-starch-functionality.

https://pancreaticcanceraction.org/help-and-support/living-with-pancreatic-cancer/type-3c-diabetes/

https://www.diabetes.org.uk/diabetes-the-basics/types-of-diabetes

https://www.frontiersin.org/articles/10.3389/fendo.2020.00064/full

https://www.healthline.com/health/diabetes/type-4-diabetes#causes

https://www.hopkinsmedicine.org/health/conditions-and-diseases/diabetes/gestational-diabetes

https://www.nationwidechildrens.org/conditions/neonatal-diabetes

https://www.nature.com/articles/nm.4432

https://www.ncbi.nlm.nih.gov/books/NBK545192/ 11.

https://www.ncbi.nlm.nih.gov/pmc/articles/PMC4852710

https://www.yourhormones.info/hormones/glucose-dependent-insulinotropic-pepti

Jain J.L., Sunjay Jain, Nitin Jain (2005). Fundamentals of Biochemistry. S.Chand & Company Ltd. New Delhi.

Life sciences. Fundamentals and Practice I. Pranav Kumar and Usha Mina (2006). Sixth Edition. Pathfinder publication. New Delhi.

Medical Biochemistry (2004). Ethiopia public health initiative.

Nauck MA (November 2006). "The incretin system: glucagon-like peptide-1 receptor agonists and dipeptidyl peptidase-4 inhibitors in type 2 diabetes". Lancet. 368 (9548): 1696–705. doi:10.1016/S0140-6736(06)69705-5. PMID 17098089. S2CID 25748028.

Objective Biochemistry. Sengar, R.S. and Chaudhary Rishu (2014). New India Publishing Agency. New Delhi.

Rang HP, Ritter JM, Flower R, Henderson G (2016). Rang and Dale›s Pharmacology (8th ed.). United Kingdom: Elsevier Churchill Livingstone. p. 385. ISBN 9780702053627. OCLC 903083639.

Serum total cholesterol, triglycerides, amylase, alkaline phosphatase, aspartate transaminase, and sugar estimates in patients with different postprandial sugar levels. A dissertation was submitted to Sardar Patel University, Vallabh Vidyanagar, Anand, Gujarat. (2012). Ms Vibhita Patel.

Index

A

Acetic acid, 53
Active transport, 12, 19, 20, 22, 23, 27, 28
Aerobic metabolism, 129, 152, 156
Agar, 118, 119, 124
Agarobiose, 119
Agaropectin, 118, 119, 124
Agarose, 118, 119, 124
Aldehyde, 32, 33, 40, 41, 45, 48, 53, 55, 59, 62, 67, 68, 69, 70, 71, 72, 73, 74, 76, 79, 80, 87, 89, 92, 163, 164, 207, 208, 215, 220
Alginate, 122
Allose, 41, 46
Alpha- amylase, 104
Alpha linkages, 86
Alpha sugar, 47
Altrose, 38, 41, 42, 46, 71
Amino sugars, 76
Amphipathic, 5, 8
Amylopectin, 89, 104, 105, 106, 107, 109, 123, 126
Amylose, 74, 97, 104, 105, 106, 123, 126
Anaerobic metabolism, 129
Anomeric carbon, 35, 46, 56, 57, 72, 76, 86, 92, 94
Anomers, 34, 47, 49, 55, 72
Antiport, 18, 19, 27
Aquaporin, 16
Arabinose, 41, 42, 45, 46, 64, 71, 118, 119
Asymmetric, 33, 45, 48, 49, 54, 59, 63, 79
ATP, 7, 17, 20, 21, 22, 24, 25, 26, 39, 129, 131, 132, 133, 134, 136, 137, 138, 141, 142, 143, 144, 145, 146, 147, 153, 157, 161, 163, 164, 166, 167, 168, 169, 170, 171, 173, 174, 175, 177, 180, 181, 182, 183, 184, 186, 190, 191, 192, 195, 196, 197, 200, 203, 204, 206, 211, 214, 215, 219, 225, 228, 229, 231, 234, 238, 239, 240, 241, 244, 245, 254, 256, 259, 261, 263, 271, 273, 281, 283

B

Beta amylase, 104, 107
Beta linkages, 86
Beta sugar, 47
Beta-blockers, 21
Biomolecules, 2, 3, 5, 15, 130, 153, 156, 219, 234

C

Calcium pump, 21
cAMP, 16, 251, 258, 261, 262, 265, 286
Caramelization, 68, 91, 100
Carbohydrate, 2, 3, 4, 8, 9, 15, 23, 32, 78, 111, 130, 147, 148, 149, 150, 155, 159, 196, 219, 234, 240, 274, 281, 284, 285, 290, 292, 293, 296, 297
Carbonyl carbon, 47, 59
Carrier proteins, 15, 17, 19, 20
Caveolin, 24
Cell membrane, 5, 6, 7, 8, 9, 10, 11, 12, 13, 14, 15, 20, 21, 22, 23, 24, 26, 27, 118, 146
Cellobiose, 42, 87, 90, 100
Cellulose, 31, 32, 39, 43, 48, 49, 60, 90, 104, 109, 110, 112, 123, 124, 125, 148, 149, 154, 155
Chemical bonds, 3, 156
Chiral carbon, 33, 35, 45, 46, 49
Chitin, 39, 43,77, 104, 112,123, 125, 126
Chitobiose, 91, 99
Cholesterol, 8, 9, 11, 24, 211, 214, 265, 289, 300
Chondroitin, 62, 113, 114, 116, 117, 228
Cis isomer, 48
Clathrin, 24
Conformational isomer, 46
Co-transport, 19
Coupled carrier, 19
Cows, 110
Cristae, 25, 27, 132, 153
Cuticle, 112

D

Davson-Danielli model, 11
Deoxyribose, 31, 53, 76
Dermatan sulfate, 9, 113
Dextran, 108
Dextrin, 97, 104, 107, 124
Dextrins, 106, 107, 125, 147
Dextrorotatory, 45, 47, 48, 50, 88, 89, 90, 92
Diabetes, 2, 17, 26, 150, 282, 283, 284, 286, 287, 288, 289, 290, 295, 296, 297, 298, 299, 300
Diabetes Mellitus, 22, 74, 75, 286, 289, 296, 299
Diastereomer, 46, 49
Diastereomers, 36, 43, 45, 46
Dihydroxyacetone, 41, 164, 167, 190, 191, 200, 271
Dioses, 41
Disaccharides, 49, 53, 69, 74, 87, 100, 101, 113, 127, 147, 271, 290

E

Electrochemical gradient, 12, 20, 22, 23, 134, 146
Electron, 3, 4, 25, 26, 27, 29, 129, 131, 132, 133, 134, 135, 136, 137, 138, 139, 140, 141, 143, 144, 145, 146, 147, 153, 154, 155, 157, 159, 177, 178, 181, 186, 190, 196, 199
Enantiomers, 36, 37,38, 43,45, 46, 47,48, 49,51, 72
Endocytosis, 23, 24, 25, 27, 28
Endothermic, 31, 73
Epimers, 35, 45, 46, 47, 49, 51
Erythritol, 32, 63
Erythrose, 36, 41, 63, 70, 71, 206, 207, 208, 211, 214, 216, 274
Erythrulose, 41
ETC,, 129, 136, 144, 175
Exocytosis, 23, 24, 25, 27, 28, 29

F

Facilitated diffusion, 13, 15, 19, 20, 27
Fischer projection, 53, 54, 73, 81
Fructan, 111
Fructose, 20, 19, 32, 33, 34, 35, 39, 41, 42, 50, 51, 54, 56, 61, 62, 63, 64, 65, 66, 67, 68, 75, 80, 82, 83, 86, 87, 92, 93, 94, 95, 98, 99, 100, 101, 102, 111, 123, 125, 126, 147, 148, 156, 161, 163, 164, 167, 169, 171, 172, 182, 206, 207, 208, 210, 211, 213, 214, 230, 232, 233, 234, 238, 241, 242, 244, 254, 268, 271, 273, 276, 278, 279, 291, 292
Fucose, 9
Furanose, 55, 56, 57, 79, 82
Furanose ring, 55, 82, 92
Furfural, 60, 64, 65, 80, 291

G

Galactoheptose, 42
Galactosamine, 9, 40, 71, 77, 113, 116, 117
Galactose, , 9, 19, 20,32, 38, 39, 41, 42, 46, 47, 51, 61, 63, 67, 71, 72, 75, 76, 77, 78, 80, 82, 86, 87, 88, 93, 94, 95, 98, 100, 101, 102, 112, 113, 118, 119, 126, 148, 149, 167, 169, 217, 242, 270, 275, 278, 298
Gamma amylase, 107
Gated -ion channels, 16
Gated protein channels, 15
Gentianose, 42, 93, 94, 95, 101, 102, 298
Gentiobiose, 42, 86, 91, 101, 127
Geometric isomers, 43, 45, 47
Glucoheptose, 42
Glucoheptulose, 42
Glucopyranose, 57, 73, 88, 90, 95, 104, 110, 123
Glucosamine, 9, 31, 40, 65, 77, 78, 84, 91, 99, 102, 112, 113, 114, 115, 122, 123
Glucosan, 105
Glucose, 4, 7, 8, 9, 13, 15, 17, 19,20, 21, 22, 25, 27, 29, 32, 33, 34, 35, 36, 38, 39, 41, 42, 45, 46, 47, 49, 50, 51, 53, 54, 55, 57, 60, 61, 62, 63, 64, 65, 67, 68, 69, 70, 71, 72, 73, 74, 75, 76, 78, 79, 80, 82, 83, 86, 87, 88, 89, 90, 91, 92, 93, 94, 95, 97, 98, 99, 100, 101, 102, 104, 105, 106, 107, 108, 109, 110, 111, 119, 120, 123, 124, 125, 126, 129, 130, 131, 147, 148, 149, 150, 151,153, 154, 155, 156, 159, 161, 163, 164, 167, 168, 169, 170, 171, 172, 173, 174, 175, 178, 182, 183, 184, 191, 192, 195, 196, 199, 201, 202, 203, 204, 205, 206, 210, 211, 213, 214215, 217, 219, 220, 228, 229, 230, 231, 233, 234, 238, 239, 240, 241, 242, 244, 245, 246, 247, 248, 249, 250, 251, 252, 253, 254,256, 257, 258, 259, 261, 262, 263, 264, 265, 266, 267, 268, 270, 271, 272, 273, 274, 275, 276, 277, 281, 282, 283, 284, 285, 286, 287, 288, 289, 290, 291, 292, 293, 294, 295, 297, 298, 300

GLUT, 17
Glyceraldehyde, 41, 45, 51, 59, 63, 70, 71, 161, 164, 165, 167, 171, 173, 182, 196, 206, 207, 208, 211, 213, 214, 216, 226, 233, 271
Glycerose, 41
Glycoaldehyde, 41
Glycocalyx, 9,23, 28
Glycogen, 21, 23, 26, 28, 31, 39, 43, 48, 74, 76, 89, 102, 104, 107, 108, 109, 123, 124, 125, 126, 131, 147, 150, 153, 156, 167, 170, 238, 243, 244, 245,246, 247, 248, 249, 250, 251, 252, 253, 254, 255, 256, 257, 258, 259, 261, 262, 263, 265, 266, 267, 268, 274, 275, 276, 277, 279, 281, 283, 284, 288, 292, 295, 298
Glycogenesis, 155, 244,246, 250, 251, 252, 254, 259, 281, 295
Glycolaldehyde, 53, 59
Glycolipid, 8, 9, 12, 39, 49, 71, 75, 150
Glycoproteins, 7, 9, 12, 39, 40, 49, 71, 74, 75, 76, 77, 119
Glycosaminoglycan, 116, 127
Glycosaminoglycans, 112, 114, 115
Glycosidic bonds, 4, 42, 86, 93, 94, 97, 99, 100, 104, 123, 246, 250, 252, 253, 257
Gulose, 41, 42

H

Haworth projection, 56, 57, 73
HbA1c, 286, 296, 298, 300
Hemiacetal, 46, 57, 59, 72, 80, 87
Hemiketal, 59, 80
Heparan sulfate, 9, 113, 115, 123
Heparin, 40, 62, 113, 115, 123, 126, 228
Heptoses, 41, 42
Herbivorous animals, 95, 100
Heteroglycans, 20, 104, 112, 125
Heteropolysaccharides, 49, 112, 118
Hexoses, 41, 42
Homoglycans, 20, 104, 124, 125
Homopolysaccharide, 110, 111, 112, 123, 246
Hyaluronic acid, 40, 43, 62, 113, 123, 228
Hydrogen bond, 4, 5
Hydrogen pump, 22
Hydrophilic, 2, 5, 6, 7, 8, 9, 10, 11, 13, 16, 26, 141
Hydrophobic, 2, 4, 5, 6, 8, 10, 12, 13, 15, 26, 122, 138, 145
Hyperinsulinism, 281, 282, 295
Hyperkalemia, 21
Hypertonic, 13, 14

Hypoinsulinism, 282, 295
Hypokalemia, 21
Hyponatremia, 108
Hypotonic, 13, 14

I

Idose, 41, 46
Iduronic acid, 113, 115, 117
Indican, 72
Insulin, 17,21, 25, 111, 148, 149, 155, 168,214, 219, 238, 242, 245,251,252, 254, 256, 259,261, 281, 282, 283, 284, 285, 286, 287, 288, 289, 295, 296, 297, 298, 299
Integral proteins, 7, 15
Intermolecular, 4
Intramolecular, 4
Inulin, 43, 75, 111, 123
Iodose, 38, 42
Isomalt, 32
Isomaltose, 86, 89, 100, 106, 127, 147
Isomerism, 34, 43, 44, 45, 47, 48
Isomers, 19, 33, 34, 35, 43, 44, 46, 47, 48, 49, 50, 51, 73, 74

K

Keratan sulfates, 113
Ketone, 32, 33, 40, 41, 48, 53, 56, 59, 62, 68, 69, 75, 87, 92, 163, 164, 206, 207, 215, 271, 287, 291, 292, 293, 294
Ketoses, 34, 41, 46, 51, 62, 63, 66, 71, 83, 218
Kidneys, 15, 17, 22, 76, 131, 150, 234, 240, 256, 261, 263, 267, 271, 292
Killiani Fischer's synthesis, 80

L

Lactose, 32, 42, 65, 75, 86, 87, 88, 93, 100, 101, 102, 147, 149, 150, 270, 275, 279, 291, 292
Levorotatory, 45, 47, 48, 50, 92
Ligand, 16, 24
Ligand-gated ion channels, 16
Light rotations, 92
Lipid bilayer, 7, 8, 10, 11, 12, 16, 17, 135
Liver, 17, 31, 39, 43, 48, 76, 77, 78, 108, 114, 115, 123, 132, 149, 150, 153, 163, 186, 204, 211, 214, 216, 220, 228, 234, 238, 239, 240,

241, 244, 246, 244, 245, 249, 251, 252, 256, 257,258, 259, 261, 262, 263, 264, 265, 266, 267, 268, 270, 271, 273, 274, 275, 281, 283, 284, 285, 286, 287, 288, 290, 292, 293, 294, 295
Lysosome, 23
Lysozyme, 121
Lyxose, 41, 42

M

Maltose, 32, 42, 65, 86, 87, 88, 89, 90, 93, 97, 100, 101, 102, 104, 105, 106, 107, 109, 147, 271, 272, 275, 291, 292
Maltotriose, 97, 100
Mannitol, 32, 63
Mannoheptulose, 42
Mannose, 9, 35, 38, 41, 42, 46, 47, 51, 61, 62, 63, 65, 67, 68, 71, 73, 76, 80, 83, 86, 94, 96, 126
Manoheptose, 42
Matrix, 25, 26, 75, 132, 133, 134, 136, 137, 139, 141, 146, 147, 157, 158, 175, 178, 184, 186, 187, 190, 199, 200, 202, 203, 232, 241, 272
Melezitose, 42, 93, 94, 95, 101
Melibiose, 94, 95
Mitochondria, 5, 22, 25, 26, 129, 131, 132, 135, 136, 144, 145, 146, 147, 153, 155, 158, 161, 169, 170, 171, 172, 174, 178, 186, 187, 190, 193, 195, 202, 203, 211, 230, 234, 232, 240, 241, 283
Molecular forces, 2, 3
Monosaccharide, 34, 42, 43, 53, 59, 72, 73, 75, 76, 78, 86, 87, 88, 92, 98, 99, 100, 102, 104, 115, 119, 120, 123, 125, 148, 270, 271, 272, 291
Mucopolysaccharide, 114, 117, 118, 124
Mucopolysaccharides, 10, 75, 113, 115, 150, 230
Mutarotation, 34, 35, 49, 55, 87, 89, 90, 92

N

NAD, 129, 133, 134, 135, 136, 137, 146, 153, 155, 158, 165, 167, 170, 173, 176, 177, 178, 180, 182, 184, 185, 186, 190, 193, 200, 202, 204, 220, 224, 229, 231, 232, 234, 240, 241, 272
NADH, 129, 134, 135, 136, 137, 138, 141, 143, 144, 146, 147, 153, 154, 158, 161, 165, 171, 177, 178, 179, 180, 181, 182, 183, 184, 185, 186, 190, 191, 192, 193, 195, 199, 215, 219, 224, 232, 234, 240, 241, 243, 254, 272
Neurotransmitter, 25
Nitrocellulose, 110
Nonbiological use, 26
Non-polar, 4, 6, 8, 26
Nonreducing sugar, 42
Nutraceuticals, 2, 26

O

Obesity, 2, 26
Oligosaccharide, 8, 97, 100, 102, 107, 127, 248, 252, 253
Oligosaccharides, 42, 53, 86, 122, 130, 147
Optical rotation, 34, 35, 48, 92, 99
Osazone, 65, 67, 72, 79, 80, 83, 87, 89, 90, 92
Osmolality, 13, 289
Osmolarity, 13
Osmotic equilibrium, 21
Osmotic pressure, 13, 14, 15, 131

P

Pancreas, 17, 25, 163, 168, 256, 261, 283, 284, 285, 286, 288, 296, 297, 298, 299
Passive transport, 12, 15, 23, 28
Pectin, 43, 63, 104, 119
Pectins, 118, 124
Pentasaccharide, 98
Pentoses, 41, 42, 60
Peptidoglycan, 39, 120, 121, 124
Peptidoglycans, 31, 48, 120, 121, 125
Peripheral proteins, 7, 12
Permeases, 17
Phagosome, 23
Pharmaceuticals, 2, 26, 123
Phlorizin, 72
Phospholipids, 4, 8, 10, 11
Pinocytosis, 23, 24, 29
Polycarboxy sulfates, 113
Polycarboxylates, 113
Polysaccharide, 4, 105, 107, 110, 112, 113, 118, 119, 120, 121, 123, 124, 125, 126, 147, 257
Polysaccharides, 39, 43, 49, 53, 74, 104, 106, 107, 112, 118, 119, 120, 123, 124, 125, 126, 130, 147, 154, 290, 292
Polysulfates, 113
Porin, 16

Postprandial, 300
Potassium ions, 21
Powerhouse, 27
Projection formula, 54
Prokaryotic cells, 26, 178
Protein channels, 15
Proton, 3, 34, 68, 132, 133, 134, 136, 137, 138, 139, 140, 141, 144, 145, 147, 153, 157, 158, 186
Pseudopodium, 23
Pyranose, 55, 56, 57, 72, 73, 75, 79, 82, 88, 89, 91, 92, 94, 98, 100
Pyranose ring, 55, 57, 79, 92

R

Racemic mixture, 36, 48
Raffinose, 42, 93, 94, 95
Rate of diffusion, 20
Receptor proteins, 7, 32
Respiratory chain, 26, 136, 145, 158, 168, 178
Rhamninose, 42
Rhamnose, 31, 53, 87, 93, 119
Ribose, 41, 42, 46, 63, 71, 76, 150, 206, 204, 206, 207, 211, 214, 215, 217, 274
Ribulose, 41, 42, 193, 196, 197, 199, 206, 204, 206, 207, 215, 217, 218

S

Sorbose, 41, 42
Stachyose, 42, 98, 100, 102
Starch, 23, 26, 28, 31, 32, 39, 43, 48, 74, 86, 89, 97, 104, 105, 106, 107, 109, 123, 124, 125, 126, 127, 147, 149, 150, 155, 292, 300
Starch grains, 104, 109
Stereoisomers, 36, 44, 45, 46, 47, 49, 51, 54
Sucrose, 32, 50, 75, 88, 89, 92, 93, 95, 97, 99, 100, 101, 147, 148, 271, 272
Sucrose, 42, 86, 92, 93, 99, 101, 271, 275
Sugar osazones, 66
Sweetness, 2, 31, 32, 48, 50, 75, 88, 89, 104, 109
Symport, 18, 19, 27

T

Tagatose, 32, 41, 42
Talose, 38, 41, 42, 46, 71
Tautomerism, 49, 62
Tetrasaccharide, 96, 97, 98, 100, 102
Tetrsoes, 41
Threose, 41
Trans isomer, 48
Transport proteins, 7, 132
Trehalse, 32, 42, 86, 93, 99, 101, 102, 127
Trioses, 41

U

Uniport, 18, 19
Uphill transport, 20

V

Verbascose, 42, 98
Voltage-gated ion channels, 16, 22

X

Xylan, 119
Xylitol, 32, 224, 232
Xylose, 41, 42, 64, 72, 118, 119, 148
Xylulose, 41, 42, 50, 206, 228, 231